Developments in Primatology: Progress and Prospects

Series Editor
Louise Barrett, Lethbridge, AB, Canada

This book series melds the facts of organic diversity with the continuity of the evolutionary process. The volumes in this series will exemplify the diversity of theoretical perspectives and methodological approaches currently employed by primatologists and physical anthropologists. Specific coverage includes primate behavior in natural habitats and captive settings; primate ecology and conservation; functional morphology and developmental biology of primates; primate systematics; genetic and phenotypic differences among living primates and paleoprimatology.

Tracie McKinney • Siân Waters
Michelle A. Rodrigues

Editors

Primates in Anthropogenic Landscapes

Exploring Primate Behavioural Flexibility
Across Human Contexts

 Springer

Editors
Tracie McKinney
School of Applied Science
University of South Wales
Pontypridd, UK

Siân Waters
Department of Anthropology
Durham University
Durham, UK

Michelle A. Rodrigues
Department of Social and Cultural Sciences
Marquette University
Milwaukee, WI, USA

ISSN 1574-3489   ISSN 1574-3497   (electronic)
Developments in Primatology: Progress and Prospects
ISBN 978-3-031-11735-0   ISBN 978-3-031-11736-7   (eBook)
https://doi.org/10.1007/978-3-031-11736-7

This Springer imprint is published by the registered company Springer Nature Switzerland AG
The registered company address is: Gewerbestrasse 11, 6330 Cham, Switzerland

Foreword: Primatology in the Anthropocene

In 1994, Karen Strier published a groundbreaking article entitled "The Myth of the Typical Primate." In it, she criticised the then dominant models and assumptions for what primates are and do, correctly noting that lives, ecologies, and dynamics "are far less uniform across primates than the myth has implied, raising questions about the generality of models of primate social systems derived from 'typical' primates." Less than a decade later, I began a 20-year journey articulating the position that humans and other primates are often "simultaneously actors and participants in sharing and shaping mutual ecologies." In the past three decades, primatology has not only massively expanded datasets and the diversity of species and populations studied, but most practitioners have also recognised that humans have always been part of other primates' lives. There is neither a "typical" primate nor a clear line dividing the wild, the captive, and the in-between in primate ecologies. Anthropogenic ecosystems and their diversity of relations are characteristic norms for most other primates; human dynamics are a central ecological factor structuring primate niches.

Here, in the third decade of the twenty-first century, approximately 60% of primate species are threatened with extinction and approximately 75% have declining populations. This is mainly due to anthropogenic (human created and induced) pressures on other primates and their habitats, largely from global and local market demands and political realities leading to extensive habitat loss (Estrada et al., 2017, 2018). The contemporary processes of the Anthropocene also generate increased hunting, trade, and exploitation of primates as food, commodities, pets, and research subjects. The other primates' lives overlap extensively with large, widespread, and rapidly growing human populations characterised by high levels of economic, political, and ecological complexity and inequity. This is why the approach framed by the editors, and carried out by the authors in the chapters, of this book is *the* critical frame all primatological study must take, or at least be engaged with/aware of. Otherwise, there is no hope of addressing the massive risk of extinction and developing sustainable relationships between humans and the other primates. In the twenty-first century, one cannot engage in study of primate ecology and behaviour scientifically, ethically, or morally without including the relations between humans and other primates. These relations, and their resultant ecologies, are mutually co-constructive, dynamic, conflictual, complex, and central to primate lives.

But one cannot just "add" humans into the mix as another variable. To succeed in the twenty-first century, primatological practitioners, those forming the questions, doing the work, the analyses, and the publication, must create diverse and inclusive teams. Diversity of experience, perspective, and training offers the broader toolkit necessary to effectively observe, model, and analyse the dynamics of ecological and behavioural processes in the Anthropocene. The focus of primatological study must include serious engagement with more than the particular primate species of interest. Recognition (and representation) of the contexts, histories, and contemporary interests of human communities is a central facet in ecosystem dynamics. This volume moves in the direction of engaging and reflecting on this contemporary mode of primatology. The authors, the topics, and the framing of the chapters represent multiple continents and countries, backgrounds, languages, and disciplinary trainings/histories. That in and of itself is a strong and necessary statement. However, there remains much work to be done in decolonising the field and shifting the centrality of discourse, infrastructure, and funding to the people and places where the other primates and humans co-exist most. There is also a need for more extensive and explicit recognition of, and collaboration with, Indigenous and other local communities across the planet who hold knowledge(s) of relations and practices that primatology has been too slow to formally and intellectually recognise and respect (Estrada et al., 2022).

Behavioural ecology in the twenty-first century is necessarily more than late twentieth-century practice and theory. Yet, much training and formation, especially in the centres of power and funding, remain rooted in the theory and methods of that earlier time. The chapters in this volume recognise the historical importance of the theory from the 1980s, 1990s, and early 2000s but subtly (and not so subtly) push against a static and hierarchical Global-North-centric frame of analyses and practice. From community-based conservation strategies to relations between primates and dogs to urban and other shared spaces, the chapters recognise the multifarious dynamics of contemporary primate ecology. Regenerating forests, disease transmission, touristic ecologies, trade, dynamic hunting realties, rehabilitation, and zoo landscapes all act as necessary locales for primate research. Ethnoprimatological frameworks, alongside other modes of practice and theory, facilitate integrated approaches, fearlessly adopting innovative methods and collaborations, and facing the challenges of the contemporary moment. In these chapters one feels and recognises the push against constraints of tradition and sees the emergence of revolutionary entanglements of practice and experience, past and present, into new modes of investigation, reflection and action.

In his recent book, Nicholas Malone wrote "primate research and conservation activities do not take place in isolation, and attempting to tease apart the ecological and social lives of primates, amidst the context of humankind's planet altering reach, is exceedingly difficult work" (Malone, 2022). This is accurate and daunting. Today, in the third decade of the twenty-first century, it might be too late for many primate populations. But those involved in primate research and conservation must try. This book positively contributes to the efforts by making a difference of tone, frame, and intent. One cannot read this volume and not see that humans and other primates have been, are, and will always be entangled in one another's lives. At the end of the day, this volume shows us that innovative multidisciplinary approaches are the contemporary baseline for a twenty-first-century primatology.

Princeton University   Agustín Fuentes
Princeton, NJ, USA
29 March 2022

References

Estrada, A., et al. (2017). Impending extinction crisis of the world's primates: Why primates matter. *Science Advances, 3*(1), e1600946. https://doi.org/10.1126/sciadv.1600946

Estrada, A., et al. (2018). Primates in peril: The significance of Brazil, Madagascar, Indonesia and the Democratic Republic of the Congo for global primate conservation. *PeerJ, 6*, e4869. https://doi.org/10.7717/peerj.4869

Estrada, A., et al. (2022). Global importance of Indigenous Peoples, their lands, and knowledge systems for saving the world's primates from extinction. *Science Advances, 8*(32), eabn2927.

Malone, N. (2022). *The dialectical primatologist: Past, present and future life in the hominoid niche*. Rutledge.

Strier, K. (1994). The myth of the "typical" primate. *Yearbook of Physical Anthropology, 37*, 233–271.

Contents

Chapter 1
Introduction

Michelle A. Rodrigues, Siân Waters, and Tracie McKinney

Over the past two decades, new perspectives have emerged on the interactions between human and non-human primates. While early primatological research prioritised natural behaviour of non-human primates in pristine environments, our growing understanding of anthropogenic influences on primate behaviour has led to a paradigm shift. Across the contexts in which non-human primates live – from the most remote forests to primates living as pets or in laboratories – their flexible behavioural adaptations are shaped by the human-primate interface. Here, we consider the expanding recent literature on this subject and how it informs our understanding of the complex relationships between human and non-human primates. We further broadly consider these ranges as a continuum and suggest that we can understand non-human primates better by understanding these continuities across wild and captive contexts.

Part I focuses on primates in their most natural state, with "Human Influences on Primate Habitats". The chapters in this section review the rapidly growing literature on anthropogenic influences on primates in their traditional environments. We begin this section with one of the most pervasive threats to primate survival worldwide – habitat loss. In *Consequences of Habitat Loss and Fragmentation for Primate Behavioral Ecology* (Chap. 2), Malcolm S. Ramsay and colleagues review the

The original version of this chapter was revised. The correction to this chapter is available at https://doi.org/10.1007/978-3-031-11736-7_19

M. A. Rodrigues (✉)
Department of Social and Cultural Sciences, Marquette University, Milwaukee, WI, USA
e-mail: Michelle.Rodrigues@marquette.edu

S. Waters
Department of Anthropology, Durham University, Durham, UK

T. McKinney
School of Applied Science, University of South Wales, Pontypridd, UK

excellent work done on this important topic in recent years. The authors disambiguate the interrelated phenomena of habitat loss, fragmentation, and connectivity. The diverse ways primates are impacted by these landscape changes – including demographically, behaviourally, and ecologically – are summarised, reinforcing the complexity of the relationship between animals and their landscapes. The chapter highlights the species- and even site-specific responses of primates to habitat fragmentation and calls for a holistic, mixed methods approach to better understanding this global challenge.

With habitat loss and fragmentation at the forefront of primate conservation, it is important that we explore the factors that render regenerating forest suitable habitat for primates. In *The Emerging Importance of Regenerating Forests for Primates in Anthropogenic Landscapes* (Chap. 3), Lucy Millington and colleagues define regenerating forests and explain the role of seed-dispersing primates in forest succession. The authors note that while very few primate taxa are recognised as regenerating forest dwellers, a wide range of primate taxa use them. Our tendency to describe primates as simply "folivores vs. frugivores" or "primary forest specialists vs. regenerating forest specialists" suggests that significant gaps remain in our understanding of primate ecological plasticity.

Exacerbating habitat fragmentation and loss is the pervasive pressure of human hunting. *Hunting of Primates in the Tropics: Drivers, Unsustainability, and Ecological and Socio-economic Consequences* (Chap. 4), by Inza Koné and colleagues, reviews the myriad impacts of hunting on wild primate populations. The authors point out that, while hunting of primates has long been an important factor in human subsistence, current hunting technologies combined with market demand have pushed this threat to an unsustainable level. In addition to behavioural and demographic consequences on the primates themselves, hunting of primates drives another vital issue – cross-species disease transmission.

Where there are people, there will be dogs. In *Dogs, Primates, and People: A Review* (Chap. 5), Siân Waters and colleagues explore this ubiquitous but poorly studied interspecies association. The chapter describes people's relationships with their dogs and subsequently with wildlife and highlights differences in the level of responsibility assumed by humans for the actions of their animals. The authors highlight the risks to primate conservation posed by dogs, which include disease transmission, harassment, and hunting. The chapter provides two case studies from Morocco and Madagascar that demonstrate creative approaches to managing dog-primate interactions. This chapter is an excellent reminder that humans can influence whole ecological communities by introducing species as much as by eliminating them.

The final chapter in this section, *Climate Change Impacts on Non-human Primates: What Have We Modelled and What Do We Do Now?* (Chap. 6) by Isabelle C. Winder and colleagues, reviews recent modelling studies that aim to understand the effects of climate change on primate populations. While most primates are expected to respond to climate change through shifting home ranges, responses are species-specific, and therefore far more data collection is needed. The authors note the relative lack of studies to date and the unbalanced taxonomic focus, highlighting the need for more work in this vital area. With a global perspective on anthropogenic impacts on non-human primate survival, this chapter concludes Part I with a strong reminder of our responsibility for the conservation of our closest relatives.

Our second section, "Primates in Human-Dominated Landscapes", examines the diverse ways primates and humans impact each other. In *Community-Based Strategies to Promote Primate Conservation in Agricultural Landscapes: Lessons Learned from Case Studies in South America* (Chap. 7), Laura A. Abondano and colleagues explain how anthropogenic forest clearance for agriculture has destroyed large swathes of primate habitat globally. The authors review how community conservation projects fostering sustainable agriculture can improve primate habitat accessibility by connecting such forest fragments. The authors stress the importance of understanding the social, political, and historical contexts of a project setting as well as the use of participatory spaces where conservationists and community members can share knowledges and experiences to foster human-primate coexistence. The Colombian and Ecuadorian case studies clearly illustrate how participatory research and meaningful inclusion of communities can facilitate sustainable agriculture as well as improve people's livelihoods and endangered primates' prospects for the future.

As built-up centres of human population spread, some primates adapt to living in urban, peri-urban, and urbanising landscapes. In *Primates in the Urban Mosaic: Terminology, Flexibility, and Management* (Chap. 8), Harriet R. Thatcher and colleagues define the terms to describe these habitats and explain how primates such as baboons adapt and diversify their behaviours in response to such extreme changes in their environment. The authors explain the complexities of the urban mosaic with its high levels of anthropogenic disturbance, diverse urban primate ecologies, and complex human dimensions. All these aspects make management of these particular interfaces very challenging. The authors recommend careful consideration of the whole to enable effective management of urban human-primate coexistence.

The more frequently people and primates interact, the higher the risk of disease transmission from one to the other. In *Infectious Diseases in Primates in Human-Impacted Landscapes* (Chap. 9), Marina Ramon and colleagues review the current knowledge about infectious diseases, describing the myriad ways people and primates impact one another's health. The authors explain how good health for both human and non-human primates is promoted and encouraged, along with reviewing how risk analysis might prevent disease outbreaks in the first instance. Two case studies examine responses by primatologists to a yellow fever outbreak in Brazil and a holistic One Health approach in Uganda. As the COVID-19 pandemic has demonstrated, the anthropogenic effects on primate and other wildlife leading to disease transmission to people have implications for us all.

In the next *Primate Conservation in Shared Landscapes* (Chap. 10), Elena Bersacola and colleagues suggest we think carefully about primate conservation priorities when reviewing our definitions of important species because some primate species may be overlooked. As such, the authors argue that as well as protecting primate habitats with high species biodiversity and low human population, we need to consider how we can also conserve primates in anthropogenic or shared landscapes. However, shared landscapes bring their own set of problems which foster negative human-primate interactions. The three case studies from Uganda, Indonesia, and Brazil all illustrate how complex considerations of issues such as species characteristics and human-primate relations need to be made in order to assist people and primates to share a landscape.

Primate Tourism (Chap. 11) is an important conservation tool. In their review of the topic, Stefano Kaburu, Malene Friis Hansen, and colleagues provide a historical overview of primate tourism before describing the considerable costs and benefits that accompany this activity in all its forms. A large part of the appeal of primate tourism is getting close to the animals. Such proximity may have unpredictable and often negative consequences for both person and primate. The authors make suggestions for further research on a number of issues of concern such as how individual primate differences might influence tourist-primate interactions and discovering what tourists might expect from their primate viewing experience. In this chapter, the authors emphasise that there are substantial benefits for both primates and communities provided by primate tourism. However, dependence on tourism to the detriment of subsistence and other activities has proven a problem during the COVID-19 pandemic.

The final chapter of this section is *Shared Ecologies, Shared Futures: Using the Ethnoprimatological Approach to Study Human-Primate Interfaces and Advance the Sustainable Coexistence of People and Primates* (Chap. 12). Erin P. Riley and colleagues review the rise of ethnoprimatology and its now substantial contribution to our understanding of the human-primate interface. The authors examine three different settings where people and primates interact – tourism, urban environments, and agroecosystems. Although these settings have been examined in detail in previous chapters, here, the authors consider how people and primates react to and affect each other in various cultures and scenarios. As the authors argue, gaining a contextual understanding of each situation can facilitate human-primate coexistence.

In the third section, "Primates in Captivity", the book explores how human management is an anthropogenic pressure that shapes primate lives. In the first chapter in this section, *Perspectives on the Continuum of Wild to Captive Behaviour* (Chap. 13), Michelle A. Rodrigues and colleagues consider how anthropogenic influences occur as a continuum and affect primates across wild, human-sympatric, and human-managed conditions. They explore how human presence and activity shape primates' lives in environments across these contexts and how recognition of this continuum can inform our understanding of co-evolutionary relationships between humans and non-human primates, particularly behavioural flexibility. They provide a case study on captive chimpanzees and bonobos to illustrate how behavioural flexibility can be studied in zoo-housed primates.

In *The Past, Present, and Future of the Primate Pet Trade* (Chap. 14), Sherrie D. Alexander and colleagues provide an overview of the history and current state of the primate pet trade. They explain that pets are traditionally kept as companions and that pet-keeping is historically rooted in many cultural traditions. Pet-keeping today is influenced by these roots, as well as changing trends associated with status, social media, and internet usage. They provide regional and taxonomic overviews of

Asian macaques, orangutans in Southeastern Asia, Malagasy lemurs, Indian slow lorises, and platyrrhines in the Americas. They then consider how the internet has changed the pet trade, using a case study of how a photo of one of the authors and a captive zoo primate ended up being used out of context within the pet trade, and assess solutions for the future to curb the pet trade.

In *Rescue, Rehabilitation, and Reintroduction* (Chap. 15), Siobhan I. Speiran and colleagues explore the current state of knowledge on how conservation organisations deal with the rescue, rehabilitation, and reintroduction of primates across regions. Primate rescue refers to removing primates from harm, whether it may be injuries sustained in the wild or harms from illegal or biomedical trade. However, rescue and rehabilitation may overlap, and there is a lack of data on the state of primate rescue across regions, often due to poor record-keeping or organisations reticent to share details about failures publicly. Reintroductions have the most extensive literature and documentation, including multiple IUCN best practice guidelines. Nonetheless, like rescue and rehabilitation, there is still inconsistency in reporting and evaluation, as well as many projects that do not adhere to recommended best practices. They provide a case study examining the rescue, rehabilitation, and reintroduction of Costa Rican primates across sanctuaries. The authors point out that there are major gaps on the literature for this topic and consider future solutions for more comprehensive tracking of these projects, as well as ethical issues that need to be addressed. They note that many projects are guided by well-meaning intentions but may not adequately serve either individual primate welfare or conservation goals.

In *Through the Looking Glass: Effects of Visitors on Primates in Zoos* (Chap. 16), Ashley N. Edes and Katie Hall explore how visitor presence affects primates housed in zoos. Human presence is ever-present for primates in zoological facilities, through both the regular interaction with keeper staff and the presence of regular daily visitors. Most of the literature frames visitor effects as negative or neutral; however, approximately a third describe positive effects. While these relationships can be bidirectional and dynamic, with visitors and animals influencing each other, most of the literature focuses on how visitors affect zoo animals rather than vice versa. The authors provide some suggestions for broadening the scope of this research, including rethinking assumptions about solely negative impacts, consideration of individual variation and physiological measures to better understand welfare impact, and steps that can be taken to increase animal choices to ameliorate negative effects.

In our final *Primate Portrayals: Narratives and Perceptions of Primates in Entertainment* (Chap. 17), Brooke C. Aldrich and colleagues survey the use of primates within entertainment and the media, including the rising presence of social media. They highlight the long usage of primates in entertainment, from circuses and tourist attractions to movies and video games, to the rise of primate content in social media. Despite activism causing pressure on limiting or changing harmful exploitation of primates for entertainment, many avenues still exist, and the expanding world of social media has exacerbated these usages. The authors consider these usages from the lens of the Five Freedoms of animal welfare while considering the motivations for using primates in entertainment and media. While the usage of primates in media is harmful from both a welfare and conservation perspective, conservationists can reduce and counter these harms by considering the sociocultural factors that create demand for primate content.

From unhabituated primates deep in equatorial rainforest to the primates chained at a tourist attraction to pose for selfies, primates are shaped by the co-occurrence of humans and human activity. In this book, we ask our readers to consider shared dynamics across all of these contexts while recognising the particularly local and specific factors that shape each facet of the human-non-human primate interface. Recognising the importance of human actions in the lives of non-human primates can help us act to conserve primates and promote their welfare across the globe.

Part I
Human Influences on Primate Habitats

Chapter 2
Consequences of Habitat Loss and Fragmentation for Primate Behavioral Ecology

Malcolm S. Ramsay, Fernando Mercado Malabet, Keren Klass, Tanvir Ahmed, and Sabir Muzaffar

Contents

Abstract Primates are a particularly sensitive order to the negative effects of habitat loss and fragmentation due to their unique life histories and habitat requirements. Given that nearly all primate populations are in some way affected by habitat loss and fragmentation, it is important for all primatologists – even those uninterested in these processes directly – to consider the effects of habitat loss and fragmentation on the behavior and conservation of their study species. In this chapter, we review

M. S. Ramsay (✉) · F. Mercado Malabet
Department of Anthropology, University of Toronto, Toronto, ON, Canada

K. Klass
The Alexander Silberman Institute of Life Science, The Hebrew University of Jerusalem, Jerusalem, Israel

T. Ahmed
Wildlife Research and Conservation Unit, Nature Conservation Management (NACOM), Dhak, Bangladesh

S. Muzaffar
Department of Biology, United Arab Emirates University, Al Ain, United Arab Emirates

© The Author(s), under exclusive license to Springer Nature Switzerland AG 2023
T. McKinney et al. (eds.), *Primates in Anthropogenic Landscapes*, Developments in Primatology: Progress and Prospects, https://doi.org/10.1007/978-3-031-11736-7_2

some of the current knowledge of the effects of habitat loss and fragmentation on primate behavior. We begin by defining key terms and discussing issues of scale. We then review some of the major literature regarding primary and secondary effects of fragmentation, highlighting its potential impact on home range, social interactions, and group composition. Finally, we note that primate responses to habitat fragmentation are species- and sometimes even site-specific and recommend a holistic approach for future research concerning habitat loss and fragmentation.

Keywords Demography · Diet · Dispersal · Habitat fragmentation · Human-primate interactions · Landscape ecology · Scale

2.1 Introduction

The ways in which non-human primates (herein referred to as primates) respond to habitat loss and fragmentation have been an important theme of research in the modern era of primatology (Marsh, 2003; Marsh & Chapman, 2013). The negative effects of habitat loss and fragmentation, driven mostly by resource extraction and related anthropogenic activities, have led to a global decrease in primates; 65% of primate species are classified as threatened with extinction, and 75% have declining populations (Estrada et al., 2020). Primates are a particularly sensitive order to the negative effects of habitat loss and fragmentation due to factors such as long life histories and large home range requirements (Arroyo-Rodriguez & Fahrig, 2014). Moreover, strictly arboreal primates are often the least capable of adapting to life in fragmented landscapes, as their ecological requirement for forest habitats limits their adaptability outside these **landscapes** (Galán-Acedo et al., 2019a; Harcourt et al., 2002). The **matrix** between forest patches in fragmented landscapes such as grassland or anthropogenic habitats (e.g., farmlands or human settlements) can be very inhospitable to primates, thus restricting **dispersal** ability which can lead to long-term population viability concerns (Marshall et al., 2016). Fragmentation also has many secondary effects; in particular, primates living in fragments may come into closer contact with humans and associated anthropogenic threats (Estrada et al., 2020). In this chapter, we summarize some of the latest research on primates in fragmented landscapes. Specifically, we examine how scale and habitat vary in primates, behavioral flexibility of primates in fragments, and demographic consequences on primate populations of living in fragments. Finally, we suggest some future directions for research on primates living in an increasingly fragmented world.

2.2 Defining Habitat Loss and Fragmentation

Habitat loss and fragmentation are interconnected but distinct ecological concepts that can affect species and landscapes in different ways (Mcgarigal & Cushman, 2002). **Habitat loss** is the conversion of one habitat to another such as the change of forest to savannah or vice versa (Fahrig, 2003). Of central importance to this process is the definition of habitat. **Habitat** can be defined as any area with the necessary conditions (e.g., food, shelter, conspecifics) to promote the sustained occurrence of a species based on their ecological and behavioral requirements (Fahrig, 1997). Put differently, a habitat refers to the location where we are most likely to find a primate because they can find food and shelter and because they can locate other individuals of the same species to facilitate important group behaviors like reproduction and dispersal. For an arboreal primate species, their habitat of choice may be forest, while a terrestrial primate species may just as suitably occupy open grasslands or the forest floor in the same patch as the previously mentioned arboreal species.

After habitat loss occurs, landscapes often contain a network of isolated habitat patches surrounded by a matrix of non-habitat that may be heterogeneous in quality (Fahrig, 2003; Galán-Acedo et al., 2019b). **Habitat fragmentation** is this process of continuous habitat being reduced to isolated habitat patches or fragments (Villard et al., 2014). Researchers continue to debate how to define the degree of habitat fragmentation; what makes a landscape fragmented as opposed to continuous? In theoretical terms, given two landscapes with equal habitat amounts, the more fragmented landscape has a greater number of habitat patches, these habitat patches have a lower average habitat area, and the landscape will contain a greater total percentage of habitat edges (Fahrig, 2017). In real landscapes, these concepts are more complicated. Fahrig (2003) demonstrated that many authors considered factors such as individual patch size, isolation, and connectivity as the defining features of fragmentation, while others considered these attributes to be separate processes that occur as a result of fragmentation at the landscape scale. What constitutes habitat fragmentation does not represent simply a semantic argument. The definition of habitat loss and fragmentation, and thus their independent effects, is critically important for conservation planning (Fahrig, 2017; Keinath et al., 2017).

It is important to note that both the habitat itself and the non-habitat matrix can be variable in quality (Fahrig, 2003). In real landscapes, all area is on a spectrum of being more or less suitable to a given species; those areas that are most suitable are deemed habitat, while areas that are least suitable are deemed matrix, with spaces between known as edge zones. Generalist species occupying these landscapes may be able to exist in many different types of habitats, whereas other species may be only able to persist in certain types of habitats, such as edge-tolerant or edge-sensitive species (Fig. 2.1). The quality of the matrix, which can be determined by the degree to which matrix area resembles habitat, will also play an important role in determining a given species' persistence in a fragmented landscape. High-quality matrix (e.g., brush thicket for a forest specialist) can facilitate movement and even allow temporary refuge for populations, while low-quality matrix may lead to isolated and disjunct populations.

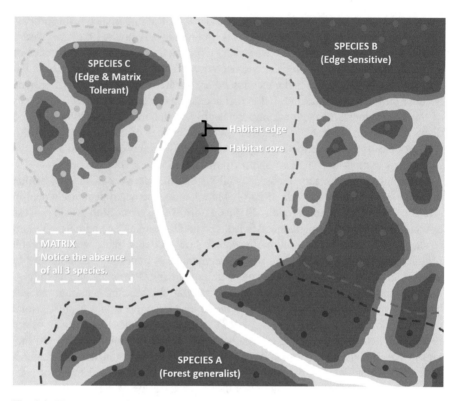

Fig. 2.1 Theoretical distribution of three ecologically distinct species across a fragmented land-scape. (Illustration by Fernando Mercado Malabet)

2.3 Habitat and Landscape Patterns

The occurrence and persistence of any primate species in a landscape of fragmented habitat are functionally correlated with the availability of adequate environmental conditions to sustain their demographic, foraging, and other behavioral needs (Fahrig, 1997; Hutchinson, 1957; Wiens et al., 1993). Species' patterns of occurrence and persistence are simultaneously affected by several interrelated characteristics of the fragmented landscapes they occupy, at both the scale of the local habitat fragments they inhabit and the much broader landscape context that surrounds those fragments (Arroyo-Rodriguez & Fahrig, 2014). For example, an arboreal primate's distribution throughout the landscape will be determined by both the species' own ecology and behavior and the conditions in the different fragments and surrounding matrix. Fragment characteristics that affect species' distributions include fragment size, when and where food is available within them, and what other species occupy these fragments. At the landscape scale, the lives of these arboreal primates are simultaneously affected by how close to different fragments they are and the composition of non-habitat matrix that separates the habitat fragments.

At the fragment-level scale, the probability of a species' occurrence and their long-term persistence are related to attributes of the fragments such as the amount of habitat (e.g., forest area for forest-dwelling species) and resources (e.g., distribution of fruits for frugivore species) available that are functionally relevant to their respective ecological and behavioral characteristics (Morris, 1987). Indeed, a number of studies have shown that the probability of forest fragments being occupied by primates is at least, in part, related to the size of the fragments – where larger fragments are more likely to exhibit higher species richness or **population density** for species whose ecology is sensitive to home range variability (Harcourt & Doherty, 2005). For example, Steffens and Lehman (2019) showed that increasing fragment area was associated with higher primate species richness. Moreover, previous research on the habitat ecology of crowned lemurs (*Eulemur coronatus*) in northern Madagascar has shown that within anthropogenically disturbed forest fragments, their distribution is influenced by fragment quality (Mercado Malabet et al., 2020). Primates respond differently to habitat characteristics depending on their ecology and behavior. This means that while a fragment can be suitable to some species, it may simultaneously restrict the occurrence of others. For example, species sensitive to edge effects may exhibit restricted distributions or absence from fragments that exhibit a large proportion of edge habitat (Lehman, 2006).

A given primate species' patterns of occurrence and persistence across a fragmented landscape are further constrained by their connectivity across the habitat mosaic. In fragmented landscapes, the connectivity of primate populations is affected by the amount and types of non-habitat matrix that are present between habitat fragments, coupled with the distance that separates any two (or more) fragments (Arroyo-Rodríguez & Fahrig, 2014; Galán-Acedo et al., 2019b). Patterns of **landscape connectivity** affect how well individuals can move across the landscape and interact with each other depending on the type of matrix that surrounds fragments, as there are inherent costs associated with movement through these non-habitat patches and corridors due to the lack of conspecifics, preferred habitat substrate for locomotion, preferred foods, and/or adequate sleeping sites (Fahrig, 1997, 2003). The matrix is not universally uninhabitable, and an organism's ability to use and move through non-habitat matrix is directly related to species-specific traits such as ecological flexibility and the degree to which matrix is similar to habitat (Fahrig, 2003).

Habitat choice patterns of primates depend on their biology (Hutchinson, 1957; Wiens et al., 1993). There is no generalized response to habitat and landscape patterns (Andriatsitohaina et al., 2020; Eppley et al., 2020; Galán-Acedo et al., 2019b; Steffens & Lehman, 2019). Habitat choice patterns are species-specific, realized by the closely synergistic relationship between the ecology and behavior of any animal and the characteristics of places that they occupy. Responses vary due to differences in correlates of any two species – such as life history, body size, group size, mating system, home range size, dietary preference, and preferred method of motility – all of which affect how a primate interacts with the habitats they occupy and how well they may be able to respond to changes in broader landscape patterns (Eppley et al., 2020; Galán-Acedo et al., 2019b). In a global-scale analysis, Keinath et al. (2017)

concluded that an organism's habitat relationships, particularly its degree of habitat specialization, were more important than other species-specific life history parameters (such as fecundity, life span, and body mass) in predicting their scale of response to fragmentation. In other words, how flexible a species' habitat needs are will affect how they respond to fragmentation, whether they can continue to inhabit these landscapes or whether their populations will be extirpated. In primates, habitat generalists will tend to tolerate fragmentation more than habitat specialists who are more prone to extinction risk, because the generalists are able to make use of fall-back resources when their preferred options are unavailable. Meanwhile, the strict habitat requirements of specialists do not grant such flexibility, forcing these species to disperse to other fragments – regardless of the inherent risk associated with dispersal (e.g., mortality or dispersal to worse quality fragments) – or risk the decline of their populations in response to the lack of resources (Gibbons & Harcourt, 2009). Thus, landscape patterns and interspecific variation in habitat use and life history traits must also be considered when examining the effects of habitat loss and fragmentation on primate populations.

2.4 Behavioral Responses to Habitat Fragmentation

2.4.1 Home Range and Density

Arboreal primate populations are often (but not always; e.g., Wong & Sicotte, 2006; DeGama-Blanchet & Fedigan, 2006) found at higher densities with decreased home ranges in forest fragments than in continuous forest (e.g., lion-tailed macaques, *Macaca silenus*, Singh et al., 2002; diademed sifakas, *Propithecus diadema*, Irwin, 2008; howler monkeys, *Alouatta* spp., Arroyo-Rodriguez & Dias, 2010; Bicca-Marques, 2003; Klass et al., 2020a). Small fragment areas can limit the number of groups and, subsequently, within-fragment population sizes; thus, population size (the number of individuals in a given population) is often positively correlated and population density (the number of individuals in a given area) negatively correlated with fragment size (e.g., lion-tailed macaques, *Macaca silenus*, Singh et al., 2002; howler monkeys, *Alouatta* spp., Arroyo-Rodriguez & Dias, 2010; brown spider monkeys, *Ateles hybridus*, de Luna & Link, 2018; ring-tailed lemurs, *Lemur catta*, Gould & Cowen, 2020). In some species, the high population densities observed in forest fragments reflect short-term, temporary crowding as a result of recent habitat loss (Debinski & Holt, 2009), in which fragments show population densities far above carrying capacity immediately post-fragmentation that then decline (e.g., Tana River red colobus, *Procolobus rufomitratus*, Decker, 1994; brown spider monkeys, *Ateles hybridus*, de Luna & Link, 2018). Other species appear able to sustain these high population densities over decades (e.g., black howler monkeys, *Alouatta pigra*, Klass et al., 2020a), although longer-term monitoring may reveal declines over more extended time periods (Kuussaari et al., 2009).

While generally primates are thought to have decreased home ranges in fragments, there appears to also be a high degree of home range size flexibility, even within individual species occupying fragmented landscapes. For example, the home range sizes of western hoolock gibbons (*Hoolock hoolock*) vary widely, with gibbons in the smallest fragments having 16–17 ha home ranges and those in medium-sized fragments having 13–24 ha home ranges (Ahsan, 2001). This variability may have many underlying causes, such as a species-specific ability to exploit resources in the matrix (Steffens et al., 2021) which would allow larger home ranges outside of the remaining fragments or ecological factors such as the presence of high-quality fallback foods which could sustain smaller home ranges in forest fragments (Irwin, 2008).

2.4.2 Dietary Shifts in Fragments

Primates in fragmented habitats often need to modify their diets in relation to the resources available, resulting in diets marked by less specialty and increased use of fallback foods (Silver & Marsh, 2003; Bicca-Marques & Calegaro-Marques, 1994). Dietary shifts to novel, often non-native food resources in fragments are widespread in many primate taxa (Singh et al., 2002; Irwin, 2008). This could also include major dietary shifts such as the introduction of novel feeding behaviors like predation (Bicca-Marques et al., 2009; Ren et al., 2009). However, if the novel foods are of lower quality or lead to an increased energy expenditure, this could negatively affect populations. For example, a population of Udzungwa red colobus (*Procolobus gordonorum*) in Tanzania inhabiting a monoculture plantation fed on lower-quality foods such as leaves and buds, and this population declined from an estimated 400 to 50 over a period of two decades (Ehardt et al., 1999; Marshall et al., 2016). Many primates in fragments also exhibit power-feeding strategies, where they increase their daily foraging path lengths, visit more food patches, and spend more time feeding per day (Harris & Chapman, 2007; Sato et al., 2016). In disturbed forest fragments (fragments with a high level of anthropogenic activity that results in a lower quality of habitat) or during seasonally lean periods, power-feeding strategies are a mechanism for primate populations to acquire enough energy in spite of reductions in the total availability of their preferred food resources. Thus, feeding strategies may be varied to cope with the availability of food resources, resulting in an altered diet, but not necessarily a negative or positive outcome (Taylor et al., 2016; Bicca-Marques & Calegaro-Marques, 1994).

2.4.3 Social Interactions in Fragments

Living in a fragmented landscape can affect the social behavior of primates in many ways (Marsh, 2003; Clarke & Young, 2000). In response to food scarcity in small fragments, increased feeding competition may result in higher rates of agonism from dominant, resident groups to displace conspecifics from food resources (Ahmed & Naher, 2021). In contrast, Asensio et al. (2007) found sympatric mantled howler monkeys (*Alouatta palliata*) and coatis (*Nasua narica*) in forest fragments had more frequent social interactions and increased levels of tolerance. Territorial primate groups are more likely to defend high-quality clumped resources in a relatively small area of habitat, while if resources are more spread out and in low-quality habitat, large home ranges or territories may be required (Lappan et al., 2017).

Average group size does not always differ in fragments compared to continuous forest (e.g., ursine colobus, *Colobus vellerosus*, Wong & Sicotte, 2006; diademed sifakas, *Propithecus diadema*, Irwin, 2008; black howler monkeys, *Alouatta pigra*, Klass et al., 2020a), although even when average size is similar, populations in fragments often show greater variability in group size. For example, lion-tailed macaque (*Macaca silenus*) groups in fragments ranged from 6 to 53 individuals, whereas in larger forest complexes, group size ranged from 8 to 18 individuals, and the variance in group size significantly increased with the degree of habitat disturbance (Singh et al., 2002). In other populations, group size is smaller in fragments than in continuous forest (e.g., black-and-white colobus, *Colobus guereza*, Onderdonk & Chapman, 2000; collared lemurs, *Eulemur collaris*, Donati et al., 2011). Population density may also shape group size (Horwich et al., 2001); in some species, this is seen specifically via changes to the number of adult males in groups. For example, Rudran and Fernandez-Duque (2003) found a preponderance of single-male groups at low densities and multi-male groups at high densities in ursine howlers (*A. arctoidea*).

Group composition in forest fragments may be altered relative to continuous forest, even when group sizes remain similar, which has consequences for sociality. In black howler monkeys (Klass et al., 2020a), the average number of adult males per group was lower, and the proportion of single-male groups was higher, in forest fragments as compared to continuous forest, despite identical mean group sizes. Klass et al., (2020a, b) posited that these patterns may result from increased male mortality during dispersal and reduced female dispersal in the fragmented landscape, resulting in an overall lower proportion of males in the fragmented landscape and an adult sex ratio that is highly skewed toward females. However, the interconnectedness of the effects of habitat loss and fragmentation makes it difficult to predict the social outcome of such changes on primate populations (Chapman et al., 2010). On the one hand, changes in sociality in fragments can have negative consequences on primate populations, including increased mortality. For example, infanticide may be more common in fragmented habitats where there is crowding of individuals as demonstrated by case studies of infanticide in populations where this behavior does not follow the sexual selection hypothesis (black-horned capuchin monkeys, *Sapajus nigritus*: Illia et al., 2021; Coquerel's sifaka, *Propithecus coquereli*: Ramsay et al., 2020). Yet other species of macaques and vervets have seemingly benefited from increased social cooperation in fragmented landscapes, resulting in increased success in food provisioning (Dhawale et al., 2020; Maurice et al., 2019; Singh, 2019).

2.5 Population-Level Effects of Fragmentation

2.5.1 Demography

Demographic studies are a necessary component for assessing the long-term viability of primate populations inhabiting fragmented landscapes (Lawler, 2011). Undisturbed wild primate populations generally have demographically stable populations (Lawler, 2011; Morris et al., 2011), with long life spans and long generation times compared to many other vertebrates (Morris et al., 2011). Population growth rates are typically close to equilibrium or slightly above (Lawler, 2011; Morris et al., 2011), and changes to adult survival rates tend to have the largest effect on population growth rates, as compared to other demographic traits, for example, age at first reproduction (Lawler, 2011). Deviations from these common patterns in populations in fragmented landscapes, or from more specific patterns in comparison to populations of the same species in continuous, undisturbed habitat, can be informative regarding the viability of the fragmented population and the ways in which the fragmented landscape may be altering population **demography** (Cristobal Azkarate et al., 2017; Klass et al., 2020a; Rudran & Fernandez-Duque, 2003).

Habitat fragmentation can alter the demographic structure in several ways. The dense, small populations often found in fragments can show increased vulnerability to random environmental and demographic fluctuations (Arroyo-Rodriguez et al., 2008; Chiarello, 2003; Umapathy & Kumar, 2003), which may cause subsequent unpredictable changes to population composition. The high primate population densities and/or reduced habitat quality often found in small forest fragments (Arroyo-Rodriguez & Dias, 2010; Laurance et al., 2017) can increase levels of nutritional and physiological stress (Cristobal-Azkarate & Dunn, 2013; Rangel-Negrín et al., 2014) caused by increased competition for resources. These stressors can in turn reduce reproductive rates (Rangel-Negrín et al., 2014) or increase individual mortality (Cristobal-Azkarate et al., 2005; Pride, 2005), causing further changes to population size and composition over time. Increased competition for limited resources in fragments can affect specific age/sex classes more than others. For example, fragment area was found to be a central variable determining the proportion of juveniles in populations of ring-tailed lemurs (*Lemur catta*) in fragments in Madagascar (Gould & Cowen, 2020), and high rates of juvenile mortality in forest fragments were recorded in lion-tailed macaques (*Macaca silenus*) in the Western Ghats, India (Umapathy et al., 2011). However, some species are also able to effectively utilize resources in the matrix, which can mitigate some of the negative demographic effects of fragmentation (Steffens et al., 2021).

2.5.2 *Dispersal*

Dispersal is at its foundation a movement behavior (Nathan et al., 2008); thus, anything that alters or restricts movement among locations/social groups, such as anthropogenic habitat fragmentation, will affect the dispersal of individuals in a population and, subsequently, population demography. Dispersal is a key individual life history stage that shapes many demographic parameters such as fertility and mortality rates and population structure, for example, via the composition of social groups and group and population sex ratios (Lawler, 2011).

The population structure of primates in fragments may be affected by changing dispersal patterns in several ways. Skewed sex ratios and reduced mate availability often characterize the small, dense populations found in fragments (Bergl & Vigilant, 2007; Ferrari et al., 2013; Oklander & Corach, 2013). Dispersal rates may therefore increase as individuals are forced to leave their natal fragment in search of reproductive opportunities elsewhere. Dispersal rates and the sex ratio of dispersing individuals may also be affected by increasing competition for resources in small, densely populated fragments (Cristobal-Azkarate & Dunn, 2013; Jones, 2005; Oklander & Corach, 2013).

Larger isolation distances between fragments or matrix types that prevent or restrict movement can reduce dispersal rates, cause increases or decreases in distances traveled during dispersal events, and lead to non-random movement paths through the matrix (e.g., howler monkeys, *Alouatta* spp., Arroyo-Rodriguez & Dias, 2010; Mandujano et al., 2004; Cross River gorillas, *Gorilla gorilla diehli*, Bergl & Vigilant, 2007; titi monkeys, *Callicebus* spp., Ferrari et al., 2013). During dispersal through non-forest matrix types, monkeys may be forced to move close to or on the ground, which can make them more vulnerable to hunting and predation, for example, from dogs (*Canis lupus familiaris*, Pozo-Montuy et al., 2011). In species with sex-biased dispersal, increased mortality rates for the dispersing sex during transition through the matrix can affect both sex ratios in the population as a whole and also within specific fragments.

Males and females may be affected in different ways by the higher costs of dispersal through the matrix. For species where both males and females disperse, some studies have shown that one sex tends to become more philopatric in fragments (e.g., increased female philopatry in black-and-gold howlers (*Alouatta caraya*) in forest fragments when compared to a population in continuous forest; Oklander & Corach, 2013). Such sex-specific changes to dispersal patterns can alter group compositions and sex ratios in fragments and the fragmented landscape overall, effects which may be further exacerbated if the dispersing sex experiences higher mortality.

2.6 Secondary Effects of Fragmentation

One of the major issues with determining the effects of fragmentation on a given species is the prevalence of secondary effects of fragmentation which can be positive, negative, or neutral (Fahrig, 2003). For example, if habitat fragmentation results in a decline of predator species that prey on a given primate, this may result in increases in primate population size. Fragmentation may also open up corridors for new predators (in particular humans and domestic species) that can have profound effects on communities in fragmented landscapes. Given that these effects are complicated and occur in dynamic systems, they must be examined in a nuanced way, not only at the species level but also site to site.

2.6.1 Predation Risk in Fragments

Large predators are some of the first species to disappear from landscapes undergoing habitat fragmentation due to large habitat area requirements (Taylor et al., 2016). A decline in predators should result in an increase of prey populations as long as the other effects of fragmentation do not lead to a net greater population decline. This positive effect has been observed in a number of primate populations (macaques, *Macaca* spp.: Singh, 2019; vervets, *Chlorocebus pygerythrus*: Maurice et al., 2019; and baboons, *Papio* spp.: Taylor et al., 2016). However, fragmentation can also increase predation via the introduction of new predators, in particular mesopredators and introduced species like domestic dogs (see Chap. 5, this volume). Primates in fragmented landscapes are often under increased predation risk by dogs (Mendes-Pontes & Soares, 2005) which can result in even higher predation rates than would be expected from a native predator. Primates themselves are often mesopredators, resulting in a competitive advantage when large predators are missing. For example, reduction of predators such as leopards (*Panthera pardus*) resulted in increasing baboon populations which are prolific omnivores that can predate on a wide variety of species (Taylor et al., 2016).

Predation risk in fragments coupled with decreased availability of preferred habitat can result in the selection of novel sleeping sites. Some primates appear to preferentially select forest-agriculture mosaics to avoid predation. For example, northern pig-tailed macaques (*Macaca leonina*) use roadside oil palm plantations as sleeping site in Bangladesh (Ahmed & Naher, 2021), hamadryas baboons (*Papio hamadryas*) use non-native palm trees rather than cliffs in Ethiopia (Schreier & Swedell, 2008), and mantled howlers (*Alouatta palliata*) in Central America inhabit coffee plantations (McCann et al., 2003). These sleeping sites may have fewer natural predators but potentially allow for easier detection by remaining predators due to lack of foliage or the need to move terrestrially because of discontinuous habitat (Chapman et al., 2006). Additionally, the movement of primates into these human-dominated zones for predator protection invariably leads to increasing human-primate interactions.

2.6.2 Human-Wildlife Interactions

Fragmentation of a primate's natural habitat can lead to the expansion of primate populations into areas of high human habitation or increased access to a primate's habitat by humans (Maurice et al., 2019; Singh, 2019; Taylor et al., 2016). The urban environment offers an abundance of anthropogenic, often high-quality, food resources, but this often leads to associated conflict with humans (see Chap. 8, this volume). Crop foraging in more rural areas and foraging in markets and homes in urban areas are common forms of primate-human interactions that can sometimes have negative outcomes (Anand & Radhakrishna, 2020; Taylor et al., 2016). Primates in forest fragments within or near urban environments frequently attack, bite, or otherwise injure people, either provoked or unprovoked (Goldberg et al., 2006; Skorupa, 1988). Agonism from humans can lead to aggressive behavioral strategies that unfortunately lead to cycles of further conflict (Nekaris et al., 2013; Uddin et al., 2020). Mitigation of these conflicts is possible, but requires a good understanding of the underlying causes of conflict and respectful cooperation with local stakeholders (Anand & Radhakrishna, 2020; Nekaris et al., 2013; Uddin et al., 2020).

In areas of where negative interactions occur, humans typically consider primates as pests which can lead to negative population outcomes (Maurice et al., 2019; Taylor et al., 2016; Uddin et al., 2020). For example, a study in India showed that most people wanted to control populations of urban primates by either translocation or sterilization (Pebsworth et al., 2021). However, these interactions are not universally negative; many primate species are associated with spirituality or taboos and are thus protected or provisioned (Uddin et al., 2020). For example, critically endangered Coquerel's sifakas (*Propithecus coquereli*) were found in high abundance in fragments near human settlements against expectations, with the authors hypothesizing that local taboos provided protection to sifaka groups ranging near the village (Ramilison et al., 2021). The complicated nature of human-primate interactions makes quantifying and interpreting these interactions difficult. Measures of anthropogenic disturbance such as the distance to the nearest settlement or the presences of trails may be uninformative or even misleading if hunters travel large distances into remote areas and avoid paths because of legality (Steffens & Lehman, 2019). Holistic approaches that involve qualitative methods, such as ethnoprimatology, may be a more effective way of measuring human-primate interactions in fragmented habitats (Nekaris et al., 2013).

Increasing levels of human-primate interactions can lead to both anthroponotic and zoonotic disease transmission (see Chap. 9, this volume). In human-dominated fragmented landscapes, the increased proximity to humans may put populations of primates at risk due to their susceptibility to human diseases like scabies, intestinal parasites, measles, metapneumovirus, and tuberculosis (Palacios et al., 2011; Wallis & Lee, 1999). Conversely, diseases transferred from primates to humans can have negative public health consequences. For example, Simian foamy viruses, a group of frequently transmitted retroviruses, have been detected in different human populations, including hunters (Gessain et al., 2013), laboratory and zoo workers (Switzer et al., 2004), and visitors to temples (Jones-Engel et al., 2007). The risk of disease transmission in both directions makes the need to understand human-primate interactions in fragmented landscapes all the more vital. However, primatologists themselves must be careful not to unknowingly put their study populations at risk. For example, genomic evidence suggests that apes, African and Asian monkeys, and some lemurs are likely to be highly susceptible to SARS-CoV-2 (Melin et al., 2020).

2.7 Conclusion and Future Directions

In this chapter, we have presented a relatively small proportion of the research done on primates in fragmented habitats. In reality, most primates, even those living in so-called pristine or undisturbed habitats, are affected by habitat loss and fragmentations at some scale. This is especially true in the context of global climate change, which is predicted to have significant negative effects on most primate habitats (Stewart et al., 2020; Chap. 6, this volume). While primates display some trends, such as the importance of habitat area and the ability to disperse between habitat patches, the overall picture of fragmentation research in primates is that it is complicated and species-specific, if not population- and/or site-specific. Some of this complication is due to theoretical questions and issues in study design and analyses. For example, are we studying primates at the right scale to identify the effects of habitat loss and fragmentation (Galán-Acedo et al., 2018; Jackson & Fahrig, 2012; Mcgarigal & Cushman, 2002; Moraga et al., 2019)? **Scale of effect** refers to the minimum size of an area at which we can observe variation in the response of an animal to changes in their environment. Issues with scale of effect occur when researchers study the ecology of an animal at a scale that is too small or too large to understand whether variation in some habitat condition can affect their ecology. For example, we may incorrectly conclude that access to fresh water is not important to the distribution of a primate species if the study was done at a small scale where all groups in the sample are located a short distance away from fresh water. To understand the effect that access to this resource has on the distribution of the species, it would be necessary to survey an area at a scale where some individuals in the sample have to put more effort into accessing water bodies than others. However, issues with scale of effect are not always considered when studying the ecology of primates in fragmented landscapes (Steffens et al., 2020). This issue is further complicated by the fact that primates also display a remarkable ecological diversity and flexibility in response to habitat loss and fragmentation, particularly in the face of a multitude of interconnected secondary effects occurring in fragmented landscapes.

Thus, their responses to these complex, interrelated secondary threats may also occur at varying scales of effect.

One area of primate fragmentation research that will likely grow in the future is the continued adoption of theoretical paradigms and mathematically rigorous modelling approaches from the general ecology literature. For example, the modelling of metapopulation dynamics in fragmented landscapes has only been applied to primates sparingly (Lawes et al., 2000; Mandujano & Escobedo-Morales, 2008; Steffens & Lehman, 2018) despite being widely used in other species for some time (Hanski, 1998). The application of methods like these to more primate species will certainly be an exciting addition to the current literature. However, rather than argue that primatologists should just be more quantitative, we suggest a more holistic approach. Primatologists should lean into the strengths of our field that was born out of the social sciences. Qualitative methods and new perspectives on quantifying human-animal interactions, such as those used in ethnoprimatology (Chap. 12, this volume), will provide many valuable insights into some of the outstanding issues in fragmentation research, especially if integrated alongside other approaches. This synthesis of multiple paradigms is not just an intellectual exercise. Given the alarming declines seen in primates across the world in increasingly fragmented landscapes, determining the effects of habitat loss and fragmentation on primates is vital for their continued survival.

Acknowledgments We wish to thank the editors for the invitation to submit this chapter. MSR was funded through a Vanier Canada Graduate Scholarship, and FMM was funded through a NSERC Postgraduate Scholarship during the writing of this manuscript.

References

Ahmed, T., & Naher, H. (2021). Population status of Northern Pig-tailed Macaque *Macaca leonina* in Satchari National Park, Bangladesh. *Asian Primates Journal, 9*(1), 32–40.

Ahsan, M. F. (2001). Socioecology of the hoolock gibbon (*Hylobates hoolock*) in two forests in Bangladesh. In Chicago Zoological Society (Eds.), *The Apes: challenges for the 21st century* (pp. 286–299). Chicago Zoological Society.

Anand, S., & Radhakrishna, S. (2020). Is human–rhesus macaque (*Macaca mulatta*) conflict in India a case of human–human conflict? *Ambio, 49*, 1685–1696.

Andriatsitohaina, B., Ramsay, M. S., Kiene, F., Lehman, S. M., Rasoloharijaona, S., Rakotondravony, R., & Radespiel, U. (2020). Ecological fragmentation effects in mouse lemurs and small mammals in northwestern Madagascar. *American Journal of Primatology, 82*(4), e23059.

Arroyo-Rodriguez, V., & Dias, P. A. D. (2010). Effects of habitat fragmentation and disturbance on howler monkeys: A review. *American Journal of Primatology, 72*, 1–16.

Arroyo-Rodriguez, V., Mandujano, S., & Benitez-Malvido, J. (2008). Landscape attributes affecting patch occupancy by howler monkeys (*Alouatta palliata mexicana*) at Los Tuxtlas, Mexico. *American Journal of Primatology, 70*, 69–77.

Arroyo-Rodriguez, V., & Fahrig, L. (2014). Why is a landscape perspective important in studies of primates? *American Journal of Primatology, 76*, 901–909.

Asensio, N., Cristóbal-Azkarate, J., Dias, P., et al. (2007). Foraging habits of *Alouatta palliata mexicana* in three forest fragments. *Folia Primatologica, 78*, 141–153.

Bergl, R. A., & Vigilant, L. (2007). Genetic analysis reveals population structure and recent migration within the highly fragmented range of the Cross River gorilla (*Gorilla gorilla diehli*). *Molecular Ecology, 16*, 501–516.

Bicca-Marques, J. C. (2003). How do howler monkeys cope with habitat fragmentation? In L. K. Marsh (Ed.), *Primates in fragments: Ecology and conservation* (pp. 283–303). Kluwer Academic/Plenum.

Bicca-Marques, J. C., & Calegaro-Marques, C. (1994). Exotic plant species can serve as stable food sources for wild howler populations. *Folia Primatologica, 63*, 209–211.

Bicca-Marques, J. C., Muhle, C. B., Prates, H. M., de Oliveira, S. G., & Calegaro-Marques, C. (2009). Habitat impoverishment and egg predation by *Alouatta caraya*. *International Journal of Primatology, 30*, 743–748.

Chapman, C. A., Wasserman, M. D., Gillespie, T. R., Speirs, M. L., Lawes, M. J., Saj, T. L., & Ziegler, T. E. (2006). Do food availability, parasitism and stress have synergistic effects on red colobus populations living in forest fragments? *American Journal of Physical Anthropology, 131*, 525–534.

Chapman, C. A., Struhsaker, T. T., Skorupa, J. P., Snaith, T. V., & Rothman, J. M. (2010). Understanding long-term primate community dynamics: Implications of forest change. *Ecological Applications, 20*, 179–191.

Chiarello, A. G. (2003). Primates of the Brazilian Atlantic forest: The influence of forest fragmentation on survival. In L. K. Marsh (Ed.), *Primates in fragments: Ecology and conservation* (pp. 99–122). Kluwer Academic/Plenum.

Clarke, G. M., & Young, A. G. (2000). Introduction: Genetics, demography and the conservation of fragmented populations. In A. G. Young & G. M. Clarke (Eds.), *Genetics, demography and viability of fragmented populations* (pp. 129–147). Cambridge University Press.

Cristobal-Azkarate, J., & Dunn, J. C. (2013). Lessons from Los Tuxtlas: 30 years of research into primates in fragments. In L. K. Marsh & C. A. Chapman (Eds.), *Primates in fragments: Complexity and resilience* (pp. 75–88). Springer.

Cristobal-Azkarate, J., Vea, J. J., Asensio, N., & Rodriguez-Luna, E. (2005). Biogeographical and floristic predictors of the presence and abundance of mantled howlers (*Alouatta palliata mexicana*) in rainforest fragments at Los Tuxtlas, Mexico. *American Journal of Primatology, 67*, 209–222.

Cristobal Azkarate, J., Dunn, J. C., Balcells, C. D., & Baro, J. V. (2017). A demographic history of a population of howler monkeys (*Alouatta palliata*) living in a fragmented landscape in Mexico. *PeerJ, 5*, e3547.

Debinski, D. M., & Holt, R. D. (2009). A survey and overview of habitat fragmentation experiments. *Conservation Biology, 14*(2), 342–355.

Decker, B. S. (1994). Effects of habitat disturbance on the behavioural ecology and demographics of the Tana River red colobus (*Colobus badius rufomitratus*). *International Journal of Primatology, 15*(5), 703–734.

DeGama-Blanchet, H. N., & Fedigan, L. M. (2006). The effects of forest fragment age, isolation, size, habitat type, and water availability on monkey density in a tropical dry forest. In A. Estrada, P. A. Garber, M. S. M. Pavelka, & L. Luecke (Eds.), *New perspectives in the study of Mesoamerican primates: Distribution, ecology, behaviour and conservation* (pp. 165–188). Springer.

Dhawale, A. K., Kumar, M. A., & Sinha, A. (2020). Changing ecologies, shifting behaviours: Behavioural responses of a rainforest primate, the lion-tailed macaque *Macaca silenus*, to a matrix of anthropogenic habitats in southern India. *PLoS One, 15*(9), e0238695.

Donati, G., Kesch, K., Ndremifidy, K., Schmidt, S. L., Ramanamanjato, J.-B., Borgognini-Tarli, S. M., & Ganzhorn, J. U. (2011). Better few than hungry: Flexible feeding ecology of collared lemurs *Eulemur collaris* in littoral forest fragments. *PLoS One, 6*(5), e19807.

Ehardt, C., Struhsaker, T., & Butynski, T. (1999). *Conservation of the endangered endemic primates of the Udzungwa Mountains, Tanzania: Surveys, habitat assessment, and long-term monitoring.* Unpublished report to the Margot Marsh Biodiversity Fund and World-Wide Fund for Nature-Tanzania.

Estrada, A., Garber, P. A., & Chaudhary, A. (2020). Current and future trends in socio-economic, demographic and governance factors affecting global primate conservation. *PeerJ, 8*, e9816.

Eppley, T. M., Santini, L., Tinsman, J. C., & Donati, G. (2020). Do functional traits offset the effects of fragmentation? The case of large-bodied diurnal lemur species. *American Journal of Primatology, 82*, e23104.

Fahrig, L. (1997). Relative effects of habitat loss and fragmentation on population extinction. The *Journal of Wildlife Management, 61*(3), 603–610.

Fahrig, L. (2003). Effects of habitat fragmentation on biodiversity. *Annual Review of Ecology, Evolution, and Systematics, 34*, 487–515.

Fahrig, L. (2017). Ecological responses to habitat fragmentation per se. *Annual Review of Ecology, Evolution, and Systematics, 48*, 1–23.

Ferrari, S. F., Santos, E. M., Jr., Freitas, E. B., Fontes, I. P., Souza-Alves, J. P., Jerusalinsky, L., Beltrão-Mendes, R., Chagas, R. R. D., Hilário, R. R., & Baião, S. A. A. (2013). Living on the edge: Habitat fragmentation at the interface of the semiarid zone in the Brazilian northeast. In L. K. Marsh & C. A. Chapman (Eds.), *Primates in fragments: Complexity and resilience* (pp. 121–136). Springer.

Galán-Acedo, C., Arroyo-Rodríguez, V., Estrada, A., & Ramos-Fernández, G. (2018). Drivers of the spatial scale that best predict primate responses to landscape structure. *Ecography, 41*(12), 2027–2037.

Galán-Acedo, C., Arroyo-Rodriguez, V., Cudney-Valenzuela, S. J., & Fahrig, L. (2019a). A global assessment of primate responses to landscape structure. *Biological Review, 94*, 1605–1618.

Galán-Acedo, C., Arroyo-Rodríguez, V., Andresen, E., Arregoitia, L. V., Vega, E., Peres, C. A., & Ewers, R. M. (2019b). The conservation value of human-modified landscapes for the world's primates. *Nature Communications, 10*, 152.

Gessain, A., Rua, R., Betsem, E., Turpin, J., & Mahieux, R. (2013). HTLV-3/4 and simian foamy retroviruses in humans: Discovery, epidemiology, cross-species transmission and molecular virology. *Virology, 435*, 187–199.

Gibbons, M. A., & Harcourt, A. H. (2009). Biological correlates of extinction and persistence of primates in small forest fragments: A global analysis. *Tropical Conservation Science, 2*(4), 388–403.

Goldberg, T. L., Gillespie, T. R., Rwego, I. B., et al. (2006). Killing of a pearl-spotted owlet (*Glaucidium perlatum*) by male red colobus monkeys (*Procolobus tephrosceles*) in a forest fragment near Kibale National Park, Uganda. *American Journal of Primatology, 68*, 1007–1011.

Gould, L., & Cowen, L. L. E. (2020). *Lemur catta* in small forest fragments: Which variables best predict population viability? *American Journal of Primatology, 82*(4), e23095.

Hanski, I. (1998). Metapopulation dynamics. *Nature, 396*(6706), 41–49.

Harcourt, A. H., & Doherty, D. A. (2005). Species-area relationships of primates in tropical forest fragments: A global analysis. *Applied Ecology, 42*(4), 630–637.

Harcourt, A. H., Coppeto, S. A., & Parks, S. A. (2002). Rarity, specialization and extinction in primates. *Journal of Biogeography, 29*(4), 445–456.

Harris, T. R., & Chapman, C. A. (2007). Variation in diet and ranging of black and white colobus monkeys in Kibale National Park, Uganda. *Primates, 48*, 208–221.

Horwich, R. H., Brockett, R. C., & Jones, C. B. (2001). Alternative male reproductive behaviours in the Belizean black howler monkey (*Alouatta pigra*). *Neotropical Primates, 8*(3), 95–98.

Hutchinson, G. E. (1957). Concluding remarks. Cold Spring Harbor Symposia on Quantitative Biology 22:415–427. Reprinted in: Classics in Theoretical Biology Series. *Bulletin of Mathematical Biology, 53*, 193–213.

Illia, G. A., Kowalewski, M., & Oklander, L. I. (2021). *Evidence of an infanticide in black-horned capuchin monkeys (Sapajus nigritus) in an Atlantic Forest remnant in Argentina*. Notas sibre Mamíferos Sudamericanos.

Irwin, M. T. (2008). Feeding ecology of *Propithecus diadema* in forest fragments and continuous forest. *International Journal of Primatology, 29*, 95–115.

Jackson, H. B., & Fahrig, L. (2012). What size is a biologically relevant landscape? *Landscape Ecology, 27*(7), 929–941.

Jones, C. B. (2005). The costs and benefits of behavioral flexibility to inclusive fitness: Dispersal as an option in heterogeneous regimes. In C. B. Jones (Ed.), *Behavioral flexibility in primates: Causes and consequences* (pp. 17–29). Springer.

Jones-Engel, L., Steinkraus, K. A., Murray, S. M., et al. (2007). Sensitive assays for simian foamy viruses reveal a high prevalence of infection in commensal, free-ranging, Asian monkeys. *Journal of Virology, 81*(14), 7330–7337.

Keinath, D. A., Doak, D. F., Hodges, K. E., Prugh, L. R., Fagan, W., Sekercioglu, C. H., Buchart, S. H. M., & Kauffman, M. (2017). A global analysis of traits predicting species sensitivity to habitat fragmentation. *Global Ecology and Biogeography, 26*, 115–127.

Klass, K., Van Belle, S., & Estrada, A. (2020a). Demographic population structure of black howler monkeys in fragmented and continuous forest in Chiapas, Mexico: Implications for conservation. *American Journal of Primatology, 82*(8), e23163.

Klass, K., Van Belle, S., Campos-Villanueva, A., Mercado Malabet, F., & Estrada, A. (2020b). Effects of variation in forest fragment habitat on black howler monkey demography in the unprotected landscape around Palenque National Park, Mexico. *PeerJ, 8*, e9694.

Kuussaari, M., Bommarco, R., Heikkinen, R. K., Helm, A., Krauss, J., Lindborg, R., Öckinger, E., Pärtel, M., Pino, J., Rodà, F., Stefanescu, C., Teder, T., Zobel, M., & Steffan-Dewenter, I. (2009). Extinction debt: A challenge for biodiversity conservation. *Trends in Ecology and Evolution, 24*(10), 564–571.

Lappan, S., Andayani, N., Kinnaird, M., Morino, L., Nuracahyo, A., & O'Brien, T. G. (2017). Social polyandry among siamangs: The role of habitat quality. *Animal Behavior, 133*, 145–152.

Laurance, W. F., Camargo, J. L. C., Fearnside, P. M., Lovejoy, T. E., Williamson, G. B., Mesquita, R. C. G., Meyer, C. F. J., Bobrowiec, P. E. D., & Laurance, S. G. W. (2017). An Amazonian rainforest and its fragments as a laboratory of global change. *Biological Review, 93*(1), 223–247.

Lawes, M. J., Mealin, P. E., & Piper, S. E. (2000). Patch occupancy and potential metapopulation dynamics of three forest mammals in fragmented Afromontane forest in South Africa. *Conservation Biology, 14*, 1088–1098.

Lawler, R. R. (2011). Demographic concepts and research pertaining to the study of wild primate populations. *Yearbook of Physical Anthropology, 54*, 63–85.

Lehman, S. M. (2006). Spatial variations in *Eulemur fulvus rufus* and *Lepilemur mustelinus* densities in Madagascar. *Folia Primatologica, 78*(1), 46–55.

de Luna, A. G., & Link, A. (2018). Distribution, population density and conservation of the critically endangered brown spider monkey (*Ateles hybridus*) and other primates of the inter-Andean forests of Colombia. *Biodiversity and Conservation, 27*, 3469–3511.

Mcgarigal, K., & Cushman, S. A. (2002). Comparative evaluation of experimental approaches to the study of habitat fragmentation effects. *Ecological Applications, 12*(2), 335–345.

Mandujano, S., Escobedo-Morales, L. A., & Palacios-Silva, R. (2004). Movements of *Alouatta palliata* among forest fragments in Los Tuxtlas, Mexico. *Neotropical Primates, 12*(3), 126–131.

Mandujano, S., & Escobedo-Morales, L. A. (2008). Population viability analysis of howler monkeys (*Alouatta palliata mexicana*) in a highly fragmented landscape in Los Tuxtlas, Mexico. *Tropical Conservation Science, 1*, 43–62.

Marsh, L. K. (Ed.). (2003). *Primates in fragments. Ecology in conservation*. Kluwer Academic/Plenum.

Marsh, L. K., & Chapman, C. (Eds.). (2013). *Primates in fragments: Complexity and resilience*. Springer.

Marshall, A. R., Rovero, F., & Struhsaker, T. T. (2016). Procolobus gordonorum. In N. Rowe & M. Myers (Eds.), *All the world's primates*. Pogonias Press.

Maurice, M. E., Fuashi, N. A., & Zeh, A. F. (2019). The effects of habitat fragmentation on the distribution of primates in the Kimbi-Fungom National Park, North West Region, Cameroon. *MOJ Ecology & Environmental Sciences, 4*, 60–68.

McCann, C., Williams-Guillen, K., Koontz, F., Roque Espinoza, A. A., Sánchez, M., & Koontz, C. (2003). Shade coffee plantations as wildlife refuge for mantled howler monkeys (*Alouatta palliata*) in Nicaragua. In L. K. Marsh (Ed.), *Primates in fragments* (pp. 321–341). Kluwer Academic.

Melin, A. D., Janiak, M. C., Marrone, F., III, Arora, P. S., & Higham, J. P. (2020). Comparative ACE2 variation and primate COVID-19 risk. *Communications Biology, 3*, 641.

Mendes-Pontes, A. R., & Soares, M. L. (2005). Sleeping sites of common marmosets in defaunated urban forest fragments: A strategy to maximize food intake. *Journal of Zoology, 266*, 55–63.

Mercado Malabet, F., Peacock, H., Razafitsalama, J., Birkinshaw, C., & Colquhoun, I. (2020). Realized distribution patterns of crowned lemurs (*Eulemur coronatus*) within a human-dominated forest fragment in northern Madagascar. *American Journal of Primatology, 82*(4), e23125.

Moraga, A. D., Martin, A. E., & Fahrig, L. (2019). The scale of effect of landscape context varies with the species' response variable measured. *Landscape Ecology, 34*, 703–715.

Morris, D. W. (1987). Ecological scale and habitat use. *Ecology, 68*(2), 362–369.

Morris, W. F., Altmann, J., Brockman, D. K., Cords, M., Fedigan, L. M., Pusey, A. E., Stoinski, T. S., Bronikowski, A. M., Alberts, S. C., & Strier, K. B. (2011). Low Demographic Variability in Wild Primate Populations: Fitness Impacts of Variation, Covariation, and Serial Correlation in Vital Rates. The American Naturalist, 177(1), E14–E28. https://doi.org/10.1086/657443

Nathan, R., Getz, W. M., Revilla, E., Holyoak, M., Kadmon, R., Saltz, D., & Smouse, P. E. (2008). A movement ecology paradigm for unifying organismal movement research. *PNAS, 105*(49), 19052–19059.

Nekaris, A. I., Boulton, A., & Nijman, V. (2013). An ethnoprimatological approach to assessing levels of tolerance between human and commensal non-human primates in Sri Lanka. *Journal of Anthropological Science, 91*, 1–14.

Oklander, L., & Corach, D. (2013). Kinship and dispersal patterns in *Alouatta caraya* inhabiting continuous and fragmented habitats of Argentina. In L. K. Marsh & C. A. Chapman (Eds.), *Primates in fragments: Complexity and resilience* (pp. 399–412). Springer.

Onderdonk, D., & Chapman, C. A. (2000). Coping with forest fragmentation: The primates of Kibale National Park, Uganda. *International Journal of Primatology, 21*(4), 587–611.

Palacios, G., Lowenstine, L. J., Cranfield, M. R., Gilardi, K. V. K., Spelman, L., Lukasik-Braum, M., Kinani, J.-F., Mudakikwa, A., Nyirakaragire, E., Bussetti, A. V., Savji, N., Hutchison, S., Egholm, M., & Lipkin, W. I. (2011). Human metapneumovirus infection in wild mountain gorillas, Rwanda. *Emerging Infectious Diseases, 17*(4), 711–713.

Pebsworth, P. A., Gawde, R., & Bardi, M. (2021). To kill or not to kill?: Factors related to people's support of lethal and non-lethal strategies for managing monkeys in India. *Human Dimensions of Wildlife, 26*(6), 541–558.

Pozo-Montuy, G., Serio-Silva, J. C., & Bonilla-Sanchez, Y. M. (2011). Influence of the landscape matrix on the abundance of arboreal primates in fragmented landscapes. *Primates, 52*, 139–147.

Pride, R. E. (2005). High faecal glucocorticoid levels predict mortality in ring-tailed lemurs (*Lemur catta*). *Biology Letters, 1*(1), 60–63.

Ramilison, M. L., Andriatsitohaina, B., Chell, C., Rakotondravony, R., Radespiel, U., & Ramsay, M. S. (2021). Distribution of the critically endangered Coquerel's sifaka (*Propithecus coquereli*) across a fragmented landscape in NW Madagascar. *African Journal of Ecology, 59*(2), 350–358.

Ramsay, M. S., Morrison, B., & Stead, S. M. (2020). Infanticide and partial cannibalism in free-ranging Coquerel's sifaka (*Propithecus coquereli*). *Primates, 61*, 575–581.

Rangel-Negrín, A., Coyohua-Fuentes, A., Chavira, R., Canales-Espinosa, D., & Dias, P. A. D. (2014). Primates living outside protected habitats are more stressed: The case of black howler monkeys in the Yucatán Peninsula. *PLoS One, 9*(11), 1–8.

Rudran, R., & Fernandez-Duque, E. (2003). Demographic changes over thirty years in a red howler population in Venezuela. *International Journal of Primatology, 24*(5), 924–947.

Ren, B., Li, D., Liu, Z., Li, B., Wei, F., & Li, M. (2009). First evidence of prey capture and meat eating by wild Yunnan snub-nosed monkeys *Rhinopithecus bieti* in Yunnan, China. *Current Zoology, 56*, 227–231.

Sato, H., Santini, L., Patel, E. R., Campera, M., Yamashita, N., Colquhoun, I. C., & Donati, G. (2016). Dietary flexibility and feeding strategies of *Eulemur*: A comparison with *Propithecus*. *International Journal of Primatology, 37*(1), 109–129.

Schreier, A., & Swedell, L. (2008). Use of palm trees as a sleeping site for hamadryas baboons (*Papio hamadryas hamadryas*) in Ethiopia. *American Journal of Primatology, 70*, 107–113.

Singh, M. (2019). Management of forest-dwelling and urban species: Case studies of the lion-tailed macaque (*Macaca silenus*) and the Bonnet macaque (*M. radiata*). *International Journal of Primatology, 40*, 613–629.

Singh, M., Singh, M., Kumar, M. A., Kumara, H. N., Sharma, A. K., & Kaumanns, W. (2002). Distribution, population structure, and conservation of lion-tailed macaques (*Macaca silenus*) in the Anaimalai Hills, Western Ghats, India. *American Journal of Primatology, 57*, 91–102.

Silver, S. C., & Marsh, L. K. (2003). Dietary flexibility, behavioral plasticity, and survival in fragments: Lessons from translocated howlers. In L. K. Marsh (Ed.), *Primates in fragments* (pp. 251–265). Kluwer Academic.

Skorupa, J. P. (1988). *The effect of selective timber harvesting on rain forest primates in Kibale Forest, Uganda* (Dissertation). University of California.

Steffens, T. S., & Lehman, S. M. (2018). Lemur species-specific metapopulation responses to habitat loss and fragmentation. *PLoS One, 13*(5), e0195791.

Steffens, T. S., & Lehman, S. M. (2019). Species-area relationships of lemurs in a fragmented landscape in Madagascar. *American Journal of Primatology, 81*(4), e22972.

Steffens, T. S., Mercado Malabet, F., & Lehman, S. M. (2020). Occurrence of lemurs in landscapes and their species-specific scale responses to habitat loss. American Journal of Primatology, 82(4), e23110. https://doi.org/10.1002/ajp.23110

Steffens, T. S., Ramsay, M. S., Andriatsitohaina, B., Radespiel, U., & Lehman, S. M. (2021). Enter the Matrix: Use of Secondary Matrix by Mouse Lemurs. Folia Primatologica, 92(1), 1–11. https://doi.org/10.1159/000510964

Stewart, B. M., Turner, S. E., & Matthews, H. D. (2020). Climate change impacts on potential future ranges of non-human primate species. *Climatic Change, 162*(4), 2301–2318.

Switzer, W. M., Bhullar, V., Shanmugam, V., Cong, M., Parekh, B., Lerche, N. W., Yee, J. L., Ely, J. J., Boneva, R., Chapman, L. E., Folks, T. M., & Heneine, W. (2004). Frequent simian foamy virus infection in persons occupationally exposed to nonhuman primates. *Journal of Virology, 78*, 2780–2789.

Taylor, R. A., Ryan, S. J., & Brashares, J. S. (2016). Hunting, food subsidies, and mesopreda-tor release: The dynamics of crop-raiding baboons in a managed landscape. *Ecology, 97*(4), 951–960.

Uddin, M. M., Ahsan, M. F., & Lingfeng, H. (2020). Human–primate conflict in Bangladesh: A review. *Journal of Animal and Plant Sciences, 30*(2), 280–287.

Umapathy, G., & Kumar, A. (2003). Impacts of forest fragmentation on lion-tailed macaque and Nilgiri Langur in Western Ghats, South India. In L. K. Marsh (Ed.), *Primates in fragments: Ecology and conservation* (pp. 163–190). Kluwer Academic/Plenum.

Umapathy, G., Hussain, S., & Shivaji, S. (2011). Impact of habitat fragmentation on the demography of lion-tailed macaque (*Macaca silenus*) populations in the rainforests of Anamalai Hills, Western Ghats, India. *International Journal of Primatology, 32*, 889–900.

Villard, M., Metzger, J. P., & Saura, S. (2014). Beyond the fragmentation debate: A conceptual model to predict when habitat configuration really matters. *Journal of Applied Ecology, 51,* 309–318.

Wallis, J., & Lee, D. R. (1999). Primate conservation: The prevention of disease transmission. *International Journal of Primatology, 20,* 803–826.

Wiens, J. A., Stenseth, N., Van Horne, B., & Ims, R. A. (1993). Ecological mechanisms and landscape ecology. *Oikos, 66*(3), 369–380.

Wong, S. N. P., & Sicotte, P. (2006). Population size and density of *Colobus vellerosus* at the Boabeng-Fiema Monkey Sanctuary and surrounding forest fragments in Ghana. *American Journal of Primatology, 68,* 465–476.

Chapter 3
The Emerging Importance of Regenerating Forests for Primates in Anthropogenic Landscapes

Lucy Millington, Onja H. Razafindratsima, Tracie McKinney, and Denise Spaan

Contents

Abstract Habitat loss is the greatest threat to primate survival. However, land altered for logging or agricultural developments is often abandoned and can regenerate after use. These regenerating forests are critical for the future of primate conservation as they provide habitats and connectivity between mature forest fragments. They can also contribute to climate change mitigation. In this chapter, we introduce what constitutes a regenerating forest, how widespread they are, and how secondary succession varies depending on disturbance history and ecological characteristics.

L. Millington (✉)
Department of Natural Sciences, Manchester Metropolitan University, Manchester, UK
e-mail: Lucy.Millington@stu.mmu.ac.uk

O. H. Razafindratsima
Department of Integrative Biology, University of California Berkeley, Berkeley, CA, USA

T. McKinney
School of Applied Sciences, University of South Wales, Pontypridd, UK

D. Spaan
Instituto de Neuroetología, Universidad Veracruzana, Xalapa, Mexico

ConMonoMaya, A.C., Chemax, Mexico

We also examine the role primate seed-dispersal plays in forest regeneration: from the transportation of seeds to changes that occur within a primate's gut that facilitate germination and impacts on plant communities. We consider how primates might cope with living in a regenerating forest, in terms of behavioral plasticity, from changes in diet to ranging patterns or group cohesion. We argue that the study of primates in regenerating forests is currently lacking and will be pivotal for future primate conservation planning.

Keywords Behavioral flexibility · Conservation · Forest regeneration · Reforestation · Secondary succession · Seed dispersal

3.1 Introduction to Regenerating Forests

Deforestation is converting forests into anthropogenically-modified landscapes at alarming rates, and the associated habitat loss is one of the greatest threats to primate populations (Estrada et al., 2017). Examples of anthropogenic drivers of deforestation in the tropics include clearance of land for agriculture and cattle ranching and logging for timber or mining (Coomes, 1995). While the global rate of deforestation has slowed over the last 30 years, rates continue to rise in many primate range countries such as Madagascar, Indonesia, and Brazil (Estrada et al., 2018; Kalbitzer & Chapman, 2018). The sheer volume of past and projected forest loss is staggering – 178 million ha of forest has been lost globally between 1990 and 2019 (FAO, 2020). This loss of tropical forests is detrimental to forest-dependent species, including the 78% of primate species that are strictly arboreal (Galán-Acedo et al., 2019a). These species provide important ecosystem services (e.g., seed dispersal, pollination; Arroyo-Rodríguez et al., 2017a) in addition to the services provided by the continued presence of standing forests (e.g., climate regulation; Diaz et al., 2006). The synergistic effects of losing standing forests and key species will affect the resilience of these ecosystems to climate change (Sales et al., 2020).

Regenerating forests, which are forests that grow on land that was previously deforested, are fast becoming important refuges for primates in the twenty-first century as they are the dominant land cover in anthropogenically-modified landscapes (Arroyo-Rodríguez et al., 2017a). In fact, many primate species use regenerating forests across the different primate-habitat regions (Fig. 3.1). Although 90% of remaining forests worldwide are classified as "naturally regenerating" (FAO, 2020), this is a misleading statistic as "naturally regenerating" is a broad term referring to a wide range of forests in different stages of succession. Some are highly homogenous and unable to sustain high levels of biodiversity, whereas others may be structurally and compositionally heterogeneous with relatively high levels of biodiversity, although often not as high as mature forests (Barlow et al., 2007).

Many terms are used interchangeably for regenerating forests including secondary, successional, second growth, and regrowth (Chokkalingam & de Jong, 2003). Similarly, mature forest has been referred to as primary, old growth, mature, and

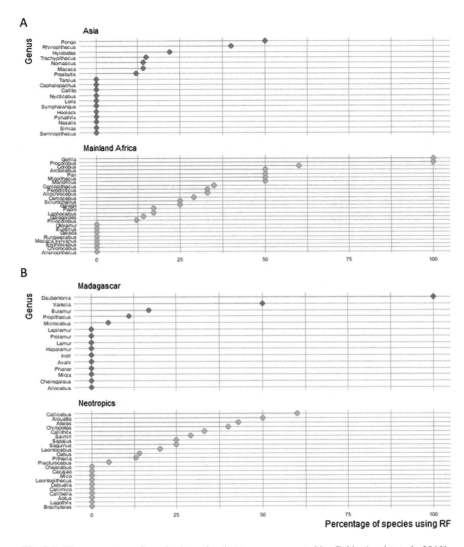

Fig. 3.1 The percentage of species in each primate genus reported by Galán-Acedo et al., 2019b, to use regenerating forests (RF) by region: (**a**) Asia and mainland Africa and (**b**) Madagascar and the Neotropics. (Taxonomy following Estrada et al. 2017)

virgin (Clark, 1996). For the purposes of this chapter, we will be using "regenerating forest" because other terms such as "secondary forest" imply poor quality. We will also use "**mature forest**" because terms like "primary/virgin" do not accurately reflect the status of many forests that have been inhabited by Indigenous people for millennia, who may have influenced the structure of tropical forests (White & Oates, 1999).

The structural and compositional complexity of regenerating forests, along with their abundance, makes them important areas that should be prioritized for biodiversity conservation (Chapman et al., 2020). Although the importance of studying primates in regenerating forests has been highlighted recently (Chapman et al., 2020), it is still an emerging field. In this chapter, we summarize the available information and identify gaps in our current knowledge on this topic to encourage future research in this field. In particular, we distinguish between active and passive regeneration practices and place forest regeneration into the context of landscape-scale ecology. We then focus on the process of forest regeneration through secondary succession and explore how primates contribute to forest regeneration through seed dispersal. We then examine the behavioral and dietary modifications that primates employ to survive in regenerating forests. Finally, we conclude with why regenerating forests are important habitats and essential for the conservation of primate populations.

Forests can regenerate on lands that have been deforested as a result of natural disturbances (e.g., hurricanes or forest fires) or anthropogenic activities (e.g., abandonment of cattle pasture or cropland; FAO, 2020), and this can happen at different spatial scales. For instance, an area cleared for agriculture may become less fertile and, therefore, lose economic value over time. Once abandoned, this area may be able to regenerate naturally through the process of secondary succession (see Sect. 3.3). On the other hand, small-scale disturbances such as tree-fall gaps, occurring when trees naturally fall, allow light to penetrate the canopy, leading to changes in plant species composition in small areas (Whitmore, 1989). These are examples of passive forest regeneration. In addition to passive regeneration, seeds can be planted to restore an area through **reforestation**, also referred to as active regeneration or active reforestation. Unfortunately, active reforestation can be costly (Corbin & Holl, 2012) and time-consuming and is met with mixed success, as levels of biodiversity in patches may remain low for decades (Wheeler et al., 2016). Active reforestation often takes place within an agroforestry context in the form of live fences and/or tree plantations (Lazos-Chavero et al., 2016) and may include exotic species that are economically beneficial to local communities.

3.2 Regenerating Forest as a Part of the Matrix

In this section, we will examine regenerating forests within the context of landscape ecology. Anthropogenically-modified landscapes are matrices comprised of different land covers, containing important habitats and resources for primates (Galán-Acedo et al., 2019b). These may include logged areas, agricultural fields, plantations, urban areas, waterbodies, stands of regenerating forests, and mature forest patches, among others.

At the landscape scale, regenerating forests can act as corridors bridging the gap between mature forest patches that have become isolated following habitat fragmentation (Raboy et al., 2004). These corridors facilitate dispersal, offer access to seasonal food resources not available in remnant mature forest patches, and provide sleeping sites (Asensio et al., 2009; Baguette & Van Dyck, 2007). Planted corridors are often comprised of both exotic and native plant species, and primates use them both to travel between patches and as food sources (lion-tailed macaques, *Macaca silenus*, Singh et al., 2001; lemurs, Ganzhorn, 1987; Ganzhorn et al., 1999; platyr-rhines, Luckett et al., 2004; slow lorises, *Nycticebus javanicus*; Nekaris et al., 2017). Ideally, for these corridors to be used by primates, they should contain fast-growing tree species and species that provide food resources (Ganzhorn, 1987; Ganzhorn et al., 1999; Singh et al., 2001). Enriching the corridor with trees of economic value to local communities (e.g., trees that can be used as timber) can ensure that corridors provide use to both humans and primates (Estrada, 2013).

Regenerating forests are important habitats for primates within anthropogenically-modified landscapes as they may provide an immediate refuge following habitat disturbance/modification. They may also contribute to the long-term maintenance of populations, particularly in landscapes composed largely of regenerating forests. For instance, many primate-habitat regions in Mesoamerica have a long-standing history of slash-and-burn agriculture, and the remaining habitats are a mosaic of mature forest patches surrounded by forests in different stages of succession (DeClerck et al., 2010). Primates in these regions use regenerating forests not only as corridors but also for feeding and other activities (Ramos-Fernández & Ayala-Orozco, 2003; Rodrigues, 2017). It is important to study primates in such landscapes to establish whether species and populations can persist without or with minimal access to mature forests as their core habitat.

By facilitating forest regeneration, we ensure that these forests will survive for both primates and humans alike. Sometimes, regenerating forests are not protected from degradation as they are considered of low conservation concern and may exist outside of protected areas. While protected areas are at the forefront of conservation planning, they often fall short in terms of their size and level of protection. For instance, currently, only 18% of all forests fall within protected areas (FAO, 2020). More needs to be done to protect forests for primates and biodiversity as a whole, including promoting forest regeneration and protecting regenerating forests from degradation. Forest regeneration is a fast way to regain the properties that make forests so important – as habitats for wildlife and for climate change mitigation through carbon storage.

3.3   Successional Pathways

In this section, we will introduce the ecological processes that underpin forest regeneration, discussing the effect that various pathways have on the structure and composition of regenerating forests. **Secondary succession** in tropical forests refers to the change in dominant species occupying an area over time, after that area has experienced a disturbance (Chazdon, 2014). It is therefore the process by which a deforested area regenerates. Directly after an anthropogenic or natural disturbance clears vegetation from the land, fast-growing, light-demanding (shade-intolerant) tree species, grasses, vines, and shrubs establish in the area (Guariguata & Ostertag, 2001). This stage of secondary succession is often referred to as the "pioneer stage" (Chazdon, 2014). Throughout the early and late stages of secondary succession (5–20 years after disturbance), pioneer species grow in height and start closing the forest canopy. As tree canopies expand, less light reaches the forest floor and prevents seedlings of shade-intolerant species from establishing (Chazdon, 2014). As such, pioneer species start being replaced by shade-tolerant species, and the regenerating forest contains both shade-tolerant and shade-intolerant species (Chazdon, 2014). At this stage of secondary succession, the regenerating forest patch is structurally similar to mature forests with tree diameter, height, and biomass assimilating that of mature forests (Chazdon, 2014). Although species richness (the number of species) may be similar to mature forests 40 years after a disturbance (Aide et al., 2000), species composition can take hundreds of years to approach mature forest levels (Dent & Wright, 2009; Guariguata & Ostertag, 2001). For example, the density of *Brosimum alicastrum*, a tree species important in the diet of Geoffroy's spider monkeys (*Ateles geoffroyi*), occurred at 288 individuals per hectare in mature forests compared to <1 individual per hectare in regenerating forest <40 years old (Ramos-Fernández & Ayala-Orozco, 2003). During the "climax" stage of succession, large-seeded tree species increase in abundance, and the canopy becomes dominated by shade-tolerant species. A regenerating forest patch can be considered "old growth" when its structure and species composition are relatively stable (Chazdon, 2014).

The process by which plant species establish in a disturbed area can follow many different successional pathways (Arroyo-Rodríguez et al., 2017a). These pathways are influenced by a mixture of disturbance history (type, frequency, and duration of the disturbance), biotic (e.g., presence of vertebrate seed dispersers, quantity of seeds in the seed bank) and abiotic factors (e.g., light availability, climate, soil quality) (Chazdon, 2014; Guariguata & Ostertag, 2001). These factors are particular to each forest stand (Guariguata & Ostertag, 2001), and, as such, no two regenerating forest plots will follow exactly the same successional pathway (Arroyo-Rodríguez et al., 2017a). Forest patches of the same age may therefore differ in their composition. Changes in plant species composition during forest succession may affect primates through changes in the availability of feeding trees and resting sites, favoring some species and providing a challenge for others. For instance, tamarins (*Saguinus mystax* and *Leontocebus nigrifrons*) increased both the time spent foraging on fruits and the number of species consumed in a regenerating forest over an almost 20-year period, likely the result of increasing habitat quality as the forest aged (Heymann et al., 2019). Similarly, Sorensen and Fedigan (2000) found that the biomass of primate food sources increased as regenerating forest aged.

It is important to consider spatial scale in studies of the effect of forest regeneration on primates. Successional pathways in a particular forest patch depend on landscape-scale factors, such as the amount of mature forest in the landscape, the connectivity between forest patches, and the type of landscape **matrix** – i.e., the different land covers surrounding forest patches, such as cropland, urban settlements, and waterbodies (Arroyo-Rodríguez et al., 2017a). These factors will determine the likelihood of seed dispersers, including primates, arriving at a regenerating forest patch; some primate species that frequent degraded habitats promote the establishment of diverse plant species in these areas by carrying seeds from mature forest patches (Kaplin & Lambert, 2002; Martinez & Razafindratsima, 2014; Wunderle Jr, 1997). Regenerating forest patches close to or connected to mature forest are, thus, more likely to contain mature forest species than isolated patches (Dent & Wright, 2009). It is vitally important to study forest regeneration and its effects on primates at the landscape scale (Arroyo-Rodríguez et al., 2017a), as primates depend on and contribute to the structure and composition of regenerating forests in their environments, yet studies of this nature are lacking.

3.4 The Importance of Primate Seed Dispersal in Forest Regeneration

In this section, we introduce the mechanisms of seed dispersal and how primates contribute to forest regeneration. Seed dispersal is a fundamental process in a plant's lifecycle and is vital for both ecosystem resilience and functioning (Howe & Smallwood, 1982; Wang & Smith, 2002). It plays a key role in influencing plant fitness and demography dynamics. Seeds dispersed away from the parent plant can escape factors that may increase mortality in the vicinity of the parent, such as increased intraspecific competition and interactions with natural enemies (Howe & Smallwood, 1982; Razafindratsima & Dunham, 2015; Wang & Smith, 2002).

Seed dispersal also plays an important role in natural succession and the regeneration of newly formed habitats. It helps plants establish in new sites and influences the composition of the future vegetation community (Howe & Miriti, 2004; Razafindratsima & Dunham, 2016; Schupp & Fuentes, 1995). Thus, it can facilitate the natural regeneration of degraded landscapes (Chazdon & Guariguata, 2016), consequently reducing management costs associated with active reforestation projects (Farwig & Berens, 2012). While some regenerating forests may benefit from seeds that are already within the soil (seed bank), in most cases, succession depends on the arrival of seeds from mature forests that are often dispersed by frugivores (Alvarez-Buylla & Martínez-Ramos, 1990; Duncan & Chapman, 1999; Guariguata & Ostertag, 2001). Some plant species are even unable to reach and establish in certain sites without animal seed dispersers (Albert et al., 2015; George & Bazzaz, 1999; Myers & Harms, 2011). Often, large-seeded plant species rely on large primate frugivores for their dispersal (Balcomb & Chapman, 2003; Kitamura et al., 2002; Peres et al., 2016). For example, in Kibale National Park, Uganda, the seeds of the plant species *Monodora myristica* are primarily dispersed by three large-bodied primates (chimpanzees, *Pan troglodytes*; baboons, *Papio anubis*; and gray-cheeked mangabeys, *Lophocebus albigena*) because these are the only frugivores able to open their hard-husked fruits (Balcomb & Chapman, 2003). The absence of large-bodied frugivorous species can, therefore, result in limited dispersal and recruitment of large-seeded plants (Cordeiro & Howe, 2001; Wotton & Kelly, 2011). Regenerating fragments where large-bodied primate frugivores are absent have fewer primate-dispersed plant species and are dominated by plants that are dispersed by other means (e.g., by birds or through abiotic means such as wind or ballistic ejection; Cordeiro & Howe, 2001; Ganzhorn et al., 1999; Effiom et al., 2013). This change in plant species composition and density within the community highlights the importance of primate-mediated seed dispersal in regenerating forests to maintain similar plant communities to remnant forest patches.

Primates are critical seed dispersal agents of many angiosperms in tropical forest ecosystems (Lambert & Garber, 1998; Razafindratsima et al., 2018; Sussman, 1991). Besides playing an essential role in transporting seeds into the regenerating forest, primates can also affect the probabilities of germination and recruitment of the seeds that they disperse through **endozoochory** (in which seeds are ingested, pass through animal gut, and get dispersed via defecation). For instance, seeds ingested and defecated by frugivorous lemurs in Madagascar were found to have higher germination rates and increased seedling growth than seeds that did not pass through their guts (Dew & Wright, 1998; Ramananjato et al., 2020; Razafindratsima & Martinez, 2012). The removal of fruit pulp from the seed by a frugivore can release it from potential germination inhibitors (Fuzessy et al., 2016; Traveset et al., 2007). The mechanical and/or chemical scarification of the seed in the gut can also enhance germination and break **dormancy** – i.e., the state in which the seed is unable to germinate under normal physical environmental conditions (Fuzessy et al., 2016; Traveset et al., 2007). In some cases, if seeds are not processed by animals, they may rot or be unable to break dormancy (Wunderle Jr, 1997). In addition, the fecal material accompanying the seed can act as a fertilizer facilitating seedling growth (Fuzessy et al., 2016; Traveset et al., 2007).

Encouraging primate frugivores to bring seeds from a mature forest into regenerating forests is pivotal in accelerating secondary succession; yet, it is an area of research that has received little attention (Chapman et al., 2020). Sites to be reforested (restoration sites) should have characteristics that attract primate seed dispersers, including the availability of key food resources and structural requirements such as refuge from predators (Duncan & Chapman, 1999; Wunderle Jr, 1997). This can be achieved by developing and/or maintaining forest or plantation corridors between the restoration site and the mature forest as well as planting key species and/or establishing human-made structures in the restoration site (Howe, 2016; Wunderle Jr, 1997). Restoration sites with remnant trees or early establishing species can receive more **zoochorous seed rain** (i.e., seeds dispersed by animals) than sites without trees because they offer resources for seed dispersers (Duncan & Chapman, 1999; Guariguata & Ostertag, 2001; Reid et al., 2015). These remnant and early establishing trees can also serve as a seed source and facilitate the movement of seed-dispersing animals (Chazdon et al., 2009a; Holl & Aide, 2011). For example, a restoration program established in the Masoala National Park (northeastern Madagascar) successfully attracted red-ruffed lemurs (*Varecia rubra*) to disperse seeds from the mature forest into regenerating forest patches by planting key food sources (Holloway, 2000, 2004; Martinez & Razafindratsima, 2014).

To identify how to incorporate primate seed dispersal into restoration efforts, Chapman and Dunham (2018) posed five fundamental questions to guide and direct these efforts. These questions included identifying which primate species use regenerating forests and for what purpose, how to encourage them to use regenerating forests, which seeds are dispersed, and what the fate of these seeds is at various stages in their lifecycle. Identifying these variables is important as restoration strategies are often context-dependent (Chapman & Dunham, 2018; Chazdon et al., 2009a; Chazdon & Guariguata, 2016). Additionally, understanding which aspects of regenerating forests may attract primates to visit them is important to increase seed rain and diversify the plant communities of these forests.

3.5 Primate Behavioral Ecology in Regenerating Forests

The vast ecological diversity in the order Primates makes describing the varied ways in which regenerating forests are used a challenge. In many cases, regenerating forests form part of the landscape matrix through which primates travel between mature forest patches, where they sleep or forage (Gascon et al., 1999). Regenerating forests can provide a more suitable substrate for travel than pastures or agroecosystems, allowing for better connectivity in a heterogeneous landscape (Anderson et al., 2007; Galán-Acedo et al., 2019b). Some taxa make greater use of this regenerating forest matrix than others through **landscape supplementation** – or foraging in the spaces surrounding their habitat patch – which can constitute a substantial amount of an animal's diet (Asensio et al., 2009). Finally, some primates, including howler monkeys (genus *Alouatta*) and some chimpanzee (genus *Pan*) populations, spend the majority of their time in regenerating forests (Arroyo-Rodríguez et al., 2017b; Bicca-Marques, 2003; Bryson-Morrison et al., 2016).

The howlers and colobus monkeys (genus *Colobus*) are informally perceived as regenerating forest specialists, but the literature for these genera usually describe them using a complex, heterogeneous landscape that includes mature forests, farmlands, and other modified landscapes in addition to regenerating forests; roughly 50% of species from each genus are known to live in regenerating forests (Fig. 3.1; Galán-Acedo et al. 2019b). Both groups are adapted for folivory but can use a range of food resources according to seasonal availability (Lambert, 2007; Nowak & Lee, 2013). Despite the broad trend of success in these genera, survival in anthropogenic landscapes is not guaranteed, nor is it exclusive to these highly folivorous primates. The black-and-white colobus (*Colobus guereza*), for example, are more abundant in regenerating forests and edge habitats (Harris and Chapman 2007) and survive well in logged forests (Chapman et al., 2017). However, where black-and-white colobus populations show increased densities in logged forests, red colobus (*Piliocolobus tephrosceles*) populations in the same forests decline, even though the two species rely on many of the same food resources (Isabirye-Basuta & Lwanga, 2008). Longitudinal records at Kibale National Park, Uganda, make it clear that primate abundance patterns are complex and dependent on a wide variety of variables

(Chapman et al., 2017). In the Americas, howler monkeys (genus *Alouatta*) survive well in regenerating forests; most studies of howler monkeys are in regenerating forest or other anthropogenically-modified landscapes (Bicca-Marques, 2003), and they are among the first primates to return to regenerating forests following deforestation events (Sorensen & Fedigan, 2000). However, other platyrrhines also frequent regenerating forests, including capuchins (genus *Cebus*) (Fedigan & Jack, 2001; Sorensen & Fedigan, 2000), titi monkeys (genus *Callicebus*) (Heiduck, 2002), and even spider monkeys (genus *Ateles*) – large frugivores that are often perceived as ecologically sensitive (Arroyo-Rodríguez et al., 2017b; Chaves et al., 2011). In fact, at least 30% of the spider monkey home range in Punta Laguna, Mexico, is made up of regenerating forest <26 years of age (Ramos-Fernández et al., 2013).

Whether a primate species can survive in regenerating forests is often attributed to the species' perceived ecological flexibility. It is widely assumed that generalist species will survive better than specialist species in changing habitats, but in fact there is no clear link between specialization and extinction risk (Nowak & Lee, 2013). Labels such as "folivore" or "frugivore" may mask a primate's true dietary flexibility, and specialist primates are probably less constrained by dietary preferences than we think – most primates switch their primary foods as resource availability fluctuates throughout the year (Lambert, 2007; Nowak & Lee, 2013). Likewise, behavioral flexibility may be overlooked in less-studied species; animals considered to be strict habitat specialists, for example, often prove to be more flexible than previously believed upon further study (Hansen et al., 2020; Nowak & Lee, 2013). When discussing primate plasticity, the most commonly reported behavioral variation is a shift in diet, followed by sociological adjustments such as changing ranging patterns or habitat use (McLennan et al., 2017; Nowak & Lee, 2013). With such flexibility, it is no surprise that many primates can alter their dietary intake to exploit plant species commonly found in regenerating forests (Bicca-Marques, 2003; Bryson-Morrison et al., 2016).

Our understanding of the effects of regenerating forest on diet or behavioral patterns is limited to comparative studies between animals in "disturbed" landscapes and their nearby conspecifics in intact mature forests. Primates in anthropogenically-modified landscapes often show a decrease in travel and foraging time and increase in resting time due to patchily distributed resources (McKinney, 2019). Primates living in forest fragments must reduce their ranging patterns in response to available space, but long-term success of these populations seems dependent on their ability to use the matrix surrounding fragments (Bicca-Marques, 2003). While comparative studies are immensely useful for our understanding of behavioral and ecological flexibility, it is worth noting that their findings are likely site-specific, and patterns highlighted may be confounded by other factors like climate change and not influenced directly by forest structure (Isabirye-Basuta & Lwanga, 2008; Nowak & Lee, 2013; Chap. 6, this volume). We are also beginning to look at landscapes in a more nuanced manner, recognizing that there is a continuum of forest successional stages and anthropogenic modification (Chazdon, Peres, et al., 2009b). Landscape research has been heavily influenced by the "island biogeography" models of the 1960s and has, therefore, focused primarily on forest patches; today, an integration of matrix landscapes and varying forest structures is sorely needed (Galán-Acedo et al., 2019c). Long-term studies of primates in regenerating forests are rare (Chazdon, Peres, et al., 2009b) but are essential for understanding the interactions of complex ecological communities (Chapman et al., 2017; Isabirye-Basuta & Lwanga, 2008). Determining whether primate presence in regenerating forests is due to a preference or adaptation for this forest type, or simply because the primates have been pushed there through habitat loss, is an important goal that will require long-term study of multiple populations of diverse primate taxa.

3.6 Conclusion

This chapter highlights the gaps in our knowledge about how and when primates use regenerating forests. At present, only a few genera are regularly described as using regenerating forests, mostly African and Neotropical species, although a recent study highlighted that a diverse array of primate taxa use them (Galán-Acedo et al., 2019b). There are, however, large gaps in our knowledge on the use of regenerating forests by Asian species, lemurs, and nocturnal primates (but see Bersacola et al., 2019; Ganzhorn et al., 1999; Martinez & Razafindratsima, 2014). Although maintaining mature forests and halting deforestation should remain the priority for conservation management plans, greater attention must be paid to the potential of regenerating forests in primate conservation (Chapman et al., 2020).

The literature on animals outside of mature forests typically considers a range of habitat types, collectively labelled the matrix, and is therefore not specific to regenerating forests. This chapter highlights the importance of exploring subtle gradients of landscape types and moving beyond the dichotomy of "natural" versus "unnatural" primate habitats. It is clear that regenerating forests are a dominant land cover globally and, as such, will become increasingly essential for primate conservation; forest regeneration on abandoned lands can create corridors and habitats and contribute to climate change mitigation. Forest regeneration to levels of diversity and structure similar to mature forests is possible given the right circumstances (e.g., the presence of seed dispersers, soil that is still viable). However, without a concerted effort to protect these later successional forests from further degradation, their ability to sustain primate populations will be diminished. This limits the influx of seed dispersers, affecting future secondary succession as well as reducing the overall habitat available for primates within anthropogenically-modified landscapes. While primate seed dispersal can facilitate forest regeneration, this process is slow, and, in some cases, a combined approach of passive and active regeneration may be necessary. In cases where the matrix is impermeable to large seed-dispersing primates, active planting interventions may be necessary to facilitate movement and aid seed dispersal into abandoned areas. Further study is needed to determine the long-term success of primate-dispersed seeds in regenerating forests and abandoned plots, and under which conditions, primates are able to disperse seeds into different areas to establish which ecological factors encourage forest restoration/regeneration in abandoned areas.

With long-lived species like primates, it will take many years of study to determine the long-term stability of populations in regenerating forests. A landscape-scale approach across taxonomic and geographic boundaries is required to better understand which species frequently exploit regenerating forests, how they use them, and which species only use younger forests when absolutely necessary. Regenerating forests may provide the necessary landscape heterogeneity required to support primate populations in a rapidly changing world.

References

Aide, M. T., Zimmerman, J. K., Pascarella, J. B., Rivera, L., & Marcano-Vega, H. (2000). Forest regeneration in a chronosequence of tropical abandoned pastures: Implications for restoration ecology. *Restoration Ecology, 8*, 328–338.

Albert, A., Auffret, A. G., Cosyns, E., Cousins, S. A., D'hondt, B., Eichberg, C., Eycott, A. E., Heinken, T., Hoffmann, M., Jaroszewicz, B., & Malo, J. E. (2015). Seed dispersal by ungulates as an ecological filter: A trait-based meta-analysis. *Oikos, 124*, 1109–1120.

Alvarez-Buylla, E. R., & Martínez-Ramos, M. (1990). Seed bank versus seed rain in the regeneration of a tropical pioneer tree. *Oecologia, 84*, 314–325.

Anderson, J., Rowcliffe, J. M., & Cowlishaw, G. (2007). Does the matrix matter? A forest primate in a complex agricultural landscape. *Biological Conservation, 135*, 212–222.

Arroyo-Rodríguez, V., Melo, F. P. L., Martínez-Ramos, M., Bongers, F., Chazdon, R. L., Meave, J. A., Norden, N., Santos, B. A., Leal, I. R., & Tabarelli, M. (2017a). Multiple successional pathways in human-modified tropical landscapes: New insights from forest succession, forest fragmentation and landscape ecology research. *Biological Reviews, 92*, 326–340.

Arroyo-Rodríguez, V., Pérez-Elissetche, G. K., Ordoñez-Gómez, J. D., González-Zamora, A., Chaves, O. M., Sánchez-López, S., Chapman, C. A., Morales-Hernández, K., Pablo-Rodríguez, M., & Ramos-Fernández, G. (2017b). Spider monkeys in human-modified landscapes: The importance of the matrix. *Tropical Conservation Science, 10*, 1–13.

Asensio, N., Arroyo-Rodríguez, V., Dunn, J. C., & Cristóbal-Azkarate, J. (2009). Conservation value of landscape supplementation for howler monkeys living in forest patches. *Biotropica, 41*, 768–773.

Baguette, M., & Van Dyck, H. (2007). Landscape connectivity and animal behavior: Functional grain as a key determinant for dispersal. *Landscape Ecology, 22*, 1117–1129.

Balcomb, S. R., & Chapman, C. A. (2003). Bridging the gap: Influence of seed deposition on seedling recruitment in a primate-tree interaction. *Ecological Monographs, 73*, 625–642.

Barlow, J., Gardner, T. A., Araujo, I. S., Ávila-Pires, T. C., Bonaldo, A. B., Costa, J. E., Esposito, M. C., Ferreira, L. V., Hawes, J., Hernandez, M. I., & Hoogmoed, M. S. (2007). Quantifying the biodiversity value of tropical primary , secondary , and plantation forests. *PNAS, 104*, 18555–18560.

Bicca-Marques, J. C. (2003). How do howler monkeys cope with habitat fragmentation? In L. K. Marsh (Ed.), *Primates in fragments* (pp. 283–303). Springer.

Bersacola, E., Sastramidjaja, W. J., Rayadin, Y., Macdonald, D. W., & Cheyne, S. M. (2019). Occupancy patterns of ungulates and pig-tailed macaques across regenerating and anthropogenic forests on Borneo. *Hystrix.* https://doi.org/10.4404/hystrix-00177-2019

Bryson-Morrison, N., Matsuzawa, T., & Humle, T. (2016). Chimpanzees in an anthropogenic landscape: Examining food resources across habitat type at Bossou, Guinea, West Africa. *American Journal of Primatology, 78*, 1237–1249.

Chapman, C. A., Bortolamiol, S., Matsuda, I., Omeja, P. A., Paim, F. P., Reyna-Hurtado, R., Sengupta, R., & Valenta, K. (2017). Primate population dynamics: Variation in abundance over space and time. *Biodiversity and Conservation, 27*, 1221–1238.

Chapman, C. A., & Dunham, A. E. (2018). Primate seed dispersal and forest restoration: An African perspective for a brighter future. *International Journal of Primatology, 39*, 427–442.

Chapman, C. A., Bicca-Marques, J. C., Dunham, A. E., Fan, P., Fashing, P. J., Gogarten, J. F., Guo, S., Huffman, M. A., Kalbitzer, U., Li, B., & Ma, C. (2020). Primates can be a rallying symbol to promote tropical forest restoration. *Folia Primatologica, 91*, 669–687.

Chaves, Ó. M., Stoner, K. E., Arroyo-Rodríguez, V., & Estrada, A. (2011). Effectiveness of spider monkeys (Ateles geoffroyi vellerosus) as seed dispersers in continuous and fragmented rain forests in Southern Mexico. *International Journal of Primatology, 32*, 177–192.

Chazdon, R. L. (2014). *Second growth: The promise of tropical forest regeneration in an age of deforestation*. University of Chicago Press.

Chazdon, R. L., & Guariguata, M. R. (2016). Natural regeneration as a tool for large-scale forest restoration in the tropics: Prospects and challenges. *Biotropica, 48*, 716–730.

Chazdon, R. L., Harvey, C. A., Komar, O., Griffith, D. M., Ferguson, B. G., Martínez-Ramos, M., Morales, H., Nigh, R., Soto-Pinto, L., Van Breugel, M., & Philpott, S. M. (2009a). Beyond reserves: A research agenda for conserving biodiversity in human-modified tropical landscapes. *Biotropica, 41*, 142–153.

Chazdon, R. L., Peres, C. A., Dent, D., Sheil, D., Lugo, A. E., Lamb, D., Stork, N. E., & Miller, S. E. (2009b). The potential for species conservation in tropical secondary forests. *Conservation Biology, 23*, 1406–1417.

Chokkalingam, U., & de Jong, W. (2003). Secondary forest: A working definition and typology. *International Forest Reviews, 3*, 19–26.

Clark, D. B. (1996). Abolishing virginity. *Journal of Tropical Ecology, 12*, 735–739.

Coomes, O. T. (1995). A century of rain forest use in western Amazonia: Lessons for extraction-based conservation of tropical forest resources. *Forest & Conservation History, 39*, 108–120.

Corbin, J. D., & Holl, K. D. (2012). Applied nucleation as a forest restoration strategy. *Forest Ecology and Management, 265*, 37–46.

Cordeiro, N. J., & Howe, H. F. (2001). Low recruitment of trees dispersed by animals in African forest fragments. *Conservation Biology, 15*, 1733–1741.

DeClerck, F. A., Chazdon, R., Holl, K. D., Milder, J. C., Finegan, B., Martinez-Salinas, A., Imbach, P., Canet, L., & Ramos, Z. (2010). Biodiversity conservation in human-modified landscapes of Mesoamerica: Past, present and future. *Biological Conservation, 143*, 2301–2313.

Dent, D. H., & Wright, J. S. (2009). The future of tropical species in secondary forests: A quantitative review. *Biological Conservation, 142*, 2833–2843.

Dew, J. L., & Wright, P. (1998). Frugivory and seed dispersal by four species of primates in Madagascar's eastern rain forest. *Biotropica, 30*, 425–437.

Diaz, S., Joseph, F., Chapin, F. S., III, & Tilman, D. (2006). Biodiversity loss threatens human well-being. *PLoS Biology, 4*, 1300–1305.

Duncan, R. S., & Chapman, C. A. (1999). Seed dispersal and potential forest succession in abandoned agriculture in tropical Africa. *Ecological Applications, 9*, 998–1008.

Effiom, E. O., Nuñez-Iturri, G., Smith, H. G., Ottosson, U., & Olsson, O. (2013). Bushmeat hunting changes regeneration of African rainforests. *Proceedings of the Royal Society B: Biological Sciences, 280*, 20130246.

Estrada, A. (2013). Socioeconomic contexts of primate conservation: Population, poverty, global economic demand, and sustainable land use. *American Journal of Primatology, 75*, 30–45.

Estrada, A., Garber, P. A., Rylands, A. B., Roos, C., Fernandez-Duque, E., Di Fiore, A., Nekaris, K. A. I., Nijman, V., Heymann, E. W., Lambert, J. E., & Rovero, F. (2017). Impending extinction crisis of the world's primates: Why primates matter. *Science Advances, 3*, e1600946.

Estrada, A., Garber, P. A., Mittermeier, R. A., Serge, W., Gouveia, S., Dobrovolski, R., Nekaris, K. A. I., Nijman, V., Rylands, A. B., Maisels, F., & Williamson, E. A. (2018). Primates in peril:

The significance of Brazil, Madagascar, Indonesia and the Democratic Republic of the Congo for global primate conservation. *PeerJ, 6,* 4869.

FAO. (2020). *Global forest resources assessment 2020.*

Farwig, N., & Berens, D. G. (2012). Imagine a world without seed dispersers: A review of threats, consequences and future directions. *Basic and Applied Ecology, 13,* 109–115.

Fedigan, L. M., & Jack, K. (2001). Neotropical primates in a regenerating Costa Rican dry forest: A comparison of howler and capuchin population patterns. *International Journal of Primatology, 22,* 689–713.

Fuzessy, L. F., Cornelissen, T. G., Janson, C., & Silveira, F. A. (2016). How do primates affect seed germination? A meta-analysis of gut passage effects on Neotropical plants. *Oikos, 125,* 1069–1080.

Galán-Acedo, C., Arroyo-Rodríguez, V., Andresen, E., & Arasa-Gisbert, R. (2019a). Ecological traits of the world's primates. *Scientific Data, 6*(1), 1–5.

Galán-Acedo, C., Arroyo-Rodríguez, V., Andresen, E., Verde Arregoitia, L., Vega, E., Peres, C. A., & Ewers, R. M. (2019b). The conservation value of human-modified landscapes for the world's primates. *Nature Communications, 10,* 1–8.

Galán-Acedo, C., Arroyo-Rodríguez, V., Cudney-Valenzuela, S. J., & Fahrig, L. (2019c). A global assessment of primate responses to landscape structure. *Biological Reviews, 94,* 1605–1618.

Ganzhorn, J. U. (1987). Possible role of plantations for primate conservation in Madagascar. *American Journal of Primatology, 12,* 205–215.

Ganzhorn, J. U., Fietz, J., Rakotovao, E., Schwab, D., & Zinner, D. (1999). Lemurs and the regeneration of dry deciduous forest in Madagascar. *Conservation Biology, 13,* 794–804.

Gascon, C., Lovejoy, T. E., Bierregaard, R. O., Jr., Malcolm, J. R., Stouffer, P. C., Vasconcelos, H. L., Laurance, W. F., Zimmerman, B., Tocher, M., & Borges, S. (1999). Matrix habitat and species richness in tropical forest remnants. *Biological Conservation, 91,* 223–229.

George, L. O., & Bazzaz, F. A. (1999). The fern understory as an ecological filter: Emergence and establishment of canopy-tree seedlings. *Ecology, 80,* 833–845.

Guariguata, M. R., & Ostertag, R. (2001). Neotropical secondary forest succession: Changes in structural and functional characteristics. *Forest Ecology and Management, 148,* 185–206.

Hansen, M. F., Nawangsari, V. A., van Beest, F. M., Schmidt, N. M., Stelvig, M., Dabelsteen, T., & Nijman, V. (2020). Habitat suitability analysis reveals high ecological flexibility in a "strict" forest primate. *Frontiers in Zoology, 17,* 1–13.

Harris, T. R., & Chapman, C. A. (2007). Variation in diet and ranging of black and white colobus monkeys in Kibale National Park, Uganda. *Primates, 48*(3), 208–221.

Heiduck, S. (2002). The use of disturbed and undisturbed forest by masked titi monkeys *Callicebus personatus melanochir* is proportional to food availability. *Oryx, 36,* 133–139.

Heymann, E. W., Culot, L., Knogge, C., Smith, A. C., Tirado Herrera, E. R., Müller, B., Stojan-Dolar, M., Ferrer, Y. L., Kubisch, P., Kupsch, D., Slana, D., Koopmann, M. L., Ziegenhagen, B., Bialozyt, R., Mengel, C., Hambuckers, J., & Heer, K. (2019). Small Neotropical primates promote the natural regeneration of anthropogenically disturbed areas. *Scientific Reports, 9,* 1–9.

Holl, K. D., & Aide, T. M. (2011). When and where to actively restore ecosystems? *Forest Ecology and Management, 261,* 1558–1563.

Holloway, L. (2000). Catalysing rainforest restoration in Madagascar. *Diversité et endémisme à Madagascar,* 115–124.

Holloway, L. (2004). Ecosystem restoration and rehabilitation in Madagascar. *Ecological Restoration, 22,* 113–119.

Howe, H. F. (2016). Making dispersal syndromes and networks useful in tropical conservation and restoration. *Global Ecology and Conservation, 6,* 152–178.

Howe, H. F., & Smallwood, J. (1982). Ecology of seed dispersal. *Annual Review of Ecology and Systematics, 13,* 201–228.

Howe, H. F., & Miriti, M. N. (2004). When seed dispersal matters. *BioScience, 54,* 651–660.

Isabirye-Basuta, G. M., & Lwanga, J. S. (2008). Primate populations and their interactions with changing habitats. *International Journal of Primatology, 29,* 35–48.

Kalbitzer, U., & Chapman, C. A. (2018). Primate responses to changing environments in the Anthropocene. In U. Kalbitzer & K. M. Jack (Eds.), *Primate life histories, sex roles, and adaptability. Developments in primatology: Progress and prospects* (pp. 283–310). Springer Nature.

Kaplin, B. A., & Lambert, J. E. (2002). Effectiveness of seed dispersal by Cercopithecus Monkeys: Implications for seed input into degraded areas. In D. J. Levey, W. R. Silva, & M. Galetti (Eds.), *Seed dispersal and frugivory: Ecology and conservation* (pp. 351–364). CABI.

Kitamura, S., Yumoto, T., Poonswad, P., Chuailua, P., Plongmai, K., Maruhashi, T., & Noma, N. (2002). Interactions between fleshy fruits and frugivores in a tropical seasonal forest in Thailand. *Oecologia, 133*, 559–572.

Lambert, J. E. (2007). Primate nutritional ecology: Feeding biology and diet at ecological and evolutionary scales. In C. J. Campbell, A. Fuentes, K. C. MacKinnon, M. Panger, & S. K. Bearder (Eds.), *Primates in perspective* (pp. 482–495). Oxford University Press.

Lambert, J. E., & Garber, P. A. (1998). Evolutionary and ecological implications of primate seed dispersal. *American Journal of Primatology, 45*, 9–28.

Lazos-Chavero, E., Zinda, K., Bennett-Curry, A., Balvanera, P., Bloomfield, G., Lindell, C., & Negra, C. (2016). Stakeholders and tropical reforestation: Challenges, trade-offs, and strategies in dynamic environments. *Biotropica, 48*, 900–914.

Luckett, J., Danforth, E., Linsenbardt, K., & Pruetz, J. (2004). Planted trees as corridors for primates at El Zota Biological Field Station, Costa Rica. *Neotropical Primates, 12*, 143–146.

Martinez, B. T., & Razafindratsima, O. H. (2014). Frugivory and seed dispersal patterns of the red-ruffed lemur, *Varecia rubra*, at a forest restoration site in Masoala National Park, Madagascar. *Folia Primatologica, 85*, 228–243.

McKinney, T. (2019). Ecological and behavioural flexibility of mantled howlers (*Alouatta palliata*) in response to anthropogenic habitat disturbance. *Folia Primatologica, 90*, 456–469.

McLennan, M. R., Spagnoletti, N., & Hockings, K. J. (2017). The implications of primate behavioral flexibility for sustainable human-primate coexistence in anthropogenic habitats. *International Journal of Primatology, 38*, 105–121.

Myers, J. A., & Harms, K. E. (2011). Seed arrival and ecological filters interact to assemble high-diversity plant communities. *Ecology, 92*, 676–686.

Nekaris, K. A. I., Poindexter, S., Reinhardt, K. D., Sigaud, M., Cabana, F., Wirdateti, W., & Nijman, V. (2017). Coexistence between Javan slow lorises (*Nycticebus javanicus*) and humans in a dynamic agroforestry landscape in West Java, Indonesia. *International Journal of Primatology, 38*, 303–320.

Nowak, K., & Lee, P. C. (2013). "Specialist" primates can be flexible in response to habitat alteration. In L. K. Marsh & C. A. Chapman (Eds.), *Primates in fragments: Complexity and resilience. Developments in primatology: Progress and prospects* (pp. 199–211). Springer.

Peres, C. A., Emilio, T., Schietti, J., Desmoulière, S. J. M., & Levi, T. (2016). Dispersal limitation induces long-term biomass collapse in overhunted Amazonian forests. *Proceedings of the National Academy of Sciences, 113*, 892–897.

Raboy, B. E., Christman, M. C., & Dietz, J. M. (2004). The use of degraded and shade cocoa forests by endangered golden-headed lion tamarins *Leontopithecus chrysomelas. Oryx, 38*, 75–83.

Ramananjato, V., Rakotomalala, Z., Park, D. S., DeSisto, C. M., Raoelinjanakolona, N. N., Guthrie, N. K., Fenosoa, Z. E., Jonhson, S. E., & Razafindratsima, O. H. (2020). The role of nocturnal omnivorous lemurs as seed dispersers in Malagasy rain forests. *Biotropica, 52*, 758–765.

Ramos-Fernández, G., & Ayala-Orozco, B. (2003). Population size and habitat use of spider monkeys at Punta Laguna, Mexico. In L. D. Marsh (Ed.), *Primates in fragments: Ecology and conservation* (pp. 191–209). Kluwer Academic/Plenum Publishers.

Ramos-Fernandez, G. Smith Aguilar, S. E., Schaffner, C. M., Vic, L.G., Aureli, F. (2013). Site Fidelity in Space Use by Spider Monkeys (Ateles geoffroyi) in the Yucatan Peninsula, Mexico. *PLoS ONE, 8(5)*, e62813.

Razafindratsima, O. H., & Dunham, A. E. (2015). Assessing the impacts of nonrandom seed dispersal by multiple frugivore partners on plant recruitment. *Ecology, 96*, 24–30.

Razafindratsima, O. H., & Dunham, A. E. (2016). Frugivores bias seed-adult tree associations through non-random seed dispersal: A phylogenetic approach. *Ecology, 97*, 2094–2102.

Razafindratsima, O. H., & Martinez, B. T. (2012). Seed dispersal by red-ruffed lemurs: Seed size, viability, and beneficial effect on seedling growth. *Ecotropica, 18*, 15–26.

Razafindratsima, O. H., Sato, H., Tsuji, Y., & Culot, L. (2018). Advances and frontiers in primate seed dispersal. *International Journal of Primatology, 39*, 315–320.

Reid, J. L., Holl, K. D., & Zahawi, R. A. (2015). Seed dispersal limitations shift over time in tropical forest restoration. *Ecological Applications, 25*, 1072–1082.

Rodrigues, M. A. (2017). Female spider monkeys (*Ateles geoffroyi*) cope with anthropogenic disturbance through fission–fusion dynamics. *International Journal of Primatology, 38*, 838–855.

Sales, L., Culot, L., & Pires, M. M. (2020). Climate niche mismatch and the collapse of primate seed dispersal services in the Amazon. *Biological Conservation, 247*, 108628.

Schupp, E. W., & Fuentes, M. (1995). Spatial patterns of seed dispersal and the unification of plant population ecology. *Ecoscience, 2*, 267–275.

Singh, M., Kumara, H. N., Kumar, M. A., & Sharma, A. K. (2001). Behavioural responses of Lion-Tailed Macaques (Macaca silenus) to a changing habitat in a tropical rain forest fragment in the Western Ghats, India. *Folia Primatologica, 75*, 278–291.

Sorensen, T. C., & Fedigan, L. M. (2000). Distribution of three monkey species along a gradient of regenerating tropical dry forest. *Biological Conservation, 92*, 227–240.

Sussman, R. W. (1991). Primate origins and the evolution of angiosperms. *American Journal of Primatology, 23*, 209–223.

Traveset, A., Robertson, W., & Rodriguez-Perez, J. (2007). A review of the role of endozoochory on seed germination. In A. J. Dennis, R. J. Green, E. W. Schupp, & D. Westcott (Eds.), *Seed dispersal: Theory and its application in a changing world* (pp. 78–103). CABI.

Wang, B. C., & Smith, T. B. (2002). Closing the seed dispersal loop. *Trends in Ecology and Evolution, 17*, 379–385.

Wheeler, C. E., Omeja, P. A., Chapman, C. A., Glipin, M., Tumwesigye, C., & Lewis, S. L. (2016). Carbon sequestration and biodiversity following 18 years of active tropical forest restoration. *Forest Ecology and Management, 373*, 44–55.

White, L. J. T., & Oates, J. F. (1999). New data on the history of the plateau forest of Okomu, southern Nigeria: An insight into how human disturbance has shaped the African rain forest. *Global Ecology and Biogeography, 8*, 355–361.

Whitmore, T. C. (1989). Canopy gaps and the two major groups of forest trees. *Ecology, 70*, 536–538.

Wotton, D. M., & Kelly, D. (2011). Frugivore loss limits recruitment of large-seeded trees. *Proceedings of the Royal Society B: Biological Sciences, 278*, 3345–3354.

Wunderle, J. M., Jr. (1997). The role of animal seed dispersal in accelerating native forest regeneration on degraded tropical lands. *Forest Ecology and Management, 99*, 223–235.

Chapter 4
Hunting of Primates in the Tropics: Drivers, Unsustainability, and Ecological and Socio-economic Consequences

Inza Koné, Johannes Refisch, Carolyn A. Jost Robinson, and Adeola Oluwakemi Ayoola

Contents

Abstract Hunting of primates is an important source of nutritional and economic sustenance for many tropical rainforest inhabitants. However, this reliance has become one of the major drivers of species loss and disappearance. This therefore requires a review of the drivers of the hunting of primates, addressing unsustainable levels of off-take, as well as its multifaceted consequences. In this chapter, we discuss the commercial and nutritional value of wild meat being a great obstacle to primate conservation resulting in population decline of primate populations which negatively affects forest regeneration, human use of forest resources, and human health. Though debates are ongoing as to whether hunting should be banned or

I. Koné (✉)
Centre Suisse de Recherches Scientifiques en Côte d'Ivoire, Abidjan, Côte d'Ivoire

UFR Biosciences, Université Félix Houphouët-Boigny, Abidjan, Côte d'Ivoire
e-mail: inza.kone@csrs.ci

J. Refisch
Great Apes Survival Partnership/UN Environmental Programme, Nairobi, Kenya

C. A. Jost Robinson
Chengeta Wildlife, Lacey, WA, USA

A. O. Ayoola
Centre Suisse de Recherches Scientifiques en Côte d'Ivoire, Abidjan, Côte d'Ivoire

T. McKinney et al. (eds.), *Primates in Anthropogenic Landscapes*, Developments in Primatology: Progress and Prospects, https://doi.org/10.1007/978-3-031-11736-7_4

regulated, we do not comment on these debates as they require focused attention and representation of multiple levels of stakeholders. What is certain is that the status quo is sustainable for neither people nor primates. We argue in this chapter that there is a need to reduce hunting pressure both to conserve endangered species and to reduce the risk of cross-species transmissions of viruses.

Keywords Primate conservation · Wild meat trade · Health risks · Cross-species transmission

4.1 Introduction

Hunting of wildlife is widespread in the tropics as an important way of procuring protein and nutrients for people (Bennett & Robinson, 2000; Lee et al., 2000; Mukesh, 2010; Powell et al., 2013; Sirén & Machoa, 2008). Hunting has always been an integral part of local subsistence in tropical forests, but has increasingly become unsustainable given the increasing commercialisation of wild meat trades. This **unsustainable off-take** is considered one of the major drivers of species loss in the tropics after habitat destruction (Brodie et al., 2021; Estrada et al., 2017; Hoffmann et al., 2010; Hughes, 2017; Milner-Gulland & Bennett, 2003; Peres, 2011; Robinson & Bennett, 2000; Wilkie et al., 2011). Mammals rank top among the animals most affected by hunting, especially antelopes and primates (Braga-Pereira et al., 2020; Fa & Brown, 2009; Hegerl et al., 2017; Ripple et al., 2016; Topp-Jørgensen et al., 2009). Primates are often the numerically dominant prey items harvested by **Indigenous** groups. This is the case throughout Amazonia, where primates rank higher than any other order of mammals in **subsistence hunting** efforts (Peres & Nascimento, 2006).

When combined with the multiplicative effects of habitat loss, hunting leads to the dramatic reduction of primate populations and to the extirpation of vulnerable primate species from habitats across their historical range (Estrada et al., 2017; McGraw, 2007; Refisch & Koné, 2005). Miss Waldron's red colobus (*Piliocolobus waldroni*), a species endemic to south-eastern Côte d'Ivoire and south-western Ghana, for example, may have been exterminated due to habitat loss and, ultimately, hunting (Oates et al., 2020). If confirmed, it will be the first primate taxon to have gone extinct in over 500 years (Linder et al., 2021; McGraw, 2005; McGraw & Oates, 2002; Oates et al., 2000). The extinction of that monkey could signal the beginning of a wave of primate extinctions across Africa (McGraw, 2007). Population declines and local extinctions in relation to direct human exploitation are widely reported in South and Central America, including Guyana (Lehman, 2000), Venezuela (Urbani, 2006), French Guiana (de Thoisy et al., 2005), and Brazilian Amazonia (Peres & Palacios, 2007). Large cebids such as woolly monkeys (*Lagothrix lagotricha*), spider monkeys (*Ateles paniscus*), and red howlers (*Alouatta seniculus*) are usually the first target species and consequently are the most dramatically affected (de Thoisy et al., 2009).

The loss of primates has cascading effects on entire ecosystems, as they are crucial pollinators, seed dispersers, and browsers; therefore, their absence reduces the diversity of plant species and the ability of the forest systems to recover from disturbance (Refisch & Koné, 2001). Recent studies have demonstrated that beyond its effects on biodiversity, hunting of primates may cause serious human health issues through the transmission of viral diseases from wild meat (Calvignac-Spencer et al., 2012; Cantlay et al., 2017; Peeters & Delaporte, 2012). In this chapter, we review the drivers of the hunting of primates and highlight evidence addressing unsustainable levels of off-take, as well as its multifaceted consequences.

4.2 From Subsistence to Commercial Hunting of Primates

Hunting has always been a source of nutritional and economic sustenance for tropical rainforest inhabitants, who originally tracked and hunted game for their own subsistence and in a relatively sustainable way (Dounias, 2016). Forest dwellers continue to attribute paramount dietary, cultural, and symbolic value to wild meat, including primate meat (Dounias, 2016; Remis & Jost Robinson, 2014; Sirén & Machoa, 2008). These values are also strong among urban dwellers, who tend to prefer wild meat over domesticated meat and thus maintain a high market demand (Obioha et al., 2012). The harvesting of wildlife, in particular that of primates, provides a major source of animal protein and nutrients (America, Bodmer, 1995; Peres, 1990, 2000; Redford, 1992; Africa, Bennett & Robinson, 2000; Jenkins et al., 2011; Schulte-Herbruggen et al., 2017; intercontinental comparison, Cawthorn & Hoffman, 2016; Fa & Peres, 2001; Hoffman & Cawthorn, 2012; Sarti et al., 2015). In West Africa, 25% of protein requirements were met by wild meat in the 1990s (Bennett & Robinson, 2000); exceptionally in Liberia, a country that had experienced a long socio-political and military conflict, 75% of the meat consumed was from wild animals. In this instance, the challenges of living in a long-standing conflict zone may have inhibited the potential to identifying alternative sources of income in the country (Angelsen et al., 2014; Bennett & Robinson, 2000).

In addition to providing animal protein and nutrients, hunting finds its importance in local traditions from hunting technology to storytelling. In many communities, a hunter is well respected (Gurven & Rueden, 2006). Acquisition of animal trophies for personal adornment is a widespread practice (Secretariat of the Convention on Biological Diversity, 2011). Given its cultural importance, it is not surprising that communities in tropical countries continue to hunt, even if they have alternative sources of income and nutrition (Bennett & Robinson, 2000; Jost Robinson et al., 2017).

At present, the commercial value of wild meat has made hunting it a major source of income (Lindsey et al., 2007; Naidoo et al., 2016; Refisch & Koné, 2005; Saayman et al., 2011; Van der Merwe et al., 2014). A widespread trade has developed in many countries including in those where hunting is prohibited (Caspary et al., 2001). In the Arabuko Sokoke Forest in Kenya, local hunters earn more by selling meat than the equivalent of the average annual income per capita (Okello & Kiringe, 2004; Wato et al., 2006). Chapman and Peres (2001) estimated that 3.8 million primates are consumed annually in the Brazilian Amazon, which represents a mean market value of \$34.4 million. Caspary and Momo (1998) estimated that in 1996, 100,000 tonnes of wild meat entered the markets in Côte d'Ivoire, representing the equivalent of 1.4% of the gross domestic product. Over the past decades, wild meat commerce has undergone a dramatic acceleration in the Congo Basin (Nasi et al., 2011). The demand for wild meat is considerably intensified by densely populated settlements that are increasingly concentrated within wildlife habitats. That situation is exacerbated by political and economic instability, corruption, infrastructure building (for hydroelectric production, fossil fuel extraction, logging, and mining), and expanding urban and rural populations (Fa et al., 2002).

The profits derived from hunting vary considerably. For example, in Côte d'Ivoire, restaurant owners, market keepers, and intermediaries, rather than the hunters themselves, are the people who derive the greatest profit from wild meat markets (Caspary et al., 2001). Hunters may only make a profit if they kill multiple medium- or large-bodied species, such as duikers and larger primates. Given that many of these species are concentrated in national parks and reserves where hunting is prohibited, it means that hunters, who derive the least profit, simultaneously face the highest risk of being arrested. Further, those individuals who are hunting wild meat often come from low-income groups. They are often hired by gun owners (e.g. restaurateurs) to hunt to supply a commercial market (Refisch & Koné, 2005). In the Congo Basin, Indigenous hunters who were once the major suppliers of wild meat to markets are now vulnerable, economically, nutritionally, and culturally, as expanding commercial trades involve new actors and converge with other forms of illicit trade involving fearless and ruthless actors (Dounias, 2016; Jost Robinson, 2012).

4.3 Unsustainability of Primate Hunting

Given global population expansion, particularly in forested regions, even subsistence hunting of game meat may now exceed sustainability (Wilkie et al., 1998). **Food taboos** may play an important role in determining hunting intensity in primate communities, but with the combination of immigration of people from different regions and already declining primate populations, the protection offered by such customs has decreased (Jimoh et al., 2012; Jones et al., 2008). Professionalisation of the wild meat trade to feed urban markets (Bennett & Robinson, 2000) along with advanced in hunting technology, especially the availability of shotguns (Braga-Pereira et al., 2020), has led to more efficient hunting. Together, these factors make current hunting pressures unsustainable, leading to dramatic impacts on wild primate populations.

Hunting has several direct effects on wildlife populations and may lead to local extinction of species in many cases (Cowlishaw & Dunbar, 2000; Ripple et al.,

2016). Furthermore, hunting for larger individuals can change the demography and reduce the proportion of animals in older age classes. Because of hunting, many populations are reduced in size and fragmented into small remnants where there is little or no contact in the form of inter-group encounters or dispersal. This has several implications for the genetic make-up of the population. Random fluctuations due to **genetic drift** can accelerate the decline of wildlife populations for several reasons (Willi et al., 2006). Inbreeding, not necessarily a component of genetic drift but often associated with it in small populations, causes a reduction of average individual fitness compared to the ancestral population in a wide variety of species. Even if the effect of fitness on inbred individuals is not large, the loss of genetic variation may reduce the ability of a population to adapt to changing environments (Bijlsma & Loeschcke, 2012). Studies suggest a link between inbreeding and greater susceptibility to infectious diseases (Altizer et al., 2003; Avecedo-Whitehouse et al., 2003; Lively et al., 1990).

Hunting affects not only numbers but also primate behaviour (Koné & Refisch, 2005). Since hunters most often target large-bodied animals which can generate higher revenue, individuals with medium-size body mass may be favoured, as shown in orangutans (*Pongo pygmaeus*) (Bennett, 1998). In theory, females may also increase fecundity to offset the mortality rate caused by hunting, but this has not been demonstrated for tropical forest species (Bennett & Robinson, 2000). It was reported that faced with human predators, monkeys adopt a temporary cryptic behaviour (Bicca-Marques & Heymann, 2013; Cäsar et al., 2012). Bshary (2000) documented that Diana monkeys (*Cercopithecus diana*) adjust their behaviour in hunted areas of the Taï National Park, Côte d'Ivoire. In fact, poachers in the Taï region often imitate animal calls to feign the presence of either leopards (*Panthera pardus*) or crowned hawk eagles (*Stephanoaetus coronatus*), which cause the monkeys to react with vocalisations and approach (Shultz, 2001; Zuberbühler et al., 1999). While this works in most areas, Diana monkeys in heavily hunted areas are rarely fooled by these imitations and remain cryptic. Koné and Refisch (2005) demonstrated that hunting pressure did not cause any significant modification in the anti-predator behaviour of the western red colobus (*Piliocolobus badius*) that was, instead, dictated by the hunting pressure on the species by western chimpanzees (*Pan troglodytes verus*). Conversely, Diana monkeys responded to human predation pressure notably by spending most of their time in high forest strata, hiding systematically behind vegetation without alarm calling, reducing movements, and constantly keeping a distance from each other and from any other animal.

Some species are more vulnerable than others. Among mammals, species with low intrinsic rates of reproduction are less resilient to hunting; this pattern holds true for most primate species. Species whose mating, nesting, social, or anti-predator behaviours make them easy to hunt are especially vulnerable (McGraw, 2007). It was demonstrated that the behavioural non-responsiveness of red colobus (*Piliocolobus badius*) to human predation increased the vulnerability of these monkeys to hunting (Koné & Refisch, 2005). Most primates are group-living species, and hunters can hunt more than one animal at one time. Furthermore, species with spectacular displays or loud calls are easy to detect and preferred preys of hunters.

Koné and Refisch (2005) suggest that the fact that Diana monkeys reduced their frequency of vocalisations and foraging in poached areas to avoid detection may lead to a disruption of social life that is partly regulated by vocalisations and to a decrease in energy intake resulting in lowered reproduction rates. They concluded that beyond the number of animals killed, the vulnerability of monkeys to hunting also lies in the counterproductive modification of their behaviour.

4.4   Multifaceted Consequences of Primate Population Decline

Reduction of animal population densities has cascading effects on the ecosystem. The plant community composition may be biased towards species that can reproduce in the absence of large animal pollinators and seed dispersers at the expense of the species which cannot. Chapman and Chapman (1995) estimated that 60% of the 25 tree species they sampled in the Kibale National Park, Uganda, could potentially be lost if all frugivores were removed. However, there is little understanding as to how hunting activities alter the processes which govern the maintenance of biodiversity and the sustainability of forest ecosystems. Examples in which plant regeneration can be directly linked to the presence of a specific group of seed dispersers are restricted to relatively species-poor ecosystems (Brown & Heske, 1990). Some tight coadaptations seem to exist between elephants and larger fruits (Chapman et al., 1993; Stephen & Inkamba-Nkulu, 2004) and between lowland gorilla (*Gorilla gorilla gorilla*) and *Cola lizae*. Indeed, it has been documented that gorillas are the only important dispersers of *Cola lizae* (Voysey et al., 1999a, b).

The decline of primate populations may also affect human use of natural resources in forest-dwelling communities. Koné et al. (2008) explored the potential for monkey seed dispersers to maintain the utility of forest fragments for people through **seed dispersal** in the Taï region, western Côte-d'Ivoire. In this study, they compared the fruiting tree species dispersed by monkeys with those used by humans in the broader Taï region. Of this total set of 75 species of trees consumed by the monkeys of the Taï National Park, 52 (69%) were dispersed almost exclusively by monkeys and were also found in neighbouring forest fragments. Of the 52 fruiting forest tree species dispersed by Taï monkeys, 25 (48%) have some utility to local inhabitants for wood, food, medicine, or ritual purposes. The authors concluded that maintaining primate populations is important not only for forest regeneration but also for the people who rely on forest resources.

Hunting also plays an important role in the vertical and horizontal transfer of important local cultural traditions. For many tropical rainforest inhabitants, hunting practices are deeply rooted in their cosmologies and worldviews and directly influence their relationships to primates. For example, the Waiwai of Guyana are swidden agriculturalists whose diet is highly supplemented by animal protein, particularly that of black spider monkey (*Ateles paniscus*). For the Waiwai, these primates serve not only as a protein source but as a means through which they realise and understand the roles of all species in their forest environment (Shaffer et al., 2017). In the Central African Republic, BaAka hunter-gatherers, who once specialised in hunting arboreal primates using a traditional crossbow (*gbano*), have experienced the decline of this trade as primate populations are lost to hunting with firearms (Jost Robinson & Remis, 2018). Additionally, the decline of primate populations jeopardises important rituals linked to some primate species, including critically endangered species. Indeed, some of the rare plant or animal species occurring in the Tanoé-Ehy Forest, south-eastern Côte d'Ivoire, are very important for traditional practices (Zadou et al., 2011). For example, the scat of Geoffroy's black-and-white colobus (*Colobus vellerosus*) is used for expiatory ceremonies by the "adouvlê". In this community, the tenth child born of a woman remains cursed for life if the expiatory ceremony is not conducted. The same is true for a child born to a woman who has become pregnant without having had her period four times after a previous childbirth. The scat of *Colobus vellerosus* is also used in the purification ceremonies of women who have committed adultery.

4.5 Hunting and Zoonotic Infectious Diseases

Considering the West African Ebola outbreak of 2013–2015 and zoonotic disease emergence in general, including COVID-19, much global attention has turned to the risks associated with wild meat consumption and the risk of contracting animal-borne infectious diseases. The focus of this section is on the potential health risks that emerge through the hunting and consumption of wild meat. There are four main ways in which disease transmission to humans can occur through direct contact: (i) hunted and consumed wild animals; (ii) traded wild animals (including at markets); (iii) wild animals kept as pets at home or at restaurants/hotels or in zoos, sanctuaries, or laboratories; and (iv) domestic or peri-domestic animals. Wild vertebrates including primates can be reservoirs of a wide range of pathogens. Harvesting wild meat and trading live animals can increase the risk of zoonotic spillover. For example, hunters are exposed to disease risk if injured by an animal during its capture or if they cut themselves when butchering the animal and also when carrying their prey back home (LeBreton et al., 2006; Subramanian, 2012).

There are many examples where pathogens have crossed the species boundaries from animals to humans and vice versa. Wolfe et al. (2005) investigated the diversity of human T-lymphotropic virus (HTLV) among Central African hunters and showed that these hunters were infected with a wide variety of HTLVs associated with many human illnesses (Wolfe et al., 2005). Another study found simian foamy virus infections among Central African hunters, concluding that retroviruses can cross into human populations via contact when getting in contact with blood and body fluids through hunting and butchering (Wolfe et al., 2004). Two of the most significant zoonotic disease transmissions in recent history are the human immunodeficiency viruses, HIV-1 and HIV-2, the two infectious agents for acquired immunodeficiency syndrome (AIDS) in humans (Barre-Sinoussi et al., 1983; Clavel et al., 1986). The closest relatives of HIV-1 are simian immunodeficiency viruses (SIVs) that infect wild populations of chimpanzees (*Pan troglodytes troglodytes*) and gorillas in western and central Equatorial Africa. Chimpanzees were the original hosts of this clade of viruses. Four lineages of HIV-1 have evolved by independent **cross-species transmission** events to humans, and one or two of those transmissions from primates to humans may have been via gorillas (Sharp & Hahn, 2010). The closest relatives of HIV-2 are simian immunodeficiency viruses in sooty mangabeys (*Cercocebus atys*) which live in West Africa (Hirsch et al., 1989). Chen et al. (1997) show that cross-species transmission from SIVs in sooty mangabeys to HIVs in humans has occurred at least six times. The possible reason is that sooty mangabeys and chimpanzees are both often kept as pets or hunted for food, thus resulting in their frequent direct contact with humans (Hahn et al., 2000). More than 40 species of African monkeys are infected with their own, species-specific SIV, and some species can carry multiple SIVs (Hahn et al., 2000; Peeters & Courgnaud, 2002; Peeters et al., 2002). These viruses are of relatively low pathogenicity in their hosts and usually do not induce symptoms, suggesting that they have evolved with their hosts over an extended period of time. However, recent evidence shows that SIVcpz can induce symptoms and result in reduced fertility in eastern chimpanzees (Keele et al., 2009).

The conclusion that HIV-1 stems from a virus infecting chimpanzee is interesting, given the close phylogenetic relationship of chimpanzees and humans. This raises a number of questions: (i) "what is the origin of the chimpanzee virus?", (ii) "did adaptation of SIVcpz to infecting chimpanzees enabled the virus to infect humans?", and (iii) "is SIVcpz infection of chimpanzees of low pathogenicity?" (Sharp & Hahn, 2010). It is thought that SIV may have previously crossed the species barrier into human hosts multiple times throughout history, but it was not until relatively recently with modern means of travel and globalisation that HIV spread beyond causing mortality in local populations. This conclusion is based on the analysis of strains found in four species of monkeys from Bioko Island in Equatorial Guinea, which was isolated from the mainland by rising sea level about 11,000 years ago, indicating that SIV must have been present in monkeys and apes for at least 30,000 years and probably much longer (Worobey et al., 2010).

There are other diseases too which have had devastating impacts on both humans and great apes. Ebola virus disease (EVD), first documented in 1976 in the Democratic Republic of Congo and in South Sudan, can affect chimpanzees, gorillas, and people. Hunters in Central Africa contracted EVD when opportunistically harvesting and handling infected gorilla and chimpanzee cadavers for meat consumption (Leendertz et al., 2016). Previous outbreaks in Gabon and the Republic of Congo in the mid-1990s killed more than 90 per cent of the gorillas and chimpanzees in some areas, and further Ebola waves in the early 2000s killed thousands of great apes (Leroy et al., 2004). Walsh et al. (2003) estimated that it will take gorilla populations that experienced 95 per cent mortality more than 130 years to recover.

There is also evidence from other continents of the health risks associated with consumption of wild meat. A risk assessment of zoonotic disease in markets in Lao

People's Democratic Republic indicated that the combination of high volumes of consumed wildlife, high-risk taxa for zoonoses, and poor biosafety increases the risk for transmission into the human population (Greatorex et al., 2016). In North America, several studies have also documented the potential disease risk and transmission pathway associated with the import of live animals in trade (Can et al., 2019; Pavlin et al., 2009). The first reported case of monkey pox outside Africa, in 2003, was linked to human infection by pet prairie dogs that had become infected by African rodents imported to the United States (Bernard & Anderson, 2006). The shift towards large-scale **commercial hunting** activities has increased the frequency of human exposure to primate retroviruses and other pathogens. The reduction of hunting pressure to sustainable levels is not only indispensable to conserve endangered species, but it has also the potential to reduce the risk of cross-species transmissions of simian retroviruses.

4.6 Conclusion

Hunting has always been a source of nutritional and economic sustenance for tropical rainforest inhabitants. It has also always been a cultural practice with specific rules and systems of values in many societies. At present, the commercial value of wild meat has made hunting a major source of income leading to its unsustainability. Hunting of primates combined with habitat destruction causes alarming population decline. That decline is accelerated by the genetic drift associated with inbreeding and by negative behavioural changes including reduction of reproduction associated with increased stress.

The decline of primate populations negatively affects forest regeneration, people's use of forest resources, cultural references of many human societies, and human health. At present, large-scale development projects pose a risk at different levels. They lead to deforestation, migration of people, and increased trade in wild meat. They also increase the risk of emergence and spread of diseases as people and wildlife migrate, bringing new pathogens with them. Further changes in the environment can also facilitate the emergence of infectious diseases. This highlights the need to better understand the linkages between changing environment and disease and the need to include health indicators in development planning. Debates are going on in several spheres as to whether hunting should be banned or regulated. We do not comment on these debates as they require focused attention and representation of multiple levels of stakeholders. What is not debatable is that the status quo is not an option for either people or primates.

Acknowledgement The authors are grateful to the reviewers who contributed significantly to improving the quality of this chapter.

References

Altizer, S., Harvell, D., & Frieddle, E. (2003). Rapid evolutionary dynamics and disease threats to biodiversity. *Trends in Ecology & Evolution, 18*(11), 589–596.

Angelsen, A., Jagger, P., Babigumira, R., Belcher, B., Hogarth, N. J., Bauch, S., Börner, J., Smith-hall, C., & Wunder, S. (2014). Environmental income and rural livelihoods: A global-comparative analysis. *World Development, 64*, 12–28.

Avecedo-Whitehouse, K., Gulland, F., Greig, D., & Amos, W. (2003). Disease susceptibility in California sea lions. *Nature, 422*, 35.

Barre-Sinoussi, F., Chermann, J. C., Rey, F., Nugeyere, M. T., Chamaret, S., Gruest, J., Dauguet, C., Axler Blin, C., Vezinet-Brun, F., Rouzioux, C., Rozenbaum, W., & Montaigner, L. (1983). Isolation of a T-lymphotropic retrovirus from a patient at risk for acquired immunodeficiency syndrome (AIDS). *Science, 220*, 868–871.

Bennett, L. E. (1998). *The natural history of Orangutan*. Natural History Publications (Borneo).

Bennett, L. E., & Robinson, J. G. (2000). *Hunting of wildlife in tropical forests – Implications for biodiversity and forest peoples*. Environment Department Papers. Paper No. 76. The World Bank.

Bernard, S. M., & Anderson, S. A. (2006). Qualitative assessment of risk for monkeypox associated with domestic trade in certain animal species, United States. *Emerging Infectious Diseases, 12*(12), 1827–1833.

Bicca-Marques, J. C., & Heymann, E. W. (2013). Ecology and behavior of titi monkeys (genus Callicebus). In A. Barnett, L. M. Veiga, S. F. Ferrari, & M. A. Norconk (Eds.), *Evolutionary biology and conservation of titis, sakis and uacaris* (pp. 196–207). Cambridge University Press.

Bijlsma, R., & Loeschcke, V. (2012). Genetic erosion impedes adaptive responses to stressful environments. *Evolutionary Applications, 5*, 117–129.

Bodmer, R. E. (1995). Managing Amazonian wildlife: Biological correlates of game choice by detribalized hunters. *Ecological Applications, 5*, 872–877.

Braga-Pereira, J. F., Bogonibc, R. A., & Nóbrega, A. R. (2020). From spears to automatic rifles: The shift in hunting techniques as a mammal depletion driver during the Angolan civil war. *Biological Conservation, 249*, 108744.

Brodie, J. F., Williams, S., & Garnera, B. (2021). The decline of mammal functional and evolutionary diversity worldwide. *PNAS, 118*(3), e1921849118.

Brown, J. H., & Heske, E. J. (1990). Control of a desert grassland transition by a keystone rodent guild. *Science, 250*, 1705–1707.

Bshary, R. (2000). Diana monkeys, Cercopithecus diana, adjust their anti-predator response to human hunting strategies. *Behavioural Ecology and Sociobiology, 50*, 251–256.

Calvignac-Spencer, S. S., Leendertz, A. J., Gillespie, T. R., & Leendertz, F. H. (2012). Wild great apes as sentinels and sources of infectious disease. *Clinical Microbiology and Infection, 18*(6), 521–527.

Can, Ö. E., D'Cruze, N., & Macdonald, D. W. (2019). Dealing in deadly pathogens: Taking stock of the legal trade in live wildlife and potential risks to human health. *Global Ecology and Conservation, 17*, e00515.

Cantlay, J. C., Ingram, D. J., & Meredith, A. L. (2017). A review of zoonotic infection risks associated with the wild meat trade in Malaysia. *EcoHealth, 14*, 361–388.

Cäsar, C., Byrne, R., Young, R. J., & Zuberbühler, K. (2012). The alarm call system of wild black-fronted titi monkeys, Callicebus nigrifrons. *Behaviour Ecology Sociobiology, 66*, 653–667.

Caspary, H. U., & Momo, J. (1998). *La chasse villagoise en Côte d'Ivpoire – résultats dans le cadre de l´ étude Filière de Viande de brousse (Enquête Chasseurs)*. Rapport DPN et Banque Mondiale, Abidjan, Côte d'Ivoire.

Caspary, H. U., Koné, I., Prouot, C., & de Pauw, M. (2001). *La chasse et la filière viande de brousse dans l'espace Taï, Côte d'Ivoire*. Tropenbos-Côte d'Ivoire Série 2.

Cawthorn, D.-M., & Hoffman, L. C. (2016). The bushmeat and food security nexus: A global account of the contributions, conundrums and ethical collisions. *Food Research International, 76*(4), 906–925.

Chapman, C. A., & Chapman, L. J. (1995). Survival without dispersers: Seedling recruitment under parents. *Conservation Biology, 9*, 675–678.

Chapman, C. A., & Peres, C. A. (2001). Primate conservation in the new millennium: The role of scientists. *Evolutionary Anthropology, 10*, 16–33.

Chapman, C. A., White, F. J., & Wrangham, R. W. (1993). Defining subgroup size in fission-fusion societies. *Folia Primatologica, 61*, 31–34.

Chen, Z., Luckay, A., Sodora, D. L., Telfer, P., Reed, P., Gettie, A., Kanu, J. M., Sadek, R. F., Yee, J., Ho, D. D., Zhang, L., & Marx, P. A. (1997). Human immunodeficiency virus type 2 (HIV-2) seroprevalence and characterization of a distinct HIV-2 genetic subtype from the natural range of simian immunodeficiency virus-infected sooty mangabeys. *Journal of Virology, 71*(5), 3953–3960.

Clavel, F., Guyader, M., Guetard, D., Salle, M., Montaigner, L., & Alizon, M. (1986). Molecular cloning and polymorphism of the human immune deficiency virus type 2. *Nature, 324*, 691–695.

Cowlishaw, G., & Dunbar, R. (2000). *Primate conservation biology*. University of Chicago Press.

de Thoisy, B., Renoux, F., & Julliot, C. (2005). Hunting in northern French Guiana and its impacts on primates communities. *Oryx, 39*, 149–157.

de Thoisy, B., Richard-Hansen, C., & Peres, C. A. (2009). Impacts of subsistence game hunting on Amazonian primates. In P. A. Garber, A. Estrada, J. C. Bicca-Marques, E. W. Heymann, & K. B. Strier (Eds.), *South American primates. Developments in primatology: Progress and prospects*. Springer.

Dounias, E. (2016). From subsistence to commercial hunting: Technical shift in cynegetic practices among Southern Cameroon forest dwellers during the 20th century. *Ecology and Society, 21*(1), 23.

Estrada, A., Garber, P. A., Rylands, A. B., Roos, C., Fernandez-Duque, E., Di Fiore, A., Nekaris, K. A. I., Nijman, V., Heymann, E. W., Lambert, J. E., Rovero, F., Barelli, C., Setchell, J. M., Gillespie, T. R., Mittermeier, R. A., Arregoitia, L. V., Guinea, M., Gouveia, S., Dobrovolski, R., Shanee, S., Shanee, N., Boyle, S. A., Fuentes, A., MacKinnon, K. C., Amato, K. R., Meyer, A. L. S., Wich, S., Sussman, R. W., Pan, R., Kone, I., & Li, B. (2017). Impending extinction crisis of the world's primates: Why primates matter. *Science Advances, 3*(1), e1600946.

Fa, J. E., & Brown, D. (2009). Impacts of hunting on mammals in African tropical moist forests: A review and synthesis. *Mammal Review, 39*(4), 231–264.

Fa, J. E., & Peres, C. (2001). Game vertebrate extraction in African and Neotropical forests: An intercontinental comparison. In J. Reynolds, G. Mace, J. G. Robinson, & K. Redford (Eds.), *Conservation of exploited species* (pp. 203–241). Cambridge University Press.

Fa, J. E., Peres, C. A., & Meuuwig, J. (2002). Bushmeat exploitation in tropical forests: An intercontinental comparison. *Conservation Biology, 16*, 232–241.

Greatorex, Z. F., Olson, S. H., Singhalath, S., Silithammavong, S., Khammavong, K., Fine, A. E., Weisman, W., Douangngeun, B., Theppangna, W., Keatts, L., Gilbert, M., Karesh, W. B., Hansel, T., Zimicki, S., O'Rourke, K., Joly, D. O., & Mazet, J. A. K. (2016). Wildlife trade and human health in Lao PDR: An assessment of the zoonotic disease risk in markets. *PLoS One, 11*(3), e0150666.

Gurven, M., & Rueden, C. V. (2006). Hunting, social status and biological fitness. *Social Biology, 53*, 81–99.

Hahn, B., Shaw, G. M., de Cock, K. M., & Sharp, P. M. (2000). AIDS as a zoonosis: Scientific and public health implications. *Science, 287*, 607–613.

Hegerl, C., Burgess, N. D., Nielsen, M. R., Ciolli, E. M. M., & Overo, F. R. (2017). Using camera trap data to assess the impact of bushmeat hunting on forest mammals in Tanzania. *Oryx, 51*(1), 87–97.

Hirsch, V. M., Olmsted, R. A., Murphey-Corb, R. H., Purcell, R. H., & Johnson, P. R. (1989). An African primate lentivirus (SIVsm) closely related to HIV-2. *Nature, 339*, 389–392.

Hoffman, L. C., & Cawthorn, D.-M. (2012). What is the role and contribution of meat from wildlife in providing high quality protein for consumption? *Animal Frontiers, 2*(4), 40–53.

Hoffmann, M., Hilton-Taylor, C., Angulo, A., Böhm, M., Brooks, T. M., Butchart, S. H., Carpenter, K. E., et al. (2010). The impact of conservation on the status of the world's vertebrates. *Science, 330*(6010), 1503–1509.

Hughes, A. C. (2017). Understanding the drivers of Southeast Asian biodiversity loss. *Ecosphere, 8*(1), 1–33.

Jenkins, R. K. B., Keane, A., Rakotoarivelo, A. R., Rakotomboavonjy, V., Randrianandrianina, F. H., Razafimanahaka, H. J., Ralaiarimalala, S. R., & Jones, J. P. G. (2011). Analysis of patterns of bushmeat consumption reveals extensive exploitation of protected species in Eastern Madagascar. *PLoS One, 6*(12), e27570.

Jimoh, S. O., Ikyaagba, E. T., Alarape, A. A., Obioha, E. E., & Adeyemi, A. A. (2012). The role of traditional laws and taboos in wildlife conservation in the Oban Hill Sector of Cross River National Park (CRNP), Nigeria. *Journal of Human Ecology, 39*, 209–219.

Jones, J. P. G., Andriamarovololona, M. M., & Hockley, N. (2008). The importance of taboos and social norms to conservation in Madagascar. *Conservation Biology, 22*(4), 976–986.

Jost Robinson, C. A. (2012). *Beyond hunters and hunted: An integrative anthropology of human-wildlife dynamics and resource use in a central African forest* (Doctoral dissertation). Purdue University.

Jost Robinson, C. A., & Remis, M. J. (2018). Engaging holism: Exploring multispecies approaches in ethnoprimatology. *International Journal of Primatology, 39*, 776–796.

Jost Robinson, C. A., Daspit, L. L., & Remis, M. J. (2017). Monkeys on the menu? Reconciling patterns of primate hunting and consumption in a central African village. In M. Waller (Ed.), *Ethnoprimatology. Developments in primatology: Progress and prospects* (pp. 47–61). Springer.

Keele, B. F., Jones, J. H., Terio, K. A., Estes, J. D., Rudicell, R. S., Wilson, M. L., Li, Y., Learn, G. H., Beasley, T. M., Schumacher-Stankey, J., Wroblewski, E., Mosser, A., Raphael, J., Kamenya, S., Londsdorf, E. V., Travis, D. A., Mlengeya, T., Kinsel, M. J., Else, J. G., Silvestri, G., Goodall, J., Sharp, P. M., Shaw, G. M., Pusey, A. E., & Hahn, B. H. (2009). Increased mortality and AIDS-like immunopathology in wild chimpanzees infected with SIVcpz. *Nature, 460*, 515–519.

Koné, I., & Refisch, J. (2005). The influence of poaching on the behaviour of monkeys. In R. Noë, S. McGraw, & K. Zuberbühler (Eds.), *Monkeys of the Taï forest: An African primate community*. Cambridge University Press.

Koné, I., Lambert, J. E., Refisch, J., & Bakayoko, A. (2008). Primate seed dispersal and its potential role in maintaining useful tree species in the Taï region, Côte-d'Ivoire: Implications for the conservation of forest fragments. *Tropical Conservation Science, 1*(3), 291–304.

LeBreton, M., Prosser, A. T., Tamoufe, U., Sateren, W., & Mpoudi-Ngole, E. (2006). Patterns of bushmeat hunting and perceptions of disease risk among central African communities. *Animal Conservation, 9*(4), 357–363.

Lee, T. M., Sigouin, A., Pinedo-Vasquez, M., & Nasi, R. (2000). The harvest of tropical wildlife for bushmeat and traditional medicine. *Annual Review of Environment and Resources, 45*, 145–170.

Leendertz, S. A. J., Gogarten, J. F., Düx, A., Calvignac-Spencer, S., & Leendertz, F. H. (2016). Assessing the evidence supporting fruit bats as the primary reservoirs for Ebola viruses. *EcoHealth, 13*(1), 18–25.

Lehman, S. M. (2000). Primate community structure in Guyana: A biogeographic analysis. *International Journal of Primatology, 21*, 333–351.

Leroy, E. M., Rouquet, P., Formenty, P., Souquière, S., Kilbourne, A., Froment, J.-M., Bermejo, M., Smit, S., Karesh, W., Swanepoel, R., Zaki, S. R., & Rollin, P. E. (2004). Multiple Ebola virus transmission events and rapid decline of Central African wildlife. *Science, 303*(5656), 387–390.

Linder, J. M., Cronin, D. T., Ting, N., Abwe, E., Davenport, T., Detwiler, K., Galat, G., Galat-Luong, A., Hart, J., Ikemeh, R., Kivai, S., Koné, I., Kujirakwinja, D., Maisels, F., McGraw, S., Oates, J., & Struhsaker, T. (2021). *Red colobus (Piliocolobus) conservation action plan, 2021–2026*. IUCN.

Lindsey, P. A., Roulet, P. A., & Romanach, S. S. (2007). Economic and conservation significance of the trophy hunting industry in sub-Saharan Africa. *Biological Conservation, 134*, 455–469.

Lively, C. M., Craddock, C., & Vrijenhoek, R. C. (1990). Red Queen hypothesis supported by parasitism in sexual and clonal fish. *Nature, 344*, 864–866.

McGraw, W. S. (2005). Update on the search for Miss Waldron's red colobus. *International Journal of Primatology, 26*, 605–619.

McGraw WS (2007) Vulnerability and conservation of the Taï monkey fauna. In Noë R, McGraw S, Zuberbühler K (eds) : Cambridge University Press. Pp. 290–316.

McGraw, W. S., & Oates, J. F. (2002). Evidence for a surviving population of Miss Waldron's Red Colobus. *Oryx, 36*(3), 223.

Milner-Gulland, E. J., & Bennett, E. L. (2003). Wild meat— The bigger picture. *Trends in Ecology and Evolution, 18*, 351–357.

Mukesh, K. C. (2010). Sustainable usage of animals by the rural people in reference to Nepal. In P. K. Jha, S. B. Karmacharya, M. K. Balla, M. K. Chettri, & B. B. Shrestha (Eds.), *Sustainable use of biological resources in Nepal* (pp. 145–151). Ecological Society (ECOS).

Naidoo, R., Weaver, L. C., Diggle, R. W., Matongo, G., Stuart-Hill, G., & Thouless, C. (2016). Complementary benefits of tourism and hunting to communal conservancies in Namibia. *Conservation Biology, 30*(3), 628–638.

Nasi, R., Taber, A., & Van Vliet, N. (2011). Empty forests, empty stomachs? Bushmeat and livelihoods in the Congo and Amazon Basins. *International Forestry Review, 13*, 355–368.

Oates, J. F., Adedi-Lartey, M., Scott, M. W., Struhsaker, T. T., & Whitesides, G. H. (2000). Extinction of the West African red colobus monkey. *Conservation Biology, 14*, 1526–1532.

Oates, J. F., Koné, I., McGraw, S., & Osei, D. (2020). *Piliocolobus waldroni* (amended version of 2019 assessment). The IUCN Red List of Threatened Species2020:e.T18248A166620835.

Obioha, E. E., Isiugo, P. N., Jimoh, S. O., Ikyaagba, E., Ngoufo, R., Serge, B. K., & Waltert, M. (2012). Bush meat harvesting and human subsistence nexus in the Oban Hill communities of Nigeria. *Journal of Human Ecology, 38*(1), 49–64.

Okello, M., & Kiringe, J. (2004). Threats to biodiversity and their implications in protected and adjacent dispersal areas of Kenya. *Journal of Sustainable Tourism, 12*, 55–69.

Pavlin, B. I., Schloegel, L. M., & Daszak, P. (2009). Risk of importing zoonotic diseases through wildlife trade, United States. *Emerging Infectious Diseases, 15*(11), 1721–1726.

Peeters, M., & Courgnaud, V. (2002). Overview of primate lentiviruses and their evolution in non-human primates in Africa. In C. Kuiken, B. Foley, E. Freed, B. Hahn, B. Korber, P. A. Marx, F. McCutchan, J. W. Mellors, & S. Wolinksy (Eds.), *HIV sequence compendium 2002* (pp. 2–23). Theoretical Biology and Biophysics Group, Los Alamos National Laboratory, Los Alamos, NM, LA-UR 03-3564.

Peeters, M., & Delaporte, E. (2012). Simian retroviruses in African apes. *Clinical Microbiology and Infection, 18*, 514–520.

Peeters, M., Courgnaud, V., Abela, B., Auzel, P., Pourrut, X., Bibollet-Ruche, F., Loul, S., Liegeois, F., Butel, C., Koulagna, D., Mpoudi-Ngole, E., Shaw, G. M., Hahn, B. H., & Delaporte, E. (2002). Risk to human health from a plethora of Simian immunodeficiency viruses in primate bushmeat. *Emerging Infectious Diseases, 8*(5), 451–457.

Peres, C. A. (1990). Effects of hunting on western Amazonian primate communities. *Biological Conservation, 54*, 47–59.

Peres, C. A. (2000). Effects of subsistence hunting on vertebrate community structure in Amazonian forests. *Conservation Biology, 14*, 240–253.

Peres, C. A. (2011). Conservation in sustainable-use tropical forest reserves. *Conservation Biology, 25*(6), 1124–1129.

Peres, C. A., & Nascimento, H. S. (2006). Impact of game hunting by the Kayapo of southeastern Amazonia: Implications for wildlife conservation in Amazonian indigenous reserves. *Biodiversity and Conservation, 15*, 2627–2653.

Peres, C. A., & Palacios, E. (2007). Basin-wide effects of game harvest on vertebrate population densities in Amazonian forests: Implications for animal-mediated seed dispersal. *Biotropica, 39*, 304–315.

Powell, B., Ickowitz, A., McMullin, S., Jamnadass, R., Padoch, C., & Pinedo-Vasquez, M. (2013). The role of forests, trees and wild biodiversity for improved nutrition-sensitivity of food and agriculture systems. In *Expert background paper for the International Conference on Nutrition 2*. Rome, Italy.

Pratt, D. G., Macmillan, D. C., & Gordon, I. J. (2004). Local community attitudes to wildlife utilisation in the changing economic and social context of Mongolia. *Biodiversity and Conservation, 13*, 591–613.

Redford, K. H. (1992). The empty forest. *Bioscience, 42*, 412–422.

Refisch, J., & Koné, I. (2001). *Influence du braconnage sur les populations simiennes: un exemple tiré d'une forêt primaire à régime pluvieux en Côte-d'Ivoire.* Tropenökologischen Begleitprogrammes - GTZ.

Refisch, J., & Koné, I. (2005). The impact of market hunting on monkey populations in the Taï region, Côte-d'Ivoire. *Biotropica, 37*(1), 136–144.

Remis, M. J., & Jost Robinson, C. A. (2014). Examining short-term nutritional status among Aka foragers in transitional economies: Implications for women's health across the lifespan. *American Journal of Physical Anthropology, 154*(3), 365–375.

Ripple, W. J., Abernethy, K., Betts, M. G., Chapron, G., Dirzo, R., Galetti, M., Levi, T., Lindsey, P. A., MacDonald, D. W., Machovina, B., Newsome, T. M., Peres, C. A., Wallach, A. D., Wolf, C., & Young, H. (2016). Bushmeat hunting and extinction risk to the world's mammals. *Royal Society Open Science, 3*, 160498.

Robinson, J. G., & Bennett, E. L. (2000). *Hunting for sustainability in tropical forests.* Columbia University Press.

Saayman, M., Van der Merwe, P., & Rossouw, R. (2011). The impact of hunting for biltong purposes on the SA economy. *Acta Commercii, 11*(1), 1–12.

Sarti, F. M., Adams, C., Morsello, C., van Vliet, N., Schor, T., Yagüe, B., Tellez, L., Quiceno-Mesa, M. P., & Cruz, D. (2015). Beyond protein intake: Bushmeat as source of micronutrients in the Amazon. *Ecological Society, 20*(4), 22.

Schulte-Herbruggen, B., Cowlishaw, G., Homewood, K., & Rowcliffe, J. M. (2017). Rural protein insufficiency in a wildlife-depleted West African farm-forest landscape. *PLoS One, 12*, e0188109.

Secretariat of the Convention on Biological Diversity. (2011). *Livelihood alternatives for the unsustainable use of bushmeat.* Report prepared for the CBD Bushmeat Liaison Group. Technical Series No. 60, Montreal, SCBD, 46 pages.

Shaffer, C. A., Marawanaru, E., & Yukuma, C. (2017). An ethnoprimatological approach to assessing the sustainability of primate subsistence hunting of indigenous Waiwai in the Konashen Community Owned Conservation Concession, Guyana. In K. M. Dore, E. P. Riley, & A. Fuentes (Eds.), *Ethnoprimatology: A practical guide to research on the human-nonhuman primate interface* (pp. 219–231). Cambridge University Press.

Sharp, P. M., & Hahn, B. H. (2010). The evolution of HIV-1 and the origin of AIDS. *Philosophical Transactions of the Royal Society B: Biological Sciences, 365*, 2487–2494.

Shultz, S. (2001). Notes on interactions between monkeys and African crowned eagles in the Taï National Park, Ivory Coast. *Folia Primatolica, 72*, 248–250.

Sirén, A., & Machoa, J. (2008). Fish, wildlife, and human nutrition in tropical forests: A fat gap? *Interciencia, 33*(3), 186–193.

Stephen, B., & Inkamba-Nkulu, C. (2004). Fruit, minerals, and forest elephant trails: Do all roads lead to Rome? *Biotropica, 36*(3), 392–401.

Subramanian, M. (2012). Zoonotic disease risk and the bushmeat trade: Assessing awareness among hunters and traders in Sierra Leone. *EcoHealth, 9*, 471–482.

Topp-Jørgensen, E., Nielsen, M. R., Marshall, A. R., & Pedersen, U. (2009). Relative densities of mammals in response to different levels of bushmeat hunting in the Udzungwa Mountains, Tanzania. *Tropical Conservation Science, 2*(1), 70–87.

Urbani, B. (2006). A survey of primate populations in Northeastern Venezuelan Guyana. *Primate Conservation, 20*, 47–52.

Van der Merwe, P., Saayman, M., & Rossouw, R. (2014). The economic impact of hunting: A regional approach. *South African Journal of Economy Management Science, 17*(4), 379–395.

Voysey, B. C., McDonald, K. E., Rogers, M. E., Tutin, C. E. G., & Parnell, R. J. (1999a). Gorillas and seed dispersal in the Lopé Reserve, Gabon. I: Gorilla acquisition by trees. *Journal of Tropical Ecology, 15*, 23–38.

Voysey, B. C., McDonald, K. E., Rogers, M. E., Tutin, C. E. G., & Parnell, R. J. (1999b). Gorillas and seed dispersal in the Lopé Reserve, Gabon. II: Survival and growth of seeds. *Journal of Tropical Ecology, 15*, 39–60.

Walsh, P. D., Abernethy, K. A., Bermejo, M., Beyers, R., De Wachter, P., Akou, M. E., Juijbregts, B., Mambounga, D. I., Toham, A. K., Kilbourn, A. M., Lahm, S. A., Latour, S., Maisels, F., Mbina, C., Mihindou, Y., Obiang, S. N., Effa, E. N., Starkey, M. P., Telfer, P., Thibault, M., Tutin, C. E. G., White, L. J. T., & Wilkie, D. S. (2003). Catastrophic ape decline in western equatorial Africa. *Nature, 422*, 611–614.

Wato, Y., Wahungu, G., & Okello, M. (2006). Correlates of wildlife snaring patterns in Tsavo West National Park, Kenya. *Biological Conservation, 132*, 500–509.

Wilkie, D. S., Curran, B., Tshombe, R., & Morelli, G. A. (1998). Modelling the sustainability of subsistence farming and hunting in the Ituri Forest of Zaire. *Conservation Biology, 12*(1), 137–147.

Wilkie, D. S., Bennett, E. L., Peres, C. A., & Cunningham, A. A. (2011). The empty forest revisited. *Annals of the New York Academy of Sciences, 1223*, 120–128.

Willi, Y., Van Buskirk, J., & Hoffmann, A. A. (2006). Limits to the adaptive potential of small populations. *Annual Review of Ecology, Evolution, and Systematics, 37*, 433–458.

Wolfe, N. D., Switzer, W. M., Carr, J. K., Bhullar, V. B., Shanmugam, V., Tamoufe, U., Prosser, A. T., Torimiro, J. N., Wright, A., Mpoudi-Ngole, E., McCutchan, F. E., Birx, D. L., Folks, T. M., Burke, D. S., & Heneine, W. (2004). Naturally acquired simian retrovirus infections in Central African hunters. *The Lancet, 363*, 932–937.

Wolfe, N. D., Heneine, W., Carr, J. K., Garcia, A. D., Shanmugam, V., Tamoufe, U., Torimiro, J. N., Prosser, A. T., LeBreton, M., Mpoudi-Ngole, E., McCutchan, F. E., Birx, D. L., Folks, T. M., Burke, D. S., & Switzer, W. M. (2005). Emergence of unique primate T-lymphotropic viruses among central African bushmeat hunters. *PNAS, 102*(22), 7994–7999.

Worobey, M., Telfer, P., Souquière, S., Hunter, M., Coleman, C. A., Metzger, M. J., Reed, P., Makuwa, M., Hearn, G., Honarvar, S., Roques, P., Apetrei, C., Kazanji, M., & Marx, P. A. (2010). Island biogeography reveals the deep history of SIV. *Science, 329*(5998), 1487.

Zadou, D. A., Koné, I., Mouroufié, K. V., Adou-Yao, C. Y., Gléanou, K. E., Kablan, A. Y., Coulibaly, D., & Ibo, G. J. (2011). La valeur de la Forêt des Marais Tanoé-Ehy (sud-est de la Côte d'Ivoire) pour la conservation: Dimension socio-anthropologique. *Tropical Conservation Science, 4*(4), 373–385.

Zuberbühler, K., Jenny, D., & Bshary, R. (1999). The predator deterrence function of primate alarm calls. *Ethology, 105*, 477–490.

Chapter 5
Dogs, Primates, and People: A Review

Siân Waters, Tamlin Watson, Zach J. Farris, Sally Bornbusch,
Kim Valenta, Tara A. Clarke, Dilip Chetry, Zoavina Randriana,
Jacob R. Owen, Ahmed El Harrad, Arijit Pal, and Chandrima Home

Contents

Abstract People are assisted by dogs in many activities which may bring them into contact with primates, often leading to negative interactions and outcomes for one or other species. People's perceptions and behaviour towards dogs vary and are influenced by cultural and other factors. We present incidents of dog-primate harassment and predation found during a literature review. We found that dog-primate contact can result in negative interactions and outcomes for one or other species, and we discuss how dogs influence primate populations globally via indirect inter-

S. Waters (✉)
Barbary Macaque Awareness and Conservation, Tetouan, Morocco

Department of Anthropology, Durham University, Durham, UK

T. Watson · A. El Harrad
Barbary Macaque Awareness and Conservation, Tetouan, Morocco

Z. J. Farris
Department of Health and Exercise Science, Appalachian State University, Boone, NC, USA

S. Bornbusch
Smithsonian Institute, National Zoological Park and Conservation Biology Institute, Washington, DC, USA

K. Valenta
Mad Dog Initiative, Antananarivo, Madagascar

Department of Anthropology, University of Florida, Gainesville, FL, USA

© The Author(s), under exclusive license to Springer Nature Switzerland AG 2023   61
T. McKinney et al. (eds.), *Primates in Anthropogenic Landscapes*, Developments in Primatology: Progress and Prospects, https://doi.org/10.1007/978-3-031-11736-7_5

actions such as the transmission of disease. Observing direct interactions between dogs and primates is exceedingly rare due to the difficulty associated with observing these encounters, and we introduce single-species occupancy modelling as a method to conduct non-invasive research to investigate the effects of dogs on primates. We explore methods for mitigating dog-primate interactions. Finding effective ways to manage dog populations in collaboration with their owners and/or changing those owners' behaviour in relation to their dogs is emerging as yet another challenge for primate conservation practitioners.

Keywords Conservation · Dog-primate interactions · Disease transmission · Human-animal relations · Occupancy modelling · One Health Approach

5.1   Introduction

Dogs (*Canis familiaris*) were first domesticated around 16,000 years BP and have been closely associated with people ever since (Perri, 2016). Due to this commensal affiliation (Gompper, 2014), free-roaming domestic dogs are now ubiquitous in landscapes everywhere except Antarctica and have diverse and, sometimes, catastrophic effects on native wildlife species (Young et al., 2011). We define free-roaming dogs (hereafter dogs) as "dogs owned by one or more individuals or families that spend the majority of their time unconfined, able to roam freely away from their owner, and may acquire some or most of their food from a source other than their owner" (Kshirsagar et al., 2020:483). We differentiate between crop protection, livestock guarding and hunting dogs while appreciating that they may have multiple roles and also roam freely when not working. Dog presence brings the capacity for negative interactions, including zoonotic disease transmission, parasite exchange, a propensity for livestock and wildlife predation, as well as risk of injury to humans (Gompper, 2014; Young et al., 2011). Inclusion of the "One Health" concept acknowledges a long-held understanding of the interconnected relationship between human health, animal health, and the health of the ecosystems in which they exist (Mackenzie & Jeggo, 2019; OIE, 2021).

T. A. Clarke · Z. Randriana
Mad Dog Initiative, Antananarivo, Madagascar

D. Chetry
Primate Research and Conservation Division, Aranyak – A Scientific and Industrial Research Organization, Guwahati, Assam, India

J. R. Owen
Los Angeles Zoo and Botanical Gardens, Los Angeles, CA, USA

A. Pal
Animal Behaviour and Cognition Programme, School of Natural and Engineering Sciences, National Institute of Advanced Studies, Bengaluru, Karnataka, India

C. Home
Indian Institute of Science, Bengaluru, Karnataka, India

Despite the interest in dog-wildlife interactions for carnivore and ungulate species (e.g. Gompper, 2014), the frequency and types of interactions between dogs and primates are not well known. A review of literature prior to 1986 revealed only seven incidences of primate mortalities due to dog attacks – the majority of which were on Asian terrestrial primates although multiple incidences of dog-primate interactions were recorded (Anderson, 1986). Reports of dog-primate interaction may have been scarce due to the difficulty of observing such events in the field and a lack of awareness of the potential importance of such interactions. For the purposes of this chapter, we use Bengtson's (2002) definition of successful predation as the predator killing but not necessarily consuming its prey. We also document primates being wounded by dogs as this has ramifications for disease spillover events, as well as incidents of dogs chasing primates which may affect primate group movements.

5.2 Human-Dog Relations

Human-dog relations can be diverse, complex, and culturally and regionally specific, and understanding them is essential when investigating the conservation of wildlife species which may be affected by dog harassment or predation (Valenta et al., 2016; Waters, et al., 2018a). Dogs repeatedly traverse from domesticated to wild behaviour, and their categorisation may shift, transgressing the spatial orderings people attempt to confine them to (Brown, 2015; Philo & Wilbert, 2000). Human-animal perceptions differ across and within societies and are shaped by factors such as the perceived merit of an animal and their utilitarian value or for their intrinsic worth within an environment, irrespective of whether wild or domesticated (Waters, Watson, et al., 2018b). Numerous influencers including culture, religion, politics, economics, history, and the attribution of symbolic meanings drive perspectives which shape how animals are treated (Hughes & Macdonald, 2013; Waters, Watson, et al., 2018b). Communities or individuals may support or vilify the domestic dogs in contact with them and see their impacts on wildlife as positive or negative accordingly (Hughes & Macdonald, 2013). People are assisted by dogs in many activities which may bring them into contact with primates. Such contact can result in negative interactions and outcomes for one or other species.

Pastoralists are often accompanied by livestock guarding dogs. However, these dogs may be distracted by primate presence, harassing and sometimes injuring or killing them (Venkataraman et al., 2015; Waters et al., 2017). People may be accompanied by their dogs for reasons of company, protection, and purposeful and opportunistic hunting (Dos Santos et al., 2018). Mushroom collectors in Bouhachem forest, Morocco, were often accompanied by their dogs which – because they were not "working" – spent long periods of time harassing Barbary macaques (Waters et al., 2017). Both the mushroom collectors and the local shepherds were indifferent to their dogs' harassment of primates (Waters unpub data), which could be frequent and prolonged, while shepherds' relations with their dogs were ambiguous at best and hostile at worst as we see in Box 5.1.

Box 5.1 People, Dogs, and Barbary Macaques
by Siân Waters

In northern Morocco, the Barbary Macaque Awareness and Conservation team investigated strategies to conserve the remaining populations of Barbary macaques (*Macaca sylvanus*) in Bouhachem forest. Shepherds use dogs to protect their livestock, and, although essential, these dogs are often treated with open hostility. For instance, it is "haram" or forbidden to allow a dog inside homes. Shepherds expected their dogs to aggressively protect their livestock from predators within the forest, though any dogs exhibiting this aggression towards humans within the village environment were sometimes fatally punished. We suggest that dogs' liminality, moving easily from domestic to wild spaces, contributes to the ambiguity of the shepherds' perceptions of them. Local shepherds believed these dogs to be feral; however, our observations identified these dogs as being owned, free-roaming village dogs rather than feral dogs.

We recorded a total of 30 events of dog harassment on 11 macaque groups over 12 months. The highest rates of occurrence ($N = 17$) occurred during the months of April and May. In general, morning attacks lasted from 7 minutes to 2 hours and 15 minutes, with a mean length of 42 minutes ($N = 9$). Late afternoon attacks were much shorter and lasted 8–27 minutes with a mean of 15 minutes ($N = 21$). The mean number of dogs observed to take part in attacks was 4, with a range of 1–18. The dogs were accompanying shepherds and herds on most occasions (59%), with 19% of attacks by unaccompanied dogs and 19% by dogs accompanying mushroom collectors. Barbary macaques are particularly susceptible to prolonged harassment from livestock guarding dogs in spring when the dogs' presence impedes the macaques' terrestrial foraging. This harassment may result in increased stress and energetically costly behaviour for lactating or pregnant female macaques (Waters et al., 2017).

Large male baboons and macaques can kill or injure crop-protecting dogs deployed by farmers to protect their crops (Hyeroba et al., 2017; Waters et al., 2017). In Uganda, the third most common cause of village dog mortality was aggressive interactions with crop-feeding olive baboons (*Papio anubis*), with the open wounds documented on 18% of the dogs in the study probably caused by these primates (Hyeroba et al., 2017).

5.2.1   Hunting Dogs and Primates

People train dogs to assist in hunting primates (Steiner et al., 2003; Waters, Watson, et al., 2018b). In Cameroon, hunters could only hunt the large, terrestrial, and Endangered drill (*Mandrillus leucophaeus*) using dogs which were used to move a drill group into a tree where they presented an easy target for armed hunters (Steiner et al., 2003). In southern Tanzania, dogs chase the Vulnerable Angolan colobus (*Colobus angolensis*) into a tree so they can be easily shot, and dogs harass red colobus (*Piliocolobus kirkii*) in Zanzibar causing them to panic and descend to the ground where they are killed (T. Davenport, pers. obs). In 2019, a researcher observed hunters with numerous dogs attacking habituated chimpanzees in Kibale, Uganda. The hunters fled, but their dogs remained, attacking a spear-wounded female and her infant son who both died of their wounds (Yong, 2019).

5.3   Review of Dog-Primate Interactions

Dogs also influence primate populations globally via indirect and/or direct interactions. We present incidents of dog-primate harassment and predation found during our review of the literature in Table 5.1. All of the 31 primate species listed in Table 5.1 are categorised according to their conservation status on the IUCN Red List. 6% ($n = 2$) are Critically Endangered, 32% ($n = 10$) are Endangered, 26% ($n = 8$) are classed as Vulnerable, 3% ($n = 1$) are classed as Near Threatened, and 32% ($n = 9$) are classed as Least Concern. Unsurprisingly, most dog attacks on primates take place when the latter are on the ground. Arboreal primates are very vulnerable when descending to the ground to move between forest patches in search of food and secure sleeping places (Candelero-Rueda & Pozo-Montuy, 2010). The Endangered Central American howler (*Alouatta pigra*) in Mexico was observed moving up to a maximum of 200 m on the ground in one study (Pozo-Montuy et al., 2013) leaving them very vulnerable to dog predation. This vulnerability is illustrated by many observations ($N = 49$) of *A. pigra* being killed by dogs (Candelero-Rueda & Pozo-Montuy, 2010; Pozo-Montuy et al., 2013). Species like the southern brown howler (*A. guariba clamitans*) live in urbanised or urbanising spaces in Brazil making interactions with dogs an increased risk (see Table 5.1). Primate groups restricted to a single remaining forest fragment are also very vulnerable, and two groups in Uganda were completely extirpated by dogs (Goldberg et al., 2008). Dog predation is rated as the fourth most significant threat to Endangered golden langurs in Bhutan due to forest destruction to build infrastructure such as roads (Thinley et al., 2020) and is an important threat to the species in Assam, India. For example, Chetry et al. (2010) reported the deaths of seven golden langurs due to dog bites with a further eight golden langurs killed by dogs in three other areas of Assam (Chetry et al., In press; AK Das, pers. comm.) There are reports of dogs often attacking langurs by trying to catch their long tails, making these species generally more vulnerable to attacks while on the ground or in low forest canopy.

Table 5.1 Reports of dog disturbance, harassment, and killing of primates

Family	Species and Red List status	Location	Targets	Time frame	Circumstance	Source
Lemuridae	Southern bamboo lemurs (*Hapalemur meridionalis*) VU	Southern Madagascar	Group	12 months	Disturbed feeding	Eppley et al. (2016)
	Ring-tailed lemur *Lemur catta* EN and Verreaux's sifaka (*Propithecus verreauxi*) CR	Southern Madagascar	1 J L. *L. catta* 3 U Unspecified species	10 years	Unspecified	Brockman et al. (2008); Moresco et al. (2012)
	Perrier's sifaka (*P. perrieri*) CR	North Madagascar	1 U	4 weeks (interviews)	Unspecified	Anania et al. (2018)
Galagidae	White-tailed small-eared galago (*Otolemur garnettii lasiotis*) LC	Taita Hills, Kenya	1 M	Single observation	Unspecified	Pihlström et al. (2021)
Callitrichidae	Black-pencilled marmoset (*Callithrix penicillata*) LC	Southern Brazil	1 AF 1 U killed	30 months	Unspecified	De Azevedo Fernandes et al. (2020)
	Wied's marmoset (*C. khulii*) VU	Agroforestry, Brazil	Group	9 months	Chased once	dos Santos et al. (2018)
	Golden-headed lion tamarin (*Leontopithecus chrysomelas*) EN		7 groups	17 months	Each group attacked at least once	Oliveira and Dietz (2011)
			Group	9 months	Chased twice	dos Santos et al. (2018)
Cebidae	Black-horned capuchin (*Sapajus nigritus*) NT	Santa Genebra Reserve, Brazil	2 U killed	44 months	Crossing gaps of forest fragments	Galetti and Sazima (2006)
			1 AF killed	11 months		Galetti and Sazima (2006)
		Minas Gerais, Brazil	1 AM killed	Opportunistic observation		Oliveira et al. (2008)

Atelidae	Southern brown howler (*Alouatta guariba clamitans*) VU	Santa Genebra Reserve, Brazil	2 U killed	44 months	Crossing gaps of forest fragments	Galetti and Sazima (2006)
		Southern Brazil	3 AM; 1 AF; 10 U killed	30 months		De Azevedo Fernandes et al. (2020)
		Southern Brazil	5 U killed	48 months	Crossing gaps of forest fragments	Chaves and Bicca-Marques (2017)
		Southern Brazil	1 sub-AM	Single obs.		Ferriera Dasilva et al. (2021)
	Mantled howler (*A. palliata*) VU	Costa Rica	1 U killed	20 months	Crossing gaps of forest fragments in a peri-urban landscape	Lindshield (2016)
	Central American black howler (*A. pigra*) EN	Tabasco, Mexico	47 U killed	6 months	Crossing gaps of forest fragments	Candelero-Rueda and Pozo-Montuy (2010)
		Tabasco, Mexico	1 AF; 1 J killed	8 months		Pozo-Montuy et al. (2013)

(continued)

Table 5.1 (continued)

Family	Species and Red List status	Location	Targets	Time frame	Circumstance	Source
Cercopithecidae	Barbary macaque (*Macaca sylvanus*) EN	Ifrane NP Morocco	6 I killed	Unspecified	Terrestrial foraging	Camperio-Ciani and Mouna (2006)
		Bouhachem forest Morocco	3 I killed; 1 AF wounded	12 months		Waters et al. (2017)
	Bonnet macaque (*M. radiata*) LC Munzala macaque (*M. munzala*) EN Lion-tailed macaque (*M. silenus*) EN Stump-tailed macaque (*M. arctoides*) VU	India	Unspecified	Online survey	Unspecified	Home et al. (2018)
	Long-tailed macaque (*M. fascicularis*) VU	Singapore	1 J killed	Opportunistic	Terrestrial foraging	Riley et al. (2015)
		Piak Nam Yai island, Thailand	Groups	5 months	Chasing (13)	Gumert et al. (2013)
	Nicobar long-tailed macaque (*M. f. umbrosus*) VU	Nicobar Islands, India	1 J killed		Terrestrial foraging	A.Pal pers. comm
	Rhesus macaque (*M. mulatta*) LC	Assam, India	3–4 U killed	Annually	Village/forest interface	Chetry et al. (2005)
	Japanese macaque (*M. fuscata*) LC	Japan	Infants killed	Unspecified	Terrestrial foraging	Knight (2011)
	Olive baboon (*Papio anubis*) LC	Kibale NP Uganda	1 adult killed	Unspecified	Unspecified	J. Rothman pers. comm
	Gelada (*Theropithecus gelada*) LC	Guassa NP, Ethiopia	3 U killed	24 months	Terrestrial foraging	Venkataraman et al. (2015)
	Vervet monkey (*Chlorocebus pygerythrus*) LC	L. Nabugabo Uganda	2 U killed	46 months	Unspecified	Chapman et al. (2016)

Family	Species	Location	Killed	Duration	Context	Reference
	Black and white colobus (*Colobus guereza*) LC	Kibale forest, Uganda	Group killed	6 months	Forest fragment	Goldberg et al. (2008)
	Ashy red colobus (*Piliocolobus tephrosceles*) EN		Group killed		Crossing gaps in forest fragments	
	Zanzibar red colobus (*Procolobus kirkii*) EN	Jozani NP, Zanzibar Tanzania	1 AM; 1 I killed		Unspecified	Georgiev et al. (2019)
	Schlegel's banded langur (*Presbytis femoralis*) VU	Johor, Malaysia	1 AM killed	21 months	Terrestrial foraging	Najmuddin et al. (2019)
	Golden langur (*Theropithecus geei*) EN	Chakrashila Wildlife Sanctuary, Assam, India	1 AM; 3 AF; 5 J killed	12 months	Village/forest interface	Chetry et al. (2005); Chetry et al. (2010)
	Common langur (*Semnopithecus entellus*) LC Himalayan langur (*Semnopithecus ajax*) EN	Bhutan	13 U killed	3 months	Unspecified	Thinley et al. (2020)
		India	Unspecified	Online survey	Unspecified	Home et al. (2018)
Hylobatidae	Eastern hoolock gibbon (*Hoolock leuconedys*) VU	Eastern Arunachal Pradesh, India	20–25 U killed	7 years	Crossing gaps in forest fragments	Kumar et al. (2013)

CR critically endangered, *EN* endangered, *VU* vulnerable, *NT* near threatened, *LC* least concern, *AM* adult male, *AF* adult female, *J* juvenile, *I* infant, *U* unknown

Primate infants and juveniles are generally not vigilant towards predators, likely leaving them susceptible to fatal interactions with dogs (e.g. Chetry et al., 2005; Riley et al., 2015; Waters et al., 2017). Such a lack of vigilance in response to dogs may substantially affect Central American howler infant and juvenile survival in Mexico (Bonilla-Sanchez et al., 2010). In Laem Son National Park, Thailand, dog presence negatively correlates with a low proportion of juveniles in some groups of long-tailed macaques (*M. f. aurea*) although no actual predation events were observed (Gumert et al., 2013).

Primates living in habitats where they are at risk of harassment, injury, and predation by dogs will suffer disruption of group activity patterns and foraging behaviour and obstructed movement (Anderson, 1986; Gumert et al., 2013). In response, some primate species adapt their behaviour. For example, when blue monkeys (*Cercopithecus mitis*) in western Kenya shifted their territory and began feeding in plantation forest, they quickly learned active avoidance behaviour of people and dogs. The monkeys appeared to wait until the sounds of herders and their cattle and dogs grew faint before they crossed into the plantation to feed. When fleeing from approaching people or dogs, the monkeys did so silently without their usual alarm calling (Cords, 2014). Barbary macaques have also been observed moving silently away from the sounds of approaching shepherds and dogs in Bouhachem forest, Morocco (S. Waters, pers. obs.). There may well be as yet unrecognised implications for some species. For example, dog harassment of long-tailed macaques acts to disrupt the monkeys' unique tradition of tool use (Gumert et al., 2013).

5.4   Studying Dog-Primate Interactions

5.4.1   Interactions: Disease Transmission

Disease **spillover** occurs when pathogens are transmitted from reservoir populations to novel host populations (Power & Mitchell, 2004). Major human epidemics have largely resulted from spillover events from wildlife into humans (e.g. Ebola, COVID-19; Goldstein et al., 2020). While less is known about the risk of disease spillover to primates than to humans (Narat et al., 2017), it is increasingly apparent that exposure to novel pathogens can have detrimental outcomes and conservation implications for primate populations, including large-scale morbidity and mortality (Chap. 9, this volume; Walsh et al., 2003).

Pathogen spillover from dogs to primates is particularly complex. Dogs interact with humans, wildlife, and other domestic species on both sides of the anthropogenic and natural frontier, putting them at great risk of zoonotic infection (Ayinmode et al., 2018). Diseases that infect dogs as well as humans include viruses (e.g. rabies), bacteria (e.g. campylobacteriosis), parasites (e.g. echinococcosis), and fungi (e.g. sporotrichosis) (Chomel, 2014). Dogs may represent a particular disease risk to primates as they simultaneously have greater associations with humans and more frequent interactions with wildlife compared to other domestic species (Gompper, 2014).

The likelihood of pathogen spillover between dogs and primates depends on multiple variables: distribution and density of hosts and pathogens, immune and molecular responses, and the pathogen's replication and transmission rates (Plowright et al., 2017). Although disease sampling of dogs can provide estimated risks of dog-primate transmission (Alexander et al., 2016), the transmission mode of dog-primate pathogen spillover is similarly complex and can occur via three routes: direct transmission, environmental transmission, and vector-based transmission (Fig. 5.1). We discuss each of these modes with examples of confirmed and suspected dog-primate spillover events.

Direct transmission refers to the transfer of pathogens through direct contact with an infected individual or their bodily fluids (VanderWaal & Ezenwa, 2016). In the rare cases when dogs co-mingle with wild primates, such as with macaques or vervets in urban settings, direct transmission may be more common (Gompper, 2014; Lyngdoh et al., 2014). Outside these settings, however, the most plausible scenarios for dog-primate direct transmission would be failed predation attempts, which are rarely observed. Theoretically, however, primates are at risk of any directly transmitted dog-associated pathogen to which they are susceptible, e.g.

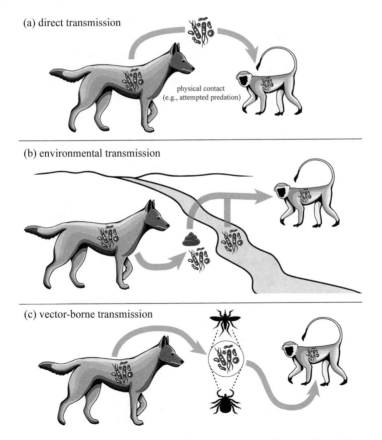

Fig. 5.1 The transmission modes of dog-primate pathogen spillover. (Illustration by Sally Bornbusch)

rabies (Kotait et al., 2019). Although direct transmission from dogs to primates may be crucial to primate conservation, the difficulty of observing direct dog-primate interactions hampers attempts to discern its importance.

Environmental transmission entails the indirect transmission of pathogens through inhalation (e.g. COVID-19) or exposure to infected water or soil (e.g. cholera or gastrointestinal parasites, respectively) (Breban et al., 2010). Because many environmentally transmitted diseases (e.g. enteric parasites) can be detected using non-invasive methods (e.g. faecal collection and microscopy; Valenta et al., 2017), they are well studied in wild primates. These pathogens can also be detected through sampling of potential environmental reservoirs, which provides information on habitat-wide disease burden and risk (Turner et al., 2016). Notably, canine parvovirus, which is commonly transmitted between dogs and soil, has high potential to spillover to and devastate native fauna, including primates (Behdenna et al., 2019; Ng et al., 2019; Rasambainarivo & Goodman, 2019).

Vector-based transmission refers to disease transfer between susceptible individuals via a third organism. Vector-borne diseases affecting primates include malaria, dengue fever, leishmaniasis, and yellow fever (De Azevedo Fernandes et al., 2017; World Health Organization, 2014). Theoretically, any disease hosted by dogs and transmitted through another organism – typically arthropods, like mosquitoes or ticks – can infect primates. For example, the transmission of *Spirocerca lupi*, a beetle-borne nematode, from dogs to lemurs has been implicated in cases of lemur mortality (Blancou & Albignac, 1976). Additionally, canine heartworm (*Dirofilaria immitis*), a mosquito-borne parasite that is fairly ubiquitous in untreated dogs, was recently discovered to have spilled over into mouse lemurs (Zohdy et al., 2019). However, because it typically requires blood samples, identifying vector-borne pathogens in primates is an invasive and challenging task.

Regarding the implications of dog-vectored diseases for primate conservation, one must consider dilution effects, whereby increased biodiversity can lead to decreased pathogen prevalence (Khalil et al., 2016). Although this is not the only process that determines pathogen prevalence, by nature of their ability to regulate host populations or interfere with transmission, diverse communities can inhibit the abundance of pathogens (Schmidt & Ostfeld, 2001). Thus, just as disease spillover can impact primate conservation and biodiversity, increasing primate biodiversity through targeted conservation may dilute the spread of novel pathogens.

5.4.2   Interactions: Non-invasive Sampling and Modelling

Studying direct interactions between dogs and primates is exceedingly rare due to the difficulty associated with observing these encounters and behaviours. **Non-invasive surveying** approaches, namely, camera trapping, for dog populations provide presence/absence data to study their spatial distribution, habitat use, and behaviours. These presence/absence data for both dogs (camera traps) and primates (line-transects, acoustic detectors) can be combined to study dog-primate space use and interactions via occupancy modelling. **Occupancy modelling** allows researchers to explore species distributions and activity, habitat suitability, inter- and intraspecific interactions, disease transmission, and others in a cost-effective, limited sampling framework.

Single-species occupancy modelling has been used to investigate the effects of dogs on primates. Farris et al. (2019) incorporated dog activity and distribution to evaluate how this invasive, exotic predator influenced the occurrence and detection of multiple lemur species at Ranomafana National Park, Madagascar. The researchers found that habitat was the most important variable for explaining the occurrence and detection of most lemur species. However, for two lemur species (red-bellied lemur *Eulemur rubriventer* and eastern lesser bamboo lemur *Hapalemur griseus*), their spatial distribution and the probability of being detected by researchers were best explained by the presence of dogs. Additionally, at least one study using two-species or conditional occupancy has been conducted to explore dog-primate interactions. Farris et al. (2014) used **two-species occupancy** to explore interspecific interactions between three lemurs and multiple potential predators, including dogs. They found that mouse lemurs (*Microcebus rufus*) showed lack of co-occurrence (occurred less than at random) with dogs in contiguous forest sites. Brown lemurs (*E. albifrons*) and dogs had a positive co-occurrence (occurred more together than at random) in fragmented sites, while woolly lemurs (*Avahi laniger*) showed no spatial relationship with dogs across either type of forest. The findings of this study highlighted the threats posed to these endemic primates, particularly as forests are altered and these invasive, exotic predators become more prevalent across the landscape. Anecdotal accounts of these interactions reported in broader studies of the population dynamics, demographics, or habitat use of primates support the importance of this issue, but lack the contextual data needed for efficient management solutions.

5.5 Managing Dog-Primate Interactions

Of the primate species appearing in Table 5.1, 64% are either Critically Endangered, Endangered, or Vulnerable to extinction according to the IUCN Red List, and for some species, dog harassment and predation is a significant threat. Dog predation on arboreal primates was not reported in Anderson's (1986) survey, but it appears to be a more common occurrence than previously thought possibly due to the ever-increasing expansion of people into primate habitat. For arboreal primates, such as howlers and gibbons, being forced to the ground to move between forest patches further stresses threatened populations both through direct mortalities and by creating a landscape of fear that restricts gene flow between small, isolated groups (Serio-Silva et al., 2019). While it may be obvious that the solution to this issue is to increase connectivity perhaps by using canopy bridges, for example (Das et al., 2009; Lindshield, 2016; Chap. 2, this volume), occupancy models could prioritise and inform where and how these connections are developed, targeting resource use and improving efficacy.

Scientists working on the issue of dog predation on primates have suggested diverse actions that might be taken to control dog populations. These include the application of fines to dog owners, constraining dogs in households, and increasing awareness about the cost of dog interactions with endangered primates (Chaves & Bicca-Marques, 2017; Oliveira et al., 2008; Thinley et al., 2020). All these suggestions involve dog owners changing their behaviour as well as that of their dogs and may be difficult to enforce in many places. Some recommend capture and removal of dogs from protected areas (Oliveira et al., 2008) or mass sterilisation (Thinley et al., 2020) which may be unacceptable to local communities or impractical and costly if the numbers of dogs are extremely high. Research into human-dog relations like that undertaken in Madagascar and Morocco (Kshirsagar et al., 2020; Valenta et al., 2016; Waters, Bell, & Setchell, 2018a) is crucial to understand the best way to collaborate with dog owners to manage dog activity in primate habitats. One such initiative is ongoing in Madagascar and described in Box 5.2.

Box 5.2 The Mad Dog Initiative: A Model in the Mitigation of Free-Roaming Dog Populations
by Tara Clarke

Strategic dog population management is critical for ensuring the health and welfare of wildlife and human communities. The Mad Dog Initiative (MDI), a US-based not-for-profit, is an example of a holistic and community-based approach working to mitigate the complex issues posed by free-roaming dogs in Madagascar. MDI employs an innovative and diverse approach that includes documenting and analysing the spread of zoonotic disease, spay/neuter/vaccination campaigns, wildlife monitoring, education, and outreach.

MDI has focused its expeditions and long-term research at Ranomafana and Andasibe-Mantadia National Parks. To date, the team has analysed thousands of dog and wildlife samples for evidence of zoonotic disease, sterilised over 1000 dogs in protected areas using mobile veterinary clinics, and vaccinated thousands. Additionally, MDI conducts camera-trap and lemur-transect surveys to assess the effect of dog populations on wildlife populations. MDI also regularly surveys people living in and around protected areas to better understand regional attitudes towards dogs and how these can be used to maximise the effectiveness of veterinary interventions. A critical part of MDI's mission is to share knowledge contributing to the inclusive development of Malagasy veterinarians and students.

MDI's holistic approach has enabled it to gain insight into human-dog relationships and dynamics, as well as of dogs' impacts on endangered and endemic wildlife. Through targeted sterilisation and vaccination campaigns, MDI has been able to address and reduce growing dog populations. Scientists and researchers might consider employing MDI's model to more successfully mitigate issues posed by dogs. While the dog dilemma is complex, and every region, community, and village is unique, strategic modifications may be necessary for achieving similar goals in different settings.

Waters and colleagues (2018b) describe how they needed to impart their findings that the dogs thought to be feral by shepherds were, in fact, owned dogs (see Box 5.1). The authors developed a One Health programme to improve the health of owned dogs to safeguard both human and animal health. They administered rabies vaccinations to owned dogs and provided their owners with brightly coloured dog collars. After observing collared dogs hunting in the forest, the shepherds realised the dogs were owned, thus avoiding a potential conflict between locals and conservationists. Many shepherds began to castrate their dogs to prevent them from roaming. Vaccination coverage was high, protecting both people and their livestock from rabies, and communities became more positive about the conservation team's presence because the programme was salient to their concerns (Waters et al., 2018b).

5.6 Discussion

Understanding the complex ways animal and human lives intertwine, their mutual embodiment within spaces and landscapes, and the ways they are categorised and conceptualised in relation to other species is essential in the development of interventions attuned to local perceptions (Brown, 2015). Whether people classify dogs as free-ranging, feral, or owned, all threaten endemic wildlife populations as agents of harassment or disease (Sparkes et al., 2016) as most live unrestrained and all may hunt with or without the presence of humans. Villatoro et al. (2018) reported that dog owners agreed unowned dogs who hunted wild animals required control, but only a limited number indicated they would actually stop their own dogs if they attacked wildlife, expressing a dichotomy in acceptable behaviour between owned and unowned dogs. This denial may be because people are unable to distinguish between owned and stray animals (Paolini et al., 2020) or because of a failure to acknowledge dogs as individuals (Waters, Watson, et al., 2018b). Thus, by focusing on "biological" behaviour and ignoring individual animal agency (Fox, 2006), dog owners effectively absolve themselves of any responsibility for the problems associated with their dogs. Dog owners in one study failed to acknowledge dog-wildlife interactions at all, perhaps indicating participants might see domestic dogs as acting purely within a "human-dominated" context and not as carnivores within wilder ecosystems (Shüttler et al., 2017).

How humans perceive and construct their relationships with animals and their environments in relation to their own needs shapes their appetite to regulate their own conduct, with potentially detrimental impacts on other species. Understanding the complex ways animal and human lives intertwine, their mutual embodiment within spaces and landscapes, and the ways they are categorised and conceptualised in relation to other species is essential in the development of interventions attuned to local perceptions (Brown, 2015). It is therefore imperative to make conservation approaches to dog management more inclusive to increase the chances of positive acceptance and strong community support (Brown, 2015; Kshirsagar et al., 2020). Our understanding of people's dynamic cultural adaptations to and beliefs and attitudes about dog-primate interactions will be fundamental in the recruitment of dog owners into programmes enabling them to manage their dogs in a more wildlife-friendly manner and to ensure owners understand the importance of maintaining dog health and population control via vaccination and sterilisation programmes under a One Health umbrella.

Continued research into the disease ecology of dogs and primates has demonstrated that dog presence within wildlife habitats facilitates the transmission of infectious diseases. As primate habitats are further eroded, pathogen spillovers are expected to accelerate. The extent of pathogen-specific primate morbidity and mortality and the associated conservation implications are important considerations for future research. Using single- and two-species occupancy that incorporates variables or covariates presents a promising approach to explore these difficult-to-study interspecific interactions between primates and invasive, exotic dog populations. These models help further clarify existing hypotheses in the literature regarding primate-dog interactions and inform conservation measures. The limited number of studies that incorporate this occupancy and/or co-occurrence modelling approach has revealed their effectiveness at exploring these relationships, and researchers should consider their use, particularly in areas where dogs or similar predators may be influencing primate spatial distribution.

Primate conservationists need to think outside the box to find innovative solutions to mitigate dog-primate interactions while working with communities to manage and maintain smaller, healthier dog populations. Including a One Health approach that considers the connections between humans, domestic animals, and wildlife will likely yield benefits when applied alongside traditional primate conservation efforts (Ellwanger & Chies, 2019; Waters, Watson, et al., 2018b). An additional benefit of a One Health approach is to convey a message to local people that conservationists are committed to them and their livestock's wellbeing as well as that of primates (Waters, Bell, & Setchell, 2018a). The threat posed to primates globally by an expanding dog population will likely increase as more primate habitat is fragmented. Finding effective ways to manage dog populations in collaboration with their owners and/or changing those owners' behaviour in relation to their dogs is emerging as yet another challenge for primate conservation practitioners.

Acknowledgements The authors are grateful to the many primatologists who generously shared their observations of dog-primate interactions for inclusion in this chapter.

References

Alexander, A. B., Poirotte, C., Porton, I. J., Freeman, K. L., Rasambainarivo, F., Olson, K. G., Iambana, B., & Dean, S. (2016). Gastrointestinal parasites of captive and free-living lemurs and domestic carnivores in eastern Madagascar. *Journal of Zoo Wildlife Medicine, 47*, 141–149.

Anania, A., Salmona, J., Rasolondraibe, E., Jan, F., Chikhi, L., Fichtel, C., Kappeler, P. M., & Rasoloarison, R. (2018). Taboo adherence and presence of Perrier's sifaka (*Propithecus perrieri*) in Andrafiamena forest. *Madagascar Conservation & Development, 13*, 1–14.

Anderson, J. R. (1986). Encounters between domestic dogs and free-ranging non-human primates. *Applied Animal Behaviour Science, 15*, 71–86.

Ayinmode, A. B., Oliveira, O., Obebe, H., Dada-Adgebola, A., & Ayede, W. G. (2018). Genotypic characterization of Cryptosporidium species in humans and peri-domestic animals in Ekiti and Oyo States, Nigeria. *Journal of Parasitology, 104*, 639–644.

Behdenna, A., Lembo, T., Calatayud, O., Cleaveland, S., Halliday, J. E. B., Packer, C., Lankester, F., Hampson, K., Craft, M. E., Czupryna, A., Dobson, A. P., Dubovi, E. J., Ernest, E., Fyumagwa, R., Hopgrant, J. G. C., Mentzel, C., Mzimbiri, I., Sutton, D., Willet, B., Haydon, D. T., & Viana, M. (2019). Transmission ecology of canine parvovirus in a multi-host, multi-pathogen system. *Proceedings of the Royal Society B: Biological Sciences, 286*, 20182772.

Bengtson, S. (2002). Origins and early evolution of predation. *The Paleontological Society Papers, 8*, 289–318.

Blancou, J., & Albignac, R. (1976). Note sur l'infestation des Lémuriens malgaches par Spirocerca lupi (Rudolphi, 1809). *Revue d'Elevage et de Medecine Veterinaire des Pays Tropicaux, 29*, 127–130.

Bonilla-Sanchez, Y. M., Serio-Silva, J. C., Pozo-Montuy, G., & Bynum, N. (2010). Population status and identification of potential habitats for the conservation of the Endangered black howler monkey (*Alouatta pigra*) in northern Chiapas, Mexico. *Oryx, 44*, 293–299.

Breban, R. J. M., Drake, M., & Rohani, P. (2010). A general multi-strain model with environmental transmission: Invasion conditions for the disease-free and endemic states. *Journal of Theoretical Biology, 264*, 729–736.

Brockman, D. K., Godrey, L. R., Dollar, L. J., & Ratsirarson, J. (2008). Evidence of invasive *Felis silvestris* predation on *Propithecus verreauxi* at Beza Mahafaly Special Reserve, Madagascar. *International Journal of Primatology, 29*, 135–152.

Brown, K. M. (2015). The role of landscape in regulating (ir)responsible conduct: Moral geographies of the 'proper control' of dogs. *Landscape Research, 40*, 39–56.

Camperio-Ciani, A., & Mouna, M. (2006). Human and environmental causes of the rapid *decline of Macaca sylvanus* in the Middle Atlas of Morocco. In J. K. Hodges & J. Cortes (Eds.), *The Barbary macaque: Biology, management and conservation* (pp. 257–275). Nottingham University Press.

Candelero-Rueda, R., & Pozo-Montuy, G. (2010). Mortalidad de monos aulladores negros (*Alouatta pigra*) en paisajes altamente fragmentados de Balancán, Tabasco. In L. Gamma, G. Pozo-Montuy, & W. Contreras-Sánchez (Eds.), *Perspectivas en Primatología de México* (pp. 65–76). Universidad Juárez Autónoma de Tabasco.

Chapman, C. A., Twinomugisha, D., Teichroeb, J. A., Valenta, K., Sengupta, R., Sarkar, D., & Rothman, J. M. (2016). How do primates survive among humans? Mechanisms employed by vervet monkeys at Lake Nabugabo, Uganda. In M. T. Waller (Ed.), *Ethnoprimatology: Primate conservation in the 21st century* (pp. 77–94). Springer.

Chaves, O. M., & Bicca-Marques, J. C. (2017). Crop feeding by brown howlers (*Alouatta guariba clamitans*) in forest fragments: The conservation value of cultivated species. *International Journal of Primatology, 38*, 263–281.

Chetry, D., Medhi, R., Bhattacharjee, P. C., & Patiri, B. N. (2005). Domestic dog (*Canis familiaris*): Threat for the golden langur *Trachypithecus geei. Journal of the Bombay Natural History Society, 102*, 220.

Chetry, D., Chetry, R., Ghosh, K., & Bhattacharjee, P. C. (2010). Status and conservation of golden langur in Chakrashila Wildlife Sanctuary, Assam, India. *Primate Conservation, 25*, 81–86.

Chetry, D., Phukan, M., Chetry, R., Baruah, B., & Bhattacharjee, P. C. (In press). Conservation status of an isolated population of golden langur *Trachypithecus geei* in an altered fragmented habitat in Assam, India. *Asian Primates Journal.*

Chomel, B. B. (2014). Emerging and re-emerging zoonoses of dogs and cats. *Animals, 4*, 434–445.

Cords, M. (2014). Blue monkeys and bridges: Transformations in habituation, habitat and people. In K. B. Strier (Ed.), *Primate ethnographies* (pp. 207–217). Pearson Education Inc.

Das, J., Biswas, J., Bhattacharjee, P. C., & Rao, S. S. (2009). Successful progression through the canopy by hoolock gibbons using canopy bridges in Borajan Reserve in the Bherjan-Borajan-

Podumoni Wildlife Sanctuary, Assam, India. In S. Lappan & J. Whittaker (Eds.), *The gibbons, developments in primatology: Progress and prospects* (pp. 467–475). Springer.

De Azevedo Fernandes, N. C. C., Cunha, M. S., Guerra, J. M., Réssio, R. A., dos Santos, C. C., Iglezias, S. D. A., de Carvalho, J., Araujo, E. L. L., Catão-Dias, J. L., & Diaz-Delgado, J. (2017). Outbreak of yellow fever among nonhuman primates, Espirito Santo, Brazil, 2017. *Emerging Infectious Diseases, 23*, 2038.

De Azevedo Fernandes, N. C. C., do Nascimento, P., Sánchez-Sarmiento, A. M., Réssio, R. A., dos Santos Cirqueira, C., Kanamura, C. T., de Carvalho, J., da Silva, S. M. P., Peruchi, A. R., de Souza, J. C., Jr., Hirano, Z. Z. M. B., & Catão-Dias, J. L. (2020). Histopathological changes and myoglobinuria in neotropical human primates attacked by dogs, Brazil. *Journal of Medical Primatology, 49*, 65–70.

Dos Santos, C. L. A., Le Pendu, Y., Gine, G. A. F., Dickman, C. R., Newsome, T. R., & Cassano, C. R. (2018). Human behaviours determine the direct and indirect impacts of free-ranging dogs on wildlife. *Journal of Mammalogy, 99*, 1261–1269.

Ellwanger, J. H., & Chies, J. A. B. (2019). The triad "dogs, conservation and zoonotic diseases" – An old and still neglected problem in Brazil. *Perspectives in Ecology and Conservation, 17*, 157–161.

Eppley, T. M., Donati, G., & Ganzhorn, J. U. (2016). Determinants of terrestrial feeding in an arboreal primate: The case of the southern bamboo lemur (*Hapalemur meridionalis*). *American Journal of Physical Anthropology, 161*, 328–342.

Farris, Z. J., Karpanty, S. M., Ratelolahy, F., & Kelly, M. J. (2014). Predator–primate distribution, activity, and co-occurrence in relation to habitat and human activity across fragmented and contiguous forests in northeastern Madagascar. *International Journal of Primatology, 35*, 859–880.

Farris, Z. J., Chan, S., Rafaliarison, R., & Valenta, K. (2019). Occupancy modeling reveals interspecific variation in habitat use and negative effects of dogs on lemur populations. *International Journal of Primatology, 40*, 706–720.

Ferriera Dasilva, A. L., Varzinczak, L. H., & Passos, F. C. (2021). Attacks of domestic dogs on common long-nosed armadillo (*Dasypus novemcinctus*) and southern brown howler monkey (*Alouatta guariba*) in fragmented Atlantic Forest and implications in a region of high priority for biodiversity conservation. *Austral Ecology, 46*, 155–158.

Fox, R. (2006). Animal behaviours, post-human lives: Everyday negotiations of the animal-human divide in pet-keeping. *Social and Cultural Geography, 7*, 525–537.

Galetti, M., & Sazima, I. (2006). Impact of feral dogs on an urban Atlantic Forest fragment in southeastern Brazil. *Natureza and Conservacao, 4*, 146–151.

Georgiev, A. V., Melvin, Z. E., Warkentin, A.-S., Winder, I. C., & Kassim, A. (2019). Two cases of dead infant carrying by female Zanzibar red colobus (*Piliocolobus kirkii*) at Jozani-Chwaka Bay National Park, Zanzibar. *African Primates, 13*, 57–60.

Goldberg, T. L., Gillespie, T. R., & Rwego, I. (2008). Health and diseases in the people, primates and domestic animals of Kibale National Park: Implications for conservation. In R. Wrangham & E. Ross (Eds.), *Science and conservation in African rain forests: The benefits of long-term research* (pp. 75–87). Cambridge University Press.

Goldstein, T., Belaganahalli, M. N., Syaluha, E. K., Lukusa, J.-P. K., Greig, D. J., Anthony, S. J., Tremeau-Bravard, A., Thakkar, T., Caciula, A., Mishra, N., Kipkin, W. I., Dhanota, J. K., Smith, B. R., Ontiveros, V. M., Randhawa, N., Cranfield, M., Johnson, C. K., Gilardi, K. V., & Mazet, J. A. K. (2020). Spillover of ebolaviruses into people in Eastern Democratic Republic of Congo prior to the 2018 Ebola virus disease outbreak. *One Health Outlook, 2*, 1–10.

Gompper, M. (2014). Introduction: Outlining the ecological influences of a subsidized predator. In M. Gompper (Ed.), *Free-ranging dogs and wildlife conservation* (pp. 1–7). Oxford University Press.

Gumert, M. D., Hamada, Y., & Malaivijitnond, S. (2013). Human activity negatively affects stone tool-using Burmese long-tailed macaques *Macaca fascicularis aurea* in Laem Son National Park, Thailand. *Oryx, 47*, 535–543.

Home, C., Bhatnagar, Y. V., & Vanak, A. T. (2018). Canine conundrum: Domestic dogs as an invasive species and their impacts on wildlife in India. *Animal Conservation, 21*, 275–282.

Hyeroba, D., Friant, S., Acon, J., Okwee-Acai, J., & Goldberg, T. L. (2017). Demography and health
of "village dogs" in rural Western Uganda. *Preventative Veterinary Medicine, 137*, 24–27.

Hughes, J., & Macdonald, D. W. (2013). A review of the interactions between free-roaming
domestic dogs and wildlife. *Biological Conservation, 157*, 341–351.

Khalil, H., Ecke, F., Evander, M., Magnusson, M., & Hörnfeldt, B. (2016). Declining ecosystem
health and the dilution effect. *Scientific Reports, 6*, 1–11.

Knight, J. (2011). *Herding monkeys to paradise: How macaque troops are managed for tourism
in Japan*. Leiden.

Kotait, I., de Novaes, O. R., Carrieri, M. L., Castilho, J. G., Macedo, C. I., Pereira, P. M. C., Boere,
V., Montebello, L., & Rupprecht, C. E. (2019). Non-human primates as a reservoir for rabies
virus in Brazil. *Zoonoses and Public Health, 66*, 47–59.

Kshirsagar, A. R., Applebaum, J. W., Randriana, Z., Rajaonarivelo, T., Rafaliarison, R. R.,
Farris, Z. J., & Valenta, K. (2020). Human-dog relationships across communities surrounding
Ranomafana and Andasibe-Mantadia National Parks, Madagascar. *Journal of Ethnobiology,
40*, 483–498.

Kumar, A., Sarma, K., Krishna, M., & Devi, A. (2013). The Eastern hoolock gibbon (*Hylobates
leuconedys*) in Eastern Arunachal Pradesh, India. *Primate Conservation, 27*, 115–123.

Lindshield, S. M. (2016). Protecting non-human primates in peri-urban environments: A case
study of Neotropical monkeys, corridor ecology, and coastal economy in the Caribe Sur of
Costa Rica. In M. T. Waller (Ed.), *Ethnoprimatology: Primate conservation in the 21st century*
(pp. 351–369). Springer.

Lyngdoh, S., Gopi, G. V., Selvan, K. M., & Habib, B. (2014). Effect of interactions among ethnic
communities, livestock and wild dogs (*Cuon alpinus*) in Arunachal Pradesh, India. *European
Journal of Wildlife Research, 60*, 771–780.

Mackenzie, J. S., & Jeggo, M. (2019). The one health approach –Why is it so important? *Tropical
Medicine and Infectious Disease, 4*, 88. https://doi.org/10.3390/tropicalmed4020088

Moresco, A., Larsen, R. S., Sauther, M. L., Cuozzo, F. P., Youssouf Jacky, I. P., & Millette,
J. B. (2012). Survival of a wild ring-tailed lemur (*Lemur catta*) with abdominal trauma in
an anthropogenically disturbed habitat. *Madagascar Conservation & Development, 7*, 49–52.

Najmuddin, M. F., Haris, H., Norazlimi, N., MD-Zain, B. M., Mohd-Ridwan, A. R., Shahrool-
Anuar, R., Husna, H. A., & Abdul-Latif, M. A. B. (2019). Predation of domestic dogs
(*Canis familiaris*) on Schlegel's banded langur (*Presbytis neglectus*) and crested hawk eagle
(*Nisaetus cirrhatus*) on dusky leaf monkey (*Trachypithecus obscurus*) in Malaysia. *Journal of
Sustainability Science and Management, 14*, 39–50.

Narat, V., Alcayna-Stevens, L., Rupp, S., & Giles-Vernick, T. (2017). Rethinking human–nonhuman primate contact and pathogenic disease spillover. *EcoHealth, 14*, 840–850.

Ng, D., Carver, S., Gotame, M., Karmasharya, D., Karmacharya, D., Man Pradhan, S., Narsingh
Rana, A., & Johnson, C. N. (2019). Canine distemper in Nepal's Annapurna Conservation
Area–Implications of dog husbandry and human behaviour for wildlife disease. *PLoS One,
14*, 0220874.

OIE. (2021). *One Health: Controlling global health risks more effectively*. World Organization for
Animal Health. https://www.oie.int/en/what-we-do/global-initiatives/one-health/

Oliveira, L. C., & Dietz, J. M. (2011). Predation risk and the interspecific association of two
Brazilian Atlantic Forest primates in *cabruca* agroforest. *American Journal of Primatology,
73*, 852–860.

Oliveira, V. B. D., Linares, A. M., Corrêa, G. L., & Chiarello, A. G. (2008). Predation on the black
capuchin monkey *Cebus nigritus* (Primates: Cebidae) by domestic dogs *Canis lupus familiaris* (Carnivora: Canidae), in the Parque Estadual Serra do Brigadeiro, Minas Gerais, Brazil.
Revista Brasileira de Zoologia, 25, 376–378.

Paolini, A., Romagnoli, S., Nardoia, M., Conte, S. R., Vulpiani, M. P., & Villa, P. D. (2020). Study
on the public perception of "community-owned dogs" in the Abruzzo Region, Central Italy.
Animals, 10, 1227. https://doi.org/10.3390/ani10071227

Perri, A. (2016). A wolf in dog's clothing: Initial dog domestication and Pleistocene wolf variation. *Journal of Archaeological Science, 68*, 1–4.

Philo, C., & Wilbert, C. (2000). *Animal spaces, beastly places: New geographies of human–animal relations.* Routledge.

Pihlström, H., Rosti, H., Benson, M., Lombo, M., & Pellikka, P. (2021). Domestic dog predation on white-tailed small-eared galago (*Otolemur garnettii lasiotis*) in the Taita Hills, Kenya. *African Primates, 15*, 31–38.

Plowright, R. K., Parrish, C. R., McCallum, H., Hudson, P. J., Ko, A. I., Graham, A. L., & Lloyd-Smith, J. O. (2017). Pathways to zoonotic spillover. *Nature Reviews Microbiology, 15*, 502–510.

Power, A. G., & Mitchell, C. E. (2004). Pathogen spillover in disease epidemics. *The American Naturalist, 164*, 79–89.

Pozo-Montuy, G., Serio-Silva, J. C., Chapman, C. A., & Bonilla-Sanchez, Y. M. (2013). Resource use in a landscape matrix by an arboreal primate: Evidence of supplementation in black howlers (*Alouatta pigra*). *International Journal of Primatology, 34*, 714–731.

Rasambainarivo, F., & Goodman, S. M. (2019). Disease risk to endemic animals from introduced species on Madagascar. *Fowler's Zoo and Wild Animal Medicine Current Therapy, 9*, 292–297.

Riley, C., Koenig, B. L., & Gumert, M. D. (2015). Observation of a fatal dog attack on a juvenile long-tailed macaque in a human-modified environment in Singapore. *Nature in Singapore, 8*, 57–63.

Schmidt, K. A., & Ostfeld, R. S. (2001). Biodiversity and the dilution effect in disease ecology. *Ecology, 82*, 609–619.

Serio-Silva, J. C., Ramírez-Julián, R., Eppley, T. M., & Chapman, C. A. (2019). Terrestrial locomotion and other adaptive behaviors in howler monkeys (*Alouatta pigra*) living in forest fragments. In R. Reyna-Hurtado & C. A. Chapman (Eds.), *Movement ecology of Neotropical Forest mammals* (pp. 125–140). Springer.

Shüttler, E., Saavedra-Aracena, L., & Jiménez, J. E. (2017). Domestic carnivore interactions with wildlife in the Cape Horn Biosphere Reserve, Chile: Husbandry and perceptions of impact from a community perspective. *PeerJ.* https://doi.org/10.7717/peerj.4124

Sparkes, J., Ballard, G., & Fleming, P. J. S. (2016). Cooperative hunting between humans and domestic dogs in eastern and northern Australia. *Wildlife Research, 43*, 20–26.

Steiner, C., Waltert, M., & Muhlenberg, M. (2003). Hunting pressure on the drill (*Mandrillus leucophaeus*) in Korup Project Area, Cameroon. *African Primates, 1-2*, 10–19.

Thinley, P., Norbu, T., Rajaratnam, R., Vernes, K., Dhendup, P., Tenzin, J., Choki, K., Wangchuk, S., Wangchuk, T., Wangdi, S., Chhetri, D. B., Powrel, R. B., Dorji, K., Rinchen, K., & Dorji, N. (2020). Conservation threats to the Endangered golden langur (*Trachypithecus geei*, Khajuria 1956) in Bhutan. *Primates, 61*, 257–266.

Turner, W. C., Kausrud, K. L., Beyer, W., Easterday, W. R., Barandongo, Z. R., Blaschke, E., Cloete, C. C., Lazak, J., Van Ert, M. N., Ganz, H. H., Turnbull, P. C. B., Sensteth, N. C., & Getz, W. M. (2016). Lethal exposure: An integrated approach to pathogen transmission via environmental reservoirs. *Scientific Reports, 6*, 1–13.

Valenta, K., Gettinger-Larson, J. A., Chapman, C. A., & Farris, Z. J. (2016). Barking up the right tree: Understanding local attitudes towards dogs in villages surrounding Ranomafana National Park, Madagascar can benefit applied conservation. *Madagascar Conservation & Development, 11*, 87–90.

Valenta, K., Twinomugisha, D., Godfrey, K., Liu, C., Schoof, V. A., Goldberg, T. L., & Chapman, C. A. (2017). Comparison of gastrointestinal parasite communities in vervet monkeys. *Integrative Zoology, 12*, 512–520.

Villatoro, F. J., Naughton-Treves, L., Sepúlveda, M. A., Stowhas, P., Mardones, F. O., & Silva-Rodríguez, E. A. (2018). When free-ranging dogs threaten wildlife: Public attitudes toward management strategies in southern Chile. *Journal of Environmental Management, 229*, 67–75.

VanderWaal, K. L., & Ezenwa, V. O. (2016). Heterogeneity in pathogen transmission: Mechanisms and methodology. *Functional Ecology, 30*, 1606–1622.

Venkataraman, V. V., Kerby, J. T., Nguyen, N., Asherafi, Z. T., & Fashing, P. J. (2015). Solitary Ethiopian wolves increase predation success on rodents when among grazing gelada herds. *Journal of Mammalogy, 96*, 129–137.

Walsh, P. D., Abernethy, K. A., Bermejo, M., Beyers, R., De Wachter, P., Akou, M. E., Huijbregts, B., Mambounga, D. I., Toham, A. K., Kilbourn, A. M., Lahm, S. A., Latour, S., Maisels, F., Mbina, C., Mihindou, Y., Obiang, S. N., Effa, E. N., Starkey, M. P., Telfer, P., Thibault, M., Tutin, C. E. G., White, L. J. T., & Wilkie, D. S. (2003). Catastrophic ape decline in western equatorial Africa. *Nature, 422*, 611–614.

Waters, S., El Harrad, A., Chetuan, M., Bell, S., & Setchell, J. M. (2017). Dogs disrupting wildlife: Domestic dogs harass and kill Barbary macaques in Bouhachem forest, North Morocco. *African Primates, 12*, 55–58.

Waters, S., Bell, S., & Setchell, J. M. (2018a). Understanding human-animal relations in the context of primate conservation: A multispecies ethnographic approach in North Morocco. *Folia Primatologia, 89*, 13–29.

Waters, S., Watson, T., Bell, S., & Setchell, J. M. (2018b). Communicating for conservation: Circumventing conflict with communities over domestic dog ownership in North Morocco. *European Journal of Wildlife Research, 64*, 69. https://doi.org/10.1007/s10344-018-1230-x

World Health Organization. (2014). *A global brief on vector-borne diseases*. Organization.

Yong, E. (2019). *A scientist witnessed poachers killing a chimp*. The Atlantic https://www.theatlantic.com/science/archive/2019/08/death-chimpanzee/595303/. Accessed 2nd February 2021.

Young, J. K., Olson, K. A., Reading, R. P., Amgalanbaatar, S., & Berger, J. (2011). Is wildlife going to the dogs? Impacts of feral and free-roaming dogs on wildlife populations. *Bioscience, 61*, 125–132.

Zohdy, S., Valenta, K., Rabaoarivola, B., Karanewsky, C. J., Zaky, W., Pilotte, N., Williams, S. A., Chapman, C. A., & Farris, Z. J. (2019). Causative agent of canine heartworm (*Dirofilaria immitis*) detected in wild lemurs. *International Journal of Parasitology: Parasites and Wildlife, 9*, 119–121.

Chapter 6
Climate Change Impacts on Non-human Primates: What Have We Modelled and What Do We Do Now?

Isabelle C. Winder, Brogan Mace, and Amanda H. Korstjens

Contents

Abstract Climate change will be a key influence on primates in the twenty-first century, potentially exacerbating the effects of habitat loss and anthropogenic activities to drive vulnerable species closer to extinction. There are many ways to assess species' vulnerability to climate change, including modelling approaches of three main types: trait-based models, species distribution models, and mechanistic models. In this chapter, we survey the literature on climate change models as applied to primates, including the type(s) of model made and the predictions obtained. Most primate genera (62 of 80) have been subject to ecological modelling, though we found no future projections for lemurs and no palaeoclimate models for lorises, tarsiers, or platyrrhines. Maximum entropy methods predominate even though direct comparisons have shown that these tend to predict more severe habitat losses when used uncritically. Most of the taxa modelled to date have been predicted substantial habitat losses by 2100, with significant variation within each taxonomic group.

Supplementary Information The online version contains supplementary material available at https://doi.org/10.1007/978-3-031-11736-7_6.

I. C. Winder (✉) · B. Mace
School of Natural Sciences, Bangor University, Bangor, UK
e-mail: i.c.winder@bangor.ac.uk

A. H. Korstjens
Bournemouth University, Poole, UK

Keywords Anthropogenic impacts · Climate change · Conservation · Ecological niches · Habitat loss · Modelling · Species distribution

6.1 Introduction

Climate change is expected to become a major factor driving human and non-human primate survival in the Anthropocene. While climate change impacts may never exceed those of land-use changes and trade, they will very likely exacerbate the effects of these other threats (Gouveia et al., 2016; Korstjens & Hillyer, 2016; Struebig et al., 2015; Titeux et al., 2017). Therefore, a discussion about the effect of anthropogenic factors on non-human primate (henceforth "primate") survival is not complete without some consideration of climate change. Primate vulnerability to climate change is determined by the level and type of changes the species will experience in its geographical range (exposure) coupled with the biological traits that determine how well the species can cope with (**sensitivity**) or adapt to (**adaptability**) the predicted changes (Foden et al., 2019; Foden & Young, 2016; Korstjens & Hillyer, 2016).

Many regions where primates occur are expected to undergo pronounced climatic changes (Carvalho et al., 2019; Graham et al., 2016; Zhang et al., 2019a), with increased maximum temperatures, reduced or excessive rainfall, increased seasonality, and an increased frequency and intensity of extreme events (cyclones and droughts) likely to be the biggest threats to primates. Across primate-occupied regions, mean temperature increases are predicted to be 10% greater than the global average (i.e. for every 1 °C increase globally, primate ranges will heat up by 1.1 °C; Graham et al., 2016). More worryingly, maximum temperatures are likely to increase by >2 °C across primate ranges, with some areas suffering up to 5–7 °C increases, taking these temperatures well above the physiological limit for primates (Carvalho et al., 2019). Greatly increased rainfall is predicted for 8% of primate species, while 4% will see large reductions (>3% increase or decrease per 1 °C mean global warming, respectively; Graham et al., 2016). Seasonality, an important predictor of species' geographical ranges (Williams et al. in press), is predicted to become more extreme across primate ranges. Temperature averages, rainfall patterns, and seasonality are major determinants of the vegetation cover and productivity of primate habitats. Climate change can also exacerbate droughts and cyclones which affect 26% and 18% of primate species, respectively (Zhang et al., 2019a), and can lead to devastating instant mortality (Campos et al., 2020). Finally, climate change may have indirect effects on primates by (for instance) changing the patterns of other anthropogenic impacts on them and their habitats. We are still far from producing an exhaustive list of the ways climate change can alter primate lives.

6.1.1 Assessing Species Vulnerability to Climate Change: Modelling Approaches

Models aimed at assessing species' vulnerability to climate change can be divided into three categories of increasing complexity, each with a different suitability for assessing sensitivity, exposure, and adaptive ability (Foden et al., 2019; Korstjens & Hillyer, 2016) (Fig. 6.1). First, **trait-based approaches** use expert opinion and published knowledge to determine how a species is likely to respond to or cope with particular climate changes based on their biological traits and inferred relationships between traits and vulnerability (e.g. Zhang et al., 2019a). Second, correlative approaches (i.e. **species distribution models**, often called niche or climate envelope models) investigate how current species distributions are influenced by climatic conditions and use these relationships to predict how species will be distributed under future climatic conditions. Finally, **mechanistic approaches** look at the underlying traits and physiological processes (mechanisms) that determine environment-species relationships to predict how species will be able to cope with and respond to changes based on their physiology, behaviour, or time budgets (e.g. Dunbar et al., 2009).

Species distribution models (SDMs) identify which areas are most suitable for a particular taxon, as identified by environmental conditions that describe the species'

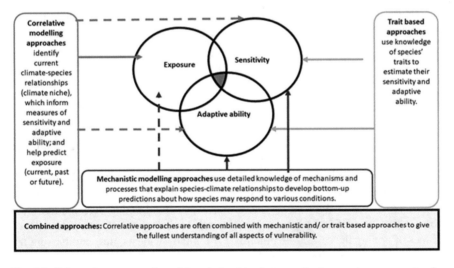

Fig. 6.1 Schematic representation of the relationships between the three main components of a species' vulnerability and the assessment approaches used to estimate species' vulnerability to climate change. Knowledge of each of these elements (sensitivity, exposure, and adaptive ability) provides the best estimate of actual vulnerability of species (dark grey overlap area in Venn diagram). Solid arrows show the most basic outcome of different approaches, and dashed lines show further more indirect outcomes or outcomes achieved by combining the approaches. Foden and Young (2016) provide an extensive overview of vulnerability indices and modelling approaches

environmental niche (Elith & Leathwick, 2009; Norberg et al., 2019). They best explain exposure but also provide important climate niche information to help understand sensitivity and adaptive ability, often used in trait-based and mechanistic approaches (Fig. 6.1). Once a taxon's environmental niche is established based on current distributions, the models can be used to identify other suitable locations elsewhere or in the past or future. SDMs require input data on the environmental conditions at locations where a species is present to compare against conditions where the species is absent. Due to lack of knowledge of absence, many models are actually developed using presence data and "pseudo-absences" or background locations, i.e. locations that are randomly selected from a predetermined area (e.g. outside the known distribution area) (Guillera-Arroita et al., 2015; Santini et al., 2021). How these pseudo-absences are selected can greatly affect the reliability of the SDM but is not always given sufficient consideration (for a critical review and test, see Santini et al., 2021). The variables that are included in establishing this environmental niche depend on the information and computational resources that are available to the researchers, as well as the research question. Recent exponential growth in accessible global datasets on environmental conditions (e.g. landscape cover, climatic conditions, human footprint) and modelling approaches can make it difficult to keep up with best practice (Araújo et al., 2019; Norberg et al., 2019; Zurell et al., 2020).

In this chapter, we systematically survey the literature covering modelling of non-human primate responses to climate change. In particular, we consider the potential presence and manifestation of taxonomic bias, explore the state of knowledge for each major primate group, and summarise predictions of habitat changes under future climate conditions.

6.2 Systematic Method

We searched the literature for modelling studies of the 80 extant non-human primate genera in the IUCN Red List of Threatened Species v. 2020.3 (IUCN, 2020) using the broadest academic database, Google Scholar (Gusenbauer, 2019). Search terms included the genus name, "species distribution model", and "ecological niche model", linked with Boolean operators. Where we could find no relevant literature for a genus, we refined the taxonomic search term (e.g. using alternative genus names or adding the rest of the binomial) and finally widened our search using the search term "climate change model".

For each genus, we sorted the resulting papers by relevance and selected up to four for detailed analysis (the modal number of relevant hits per genus was 2). Where a genus had more than four relevant papers (which affected the great apes, *Hylobates*, *Macaca*, *Rhinopithecus*, *Cercopithecus*, *Ateles*, *Aotus*, *Alouatta*, *Saguinus*, and *Sapajus*), we read abstracts and methods of all we found and selected a representative sample of four that included (1) as many species as possible; (2) a range of approaches, if several had been used; and (3) predictive and/or palaeoclimate studies if they had been done. This meant, for instance, that for the best-studied genera, we excluded some very-small-scale studies (especially where they modelled a single population or a region rather than a species' full range). We also sometimes found Masters theses and then papers deriving from the same models (we retained the peer-reviewed papers) or research teams which had produced a series of papers each considering a different species but using the same method and scenario (here we chose a representative example or, if it existed, a recent synthesis covering multiple species). For our final literature sample, we then performed a content analysis by extracting information on climate change scenario(s) and date(s) modelled, modelling approaches, focus and scale (taxon-specific, regional or larger), and the main aims and findings.

6.3   Results

Over half the studies we found dated to 2018–2021. All the major primate groups have been modelled, though coverage ranged from 30% to 100% of genera (see Fig. 6.2).

The taxonomic differences in coverage that we observed differed slightly from the bias in favour of apes and lemurs found in the wider climate change literature (Bernard & Marshall, 2020) and are less pronounced than those seen in primatological field studies (Bezanson & McNamara, 2019).

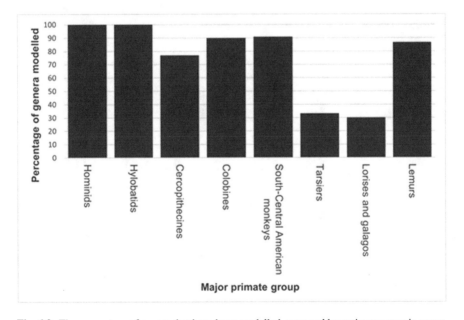

Fig. 6.2 The percentage of genera that have been modelled, arranged by major taxonomic group

Correlative species distribution models dominated our literature (Fig. 6.3). A few studies used mechanistic time-budget models (Dunbar, 1998, on *Theropithecus*) or trait-based biophysical models (Stalenberg, 2019). The majority of the correlative SDMs used maximum entropy modelling, either as a stand-alone tool (using Maxent software; Phillips et al., 2006) or in R (54/89 studies, 65.1%). Ensemble modelling, using multiple approaches (including maximum entropy, general and generalised linear models), was used in 15 studies (18.1%). Only three (3.6%) used generalised linear models, with all other methods accounting for 1–2.5% (one or

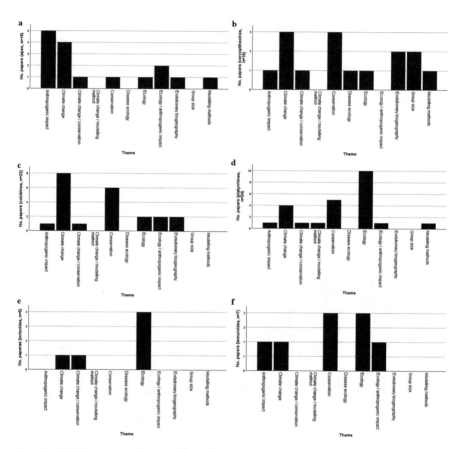

Fig. 6.3 Modelling approaches used for works on each major primate taxonomic group: (**a**) apes; (**b**) cercopithecines; (**c**) colobines; (**d**) platyrrhines; (**e**) lorises and galagos; and (**f**) lemurs. The tarsiers are excluded, because we found only one relevant paper, which used maximum entropy modelling in R. Full details of each paper can be found in Supplemental Tables 1, 2, 3, 4, 5, 6, and 7. Most studies (54.8%) only modelled the present day, while 33.3% projected into the future and 10.7% into the past. This pattern varied by group (see Fig. 6.4; we found no future projections for lemurs, and only future but no palaeoclimate models for lorises and galagos and platyrrhines). Present-only models were generally conservation orientated, looking at habitat fragmentation, primate distributions, and conservation planning. Common themes throughout the literature include climate change, ecology, anthropogenic impacts, and conservation (Fig. 6.5)

two papers) each. There were some differences among taxonomic groups, with maximum entropy modelling dominating even more where a group was understudied or subject to one or two broad works rather than numerous specific ones.

6.3.1 Apes and Monkeys

All seven ape genera have been the subject of modelling studies, but only *Hylobates* and the hominids had >4 studies each. Maximum entropy models were used in 62.5% and ensemble approaches in 25% of studies. Only Barratt et al. (2020) explored palaeoclimate models, while eight studies modelled the present (Ario et al., 2018; Bonnin et al., 2020; Etiendem et al., 2013; Rahman et al., 2019; Singh et al., 2018; Tran & Vu, 2020; Wich et al., 2012; Yuh et al., 2020), and seven predicted the future (named below).

Outcomes of future predictions varied by taxon and approach. Mwambo (2010) and Carvalho et al. (2021) predicted substantial habitat losses for *Pan* and *Gorilla* by 2090. Thorne et al. (2013), however, found that these losses were typical only when using "standard" correlative models and a limiting factor model predicted minimal change for mountain gorillas. Predictions of climate change impacts, they concluded, are sensitive to the assumptions of the selected method.

For Asian apes, predictions were more negative. Wich et al. (2016) used land-cover change scenarios to predict substantial population declines for *Pongo abelii* by 2030. Other taxa potentially at high risk included various species of *Hylobates* (Condro et al., 2021; Trisurat, 2018), *Pongo abelii*, and *Symphalangus syndactylus* (Condro et al., 2021). Potentially better off were *Pongo pygmaeus* (Condro et al., 2021) and *Hoolock hoolock* (Deb et al., 2019), both predicted range increases, while *Hylobates muelleri* and *H. abbotti* were predicted only small changes (Condro et al., 2021). The differences in apparent risk may result from different vulnerability profiles or modelling approaches.

We found models on 10/13 cercopithecine genera (all save *Erythrocebus*, *Miopithecus*, and *Allenopithecus*, though *Chlorocebus* is included only under the older name *Cercopithecus* in Willems & Hill, 2009) and 9/10 colobine genera (all but *Procolobus*). Among the 15 studies found, 2 predicted past distributions (Khanal et al., 2018a; Chala et al., 2019), 6 predicted the future (listed below), and 7 focused on the present (Willems & Hill, 2009; Green, 2012; Cronin et al., 2015, 2017, who reported the same models; Moyes et al., 2016; Korstjens et al., 2018; Fuchs et al., 2018; Greenspan et al., 2020).

Predicted outcomes of climate change varied. Dunbar's (1998) time-budget model predicted that *Theropithecus gelada* populations will fragment as climates warm, while Hill and Winder (2019) predicted that 3/6 *Papio* species might suffer substantial habitat losses. Condro et al. (2021) predicted habitat losses by 2050 for eight macaque species, but suggested *Macaca ochreata* and *M. siberu* would be less affected. *Mandrillus leucophaeus* may experience substantial habitat loss from earlier on in the twenty-first century than *Cercocebus atys* (Baker & Willis, 2015).

Finally, Korstjens (2019) suggested that habitat suitability for *Cercopithecus* would reduce only slightly by 2070. The final study that looked to the future, Ayebare et al. (2013), did not present any species-specific results.

In the colobine literature, there are proportionally more single-species or regional studies than for cercopithecines, particularly in Asia. Of 22 studies, 5 focused on the past (Ehlers-Smith, 2014; Khanal et al., 2018c; Moody, 2007; Ren et al., 2017; Windyoningrum, 2013), 9 predicted the future (see below), and 8 focused on the present (Anh et al., 2019; Atmoko et al., 2020; Cavada et al., 2017; Cronin et al., 2015, 2017; Khanal et al., 2018b; McDonald et al., 2019; Singh et al., 2018; Tran et al., 2018).

Korstjens (2019) suggested that *Colobus* may be more at risk than *Cercopithecus*. At species level, Baker and Willis (2015) predicted substantial habitat loss for *C. polykomos* (while *C. vellerosus* may be unaffected along with *Piliocolobus pennantii*). *P. badius*, however, is predicted range loss by all their models, and *P. preussi* may lose habitat by 2040.

In Asia, *Pygathrix* (Tran et al., 2020; Vu et al., 2020), *Rhinopithecus* (Luo et al., 2015; Zhang et al., 2019b), and *Semnopithecus* species (Bagaria et al., 2020) are predicted substantial habitat loss. Condro et al. (2021) predicted complete habitat losses for two species of *Trachypithecus*, *Simias concolor*, and six species of *Presbytis*. They found that two more *Presbytis* species would experience smaller habitat losses, while three would be stable or increase their habitats. Finally, Zhao et al. (2019) suggested that *Rhinopithecus bieti* might lose 8–22% of its suitable habitat by 2050 unless anthropogenic land-use changes were controlled, in which case, their suitable habitat might increase.

We found models for 20 of 22 platyrrhine genera (Platyrrhini), all save *Leontocebus* and *Brachyteles*. Of these, none modelled the past, 7 the future (Fig. 6.4), and 17 the present (Boubli & De Lima, 2009; Calixto-Pérez et al., 2018; Campos & Jack, 2013; Clément et al., 2014; De Marco et al., 2020; Garbino et al., 2015; Guy et al., 2016; Hasui et al., 2017; Helenbrook & Valdez, 2020; Holzmann et al., 2015; Howard et al., 2015; Moraes et al., 2019; Ochoa-Quintero et al., 2017; Ortega Huerta, 2007; Rezende et al., 2020; Shanee et al., 2015; Vidal-García & Serio-Silva, 2011).

For platyrrhines, as for colobines and cercopithecines, projections of climate change impacts often also include the impact of deforestation and other anthropogenic changes, and studies that expressly separate these often suggest forest loss will be more damaging. *Lagothrix lagothricha* is predicted a 13% loss of suitable habitat due to climate change, but 18% where forest loss continues apace (Linero et al., 2020). Other species predicted severe habitat losses included *Callithrix flaviceps*, *C. pencillata* and *C. aurita* (though other members of the same genus were less affected; Braz et al., 2019), *Alouatta belzebul*, *Sapajus flavius* and *S. libidinosus* (Moraes et al., 2020), *Aotus miconax*, *Lagothrix flavicauda*, and, if land-use change continues alongside climate change, *Plecturocebus oenanthe* (Shanee, 2016).

Several platyrrhine studies focused on broad biodiversity measures and how biodiversity might shift or decline as climates change. Sales et al. (2017) found that biodiversity of taxa including primates was likely to decline in the Amazon, and

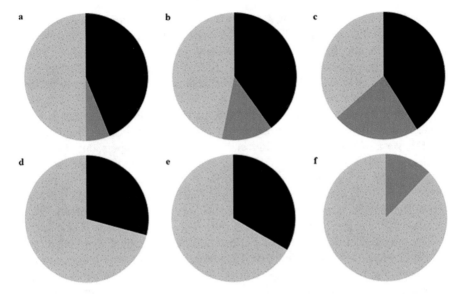

Fig. 6.4 Periods covered by modelling studies of the major primate groups: (**a**) apes; (**b**) cercopithecines; (**c**) colobines; (**d**) platyrrhines; (**e**) lorises and galagos; and (**f**) lemurs. Black denotes papers projecting the future, mid-grey includes papers examining palaeoclimates, and pale grey includes papers examining the present alone. Tarsiers are excluded because only one paper (which projected to 2050) was found

Sales et al. (2020) suggested that overall, the contribution of primates to seed dispersal in this area, based on modelled range area and dispersal potential, is likely to decline substantially. Sales et al. (2019) found that 20% of 80 primate species might expand their ranges as a result of climate change, if they can disperse as well as they can today. If fragmentation prevents dispersal of platyrrhines through the Amazon, these 80 species will lose on average 90% of their suitable habitat.

6.3.2 Prosimians

Only one of the three tarsier genera (genus *Tarsius*) has been modelled and only once. Condro et al. (2021) used maximum entropy modelling in R to explore climate change impacts on Indonesian biodiversity and included eight species. They predicted total habitat loss by 2050 for six species, near complete loss for *Tarsius dentatus*, and roughly stable habitat for *T. pumilus*.

The lorises and galagos (superfamily Lorisoidea) were also rarely studied, with only one of the six genera of Galagidae and two of the four among Lorisidae the subject of any models. We found no palaeoclimate models, two that predicted the future (see below), and four that focused on the present (Kumara et al., 2021; Nekaris & Stengel, 2013; Thorn et al., 2009; Voskamp et al., 2014).

Condro et al. (2021) predicted total habitat loss for *Nycticebus javanicus* and ~75% loss for *N. bancanus*. The remaining five taxa (*N. kayan*, *N. hilleri*, *N. borneanus*, *N. coucang*, and *N. menagensis*) were all predicted small increases of up to ~25% of the current extent (Condro et al., 2021). Erasmus et al. (2002) suggested significant shifts in overall biodiversity of South African animal communities, including *Galago moholi*, by 2050.

We found seven works specifically focusing on modelling the niches and climatic vulnerabilities of Lemuroidea, two of which (Herrera et al., 2018; Peacock, 2011) cover a wide range of taxa. In fact, beyond these two papers, only *Lepilemur* (Stalenberg, 2019) and *Eulemur*, which was by far the best studied, were modelled. *Allocebus* was not studied at all. Among these, none addressed the future, one (Stalenberg, 2019) reconstructed the past, and six (Blair et al., 2013; Herrera et al., 2018; Kamilar & Tecot, 2016; Mercado Malabet et al., 2020; Ormsby, 2019; Peacock, 2011) explored the present.

The only model of the effects of climate change on lemurs was Stalenberg's (2019) biophysical model. Biophysical models estimate a species' thermal niche and water requirements and compare physiological needs to environmental conditions to identify suitable habitats for it. Stalenberg's model found that *Lepilemur leucopus* was at higher risk of thermal stress in the 1975–2005 than the 1931–1960 time period.

6.4   Discussion and Conclusions

Climate change impact on primates is still a relatively new area of research, with over half of the studies we found published from 2018 onwards, yet it is no doubt an important one (Fig. 6.5). Our systematic review shows that there are still relatively few studies out there, and fewer than half of the species distribution models that have been built have been used to predict the future. Efforts are not equally distributed across genera, with apes and large-bodied monkeys being the best represented, a bias that we also see in the wider primatological literature (Bezanson & McNamara, 2019; Hawes et al., 2013). Bezanson and McNamara (2019) suggest that taxonomic bias may be related to the presence of long-term field sites and publication bias. Although modelling work is often desk-based, it does rely on a reasonable knowledge of the localities where the target species is found.

In the wider context of publishing in primatology, the small study counts we found for this specific topic highlight it as an understudied subject. To date, we have very limited understanding of how climate change is potentially going to affect primates, and with 2050 (one of the predicted scenarios used in climate modelling) less than 30 years away, we urgently need to make meaningful predictions and adapt subsequent conservation strategies. Those studies we do have suggest that primates will need to shift ranges to accommodate changing climatic conditions (Estrada et al., 2017; Schloss et al., 2012), with consequences for their management and conservation. Primate responses to changing climates will be species-specific

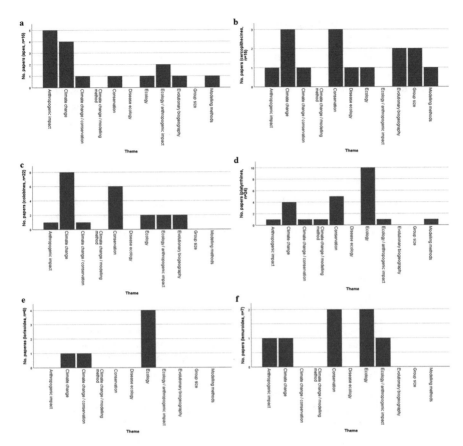

Fig. 6.5 Common themes in modelling literature for each major primate group: (**a**) apes; (**b**) cercopithecines; (**c**) colobines; (**d**) platyrrhines; (**e**) lorises and galagos; and (**f**) lemurs. Tarsiers are excluded, as only one paper was found – it explored climate change impacts on biodiversity

(Condro et al., 2021; Schloss et al., 2012), so we cannot rely on generalisations across subfamilies. The sooner we identify the potential impacts of climate change on primates, the longer we give ourselves to implement the necessary steps to mitigate effects and avoid population declines.

Modelling efforts did not seem to correspond with extinction risk, i.e. species classified by the IUCN as "critically endangered" or "endangered" were no more likely to be the focus of these studies. In fact, some endangered species (*Prolemur simus*, *Macaca pagensis*) have yet to be studied this way. Species that are already struggling may face even more challenging futures that we cannot accurately prepare for and mitigate if we do not have predictions of what the future may mean. This is yet again a pattern in the wider literature (Bezanson & McNamara, 2019) and may be a result of data scarcity. Although modelling can draw on online or published pools of data, such as the Global Biodiversity Information Facility (GBIF) or iNaturalist, sometimes datasets for a species are still missing, small, inaccurate, or of poor quality (Maldonado et al., 2015), making it difficult to build accurate, high-quality models.

Both this chapter and the literature within it emphasise the need for more foundational knowledge on many primate species. When we are unsure on current ranges, habitat preferences, and a species' ability to cope with anthropogenic factors, accurate and meaningful projections may be impossible, and we run the risk of overgeneralisation. Some species may gain incidental protection from sharing their habitats with others we have modelled, but this is not the ideal given that responses are so species-specific (Pacifici et al., 2017; Schloss et al., 2012). It is also important to note that simply modelling distributions based on perceived correlations between occurrence and climate conditions may not give us the full picture. Herrera et al.'s (2018) study found that food tree distribution was a more accurate predictor for lemur distribution than climate. Indeed, different aspects of the environment will influence different species, with arboreal species perhaps more concerned with forest connectivity than terrestrial ones, and these factors can also affect population density (Pozo-Montuy et al., 2011).

Correlative methods, and specifically maximum entropy models, were by far the most common approach we found, but climate change modellers are far from reaching consensus on methodology. With SDMs, there is ongoing disagreement on how small a sample size is appropriate and how accurate small-sample studies are (Santini et al., 2021; Wisz et al., 2008). Occurrence data used in studies of rare or cryptic species in particular is likely to have been opportunistically collected (and thus not randomly sampled or reflective of entire populations). Conducting studies on the species that have established field sites simply because they are the taxa we have sufficient high-quality data for is likely to unintentionally reinforce taxonomic biases. Additionally, animals do not exist in closed systems. Many external factors that are not commonly included in SDM studies can influence primate distributions. For example, anthropogenic and biotic factors (such as natural disasters) ultimately will play a role in shaping current and future distributions (Graham et al., 2016; Kamilar & Tecot, 2016), and, while some primatological studies included these factors, this was rare. Studies that fail to include anthropogenic layers risk overestimating primate ranges (Kamilar & Tecot, 2016). SDMs also do not take into account behavioural flexibility or the issues of sensitivity and adaptability mentioned in our introduction. As a more positive note, it is worth our recognising that some primate species have adapted fairly successfully to human-dominated landscapes, persisting in urban areas (Aguiar et al., 2014; Jaman & Huffman, 2013). Human encroachment and urbanisation may impact such species less than others. Overall, while the picture emerging from existing models of primate responses to climate change may seem grim, we would like to end by proposing that there are significant gains to be made for primatologists and conservationists if we work to develop better methods and simultaneously generate new knowledge about our models' predictive abilities.

References

Aguiar, L. M., Cardoso, R. M., Back, J. P., et al. (2014). Tool use in urban populations of capuchin monkeys *Sapajus* spp. (Primates: Cebidae). *Zoologia, 31*(5), 516–519.

Anh, N. T., Duc Minh, L., Viet Hung, P., & Thi Duyen, V. (2019). Modeling the red-shanked douc (*Pygathrix nemaeus*) distribution in Vietnam using Maxent. *VNU Journal of Science: Earth and Environmental Sciences, 35*, 61–71.

Araújo, M. B., Anderson, R. P., Barbosa, A. M., et al. (2019). Standards for distribution models in biodiversity assessments. *Science Advances, 5*(1), eaat4858.

Ario, A., Kartono, A. P., Prasetyo, L. B., & Supriatna, J. (2018). Habitat suitability of release site for Javan gibbon (*Hylobates moloch*) in Mount Malabar protected forests, West Java. *Jurnal Manajemen Hutan Tropika, 24*, 95–104.

Atmoko, T., Mardiastuti, A., Bismark, M., et al. (2020). Habitat suitability of proboscis monkey (*Nasalis larvatus*) in Berau delta, East Kalimantan, Indonesia. *Biodiversitas, 21*, 5155–5163.

Ayebare, S., Ponce-Reyes, R., Segan, D. B., et al. (2013). *Identifying climate resilient corridors for conservation in the Albertine Rift*. Research report by the Wildlife Conservation Society to the MacArthur Foundation.

Bagaria, P., Sharma, L. K., Joshi, B. D., et al. (2020). West to east shift in range predicted for Himalayan Langur in climate change scenario. *Global Ecology and Conservation, 22*, e00926.

Baker, D. J., & Willis, S. G. (2015). *Projected impacts of climate change on biodiversity in West African protected areas*. UNEP-WCMC technical report.

Barratt, C. D., Lester, J. D., Gratton, P., et al. (2020). Late quaternary habitat suitability models for chimpanzees (Pan troglodytes) since the last interglacial (120,000 BP). *bioRxiv* 2020.05.15.066662.

Bernard, A. B., & Marshall, A. J. (2020). Assessing the state of knowledge of contemporary climate change and primates. *Evolutionary Anthropology: Issues, News, and Reviews, 29*, 317–331.

Bezanson, M., & McNamara, A. (2019). The what and where of primate field research may be failing primate conservation. *Evolutionary Anthropology: Issues, News, and Reviews, 28*, 166–178.

Blair, M. E., Sterling, E. J., Dusch, M., et al. (2013). Ecological divergence and speciation between lemur (*Eulemur*) sister species in Madagascar. *Journal of Evolutionary Biology, 26*, 1790–1801.

Bonnin, N., Stewart, F. A., Wich, S. A., et al. (2020). Modelling landscape connectivity change for chimpanzee conservation in Tanzania. *Biological Conservation, 252*, 108816.

Boubli, J. P., & De Lima, M. G. (2009). Modeling the geographical distribution and fundamental niches of *Cacajao* spp. and *Chiropotes israelita* in Northwestern Amazonia via a maximum entropy algorithm. *International Journal of Primatology, 30*, 217–228.

Braz, A. G., Lorini, M. L., & Vale, M. M. (2019). Climate change is likely to affect the distribution but not parapatry of the Brazilian marmoset monkeys (*Callithrix* spp.). *Diversity and Distributions, 25*, 536–550.

Calixto-Pérez, E., Alarcón-Guerrero, J., Ramos-Fernández, G., et al. (2018). Integrating expert knowledge and ecological niche models to estimate Mexican primates' distribution. *Primates, 59*, 451–467.

Campos, F. A., & Jack, K. M. (2013). A potential distribution model and conservation plan for the critically endangered Ecuadorian capuchin, *Cebus albifrons aequatorialis*. *International Journal of Primatology, 34*, 899–916.

Campos, F. A., Kalbitzer, U., Melin, A. D., et al. (2020). Differential impact of severe drought on infant mortality in two sympatric neotropical primates. *Royal Society Open Science, 7*, 200302.

Carvalho, J. S., Graham, B., Rebelo, H., et al. (2019). A global risk assessment of primates under climate and land use/cover scenarios. *Global Change Biology, 25*(9), 3163–3178.

Carvalho, J. S., Graham, B., Bocksberger, G., et al. (2021). Predicting range shifts of African apes under global change scenarios. *Diversity and Distributions, 27*(9), 1663–1679.

Cavada, N., Ciolli, M., Rocchini, D., et al. (2017). Integrating field and satellite data for spatially explicit inference on the density of threatened arboreal primates. *Ecological Applications, 27*, 235–243.

Chala, D., Roos, C., Svenning, J. C., & Zinner, D. (2019). Species-specific effects of climate change on the distribution of suitable baboon habitats – Ecological niche modeling of current and Last Glacial Maximum conditions. *Journal of Human Evolution, 132*, 215–226.

Clément, L., Catzeflis, F., Richard-Hansen, C., et al. (2014). Conservation interests of applying spatial distribution modelling to large vagile Neotropical mammals. *Tropical Conservation Science, 7*, 192–213.

Condro, A. A., Prasetyo, L. B., Rushayati, S. B., et al. (2021). Predicting hotspots and prioritizing protected areas for endangered primate species in Indonesia under changing climate. *Biology, 10*(2), 154.

Cronin, D. T., Bovcuma Mene, D., Perella, C., et al. (2015). *The future of the biodiversity of the Gran Caldera Scientific Reserve*. Report by the Bioko Biodiversity Protection Program.

Cronin, D. T., Sesink Clee, P. R., et al. (2017). Conservation strategies for understanding and combating the primate bushmeat trade on Bioko Island, Equatorial Guinea. *American Journal of Primatology, 79*, 1–16.

De Marco, P., Villén, S., Mendes, P., et al. (2020). Vulnerability of Cerrado threatened mammals: An integrative landscape and climate modeling approach. *Biodiversity and Conservation, 29*, 1637–1658.

Deb, J. C., Phinn, S., Butt, N., & McAlpine, C. A. (2019). Adaptive management and planning for the conservation of four threatened large Asian mammals in a changing climate. *Mitigation and Adaptation Strategies for Global Change, 24*, 259–280.

Dunbar, R. I. M. (1998). Impact of global warming on the distribution and survival of the gelada baboon: A modelling approach. *Global Change Biology, 4*, 293–304.

Dunbar, R. I. M., Korstjens, A. H., & Lehmann, J. (2009). Time as an ecological constraint. *Biological Reviews, 84*(3), 413–429.

Ehlers-Smith, D. A. (2014). The effects of land-use policies on the conservation of Borneo's endemic Presbytis monkeys. *Biodiversity and Conservation, 23*, 891–908.

Elith, J., & Leathwick, J. R. (2009). Species distribution models: Ecological explanation and prediction across space and time. *Annual Review of Ecology, Evolution and Systematics, 40*, 677–697.

Erasmus, B. F. N., Van Jaarsveld, A. S., Chown, S. L., et al. (2002). Vulnerability of South African animal taxa to climate change. *Global Change Biology, 8*, 679–693.

Estrada, A., Garber, P. A., Rylands, A. B., et al. (2017). Impending extinction crisis of the world's primates: Why primates matter. *Science Advances, 3*, e1600946.

Etiendem, D. N., Funwi-Gabga, N., Tagg, N., et al. (2013). The cross river gorillas (*Gorilla gorilla diehli*) at Mawambi Hills, South-West Cameroon: Habitat suitability and vulnerability to anthropogenic disturbance. *Folia Primatologica, 84*, 18–31.

Foden, W. B., & Young, B. E. (2016). *IUCN SSC guidelines for assessing species' vulnerability to climate change*. Occasional paper of the IUCN SSC no. 59. Available at https://www.iucn.org/content/iucn-ssc-guidelines-assessing-species-vulnerability-climate-change. Accessed 5.10.2021.

Foden, W. B., Young, B. E., Akçakaya, H. R., et al. (2019). Climate change vulnerability assessment of species. *Climate Change, 10*(1), e551.

Fuchs, A. J., Gilbert, C. C., & Kamilar, J. M. (2018). Ecological niche modeling of the genus *Papio*. *American Journal of Physical Anthropology, 166*, 812–823.

Garbino, G. S. T., Semedo, T., & Pansonato, A. (2015). Notes on the western black-handed tamarin, *Saguinus niger* (É. Geoffroy, 1803) (Primates) from an amazonia-cerrado ecotone in central-western Brazil: new data on its southern limits. *Mastozoología Neotropical, 22*(2), 311–318.

Gouveia, S. F., Souza-Alves, J. P., Rattis, L., et al. (2016). Climate and land use changes will degrade the configuration of the landscape for titi monkeys in eastern Brazil. *Global Change Biology, 22*(6), 2003–2012.

Graham, T. L., Matthews, H. D., & Turner, S. E. (2016). A global-scale evaluation of primate exposure and vulnerability to climate change. *International Journal of Primatology, 37*(2), 158–174.

Green, J. M. H. (2012). *Incorporating costs and processes into systematic conservation planning in a biodiversity hotspot* (Doctoral Thesis). University of Cambridge.

Greenspan, E., Montgomery, C., Stokes, D., Wantai, S., & Moo, S. S. B. (2020). Prioritizing areas for conservation action in Kawthoolei, Myanmar using species distribution models. *Journal for Nature Conservation, 58*, 125918.

Guillera-Arroita, G., Lahoz-Monfort, J. J., Elith, J., et al. (2015). Is my species distribution model fit for purpose? Matching data and models to applications. *Global Ecology and Biogeography, 24*(3), 276–292.

Gusenbauer, M. (2019). Google Scholar to overshadow them all? Comparing the sizes of 12 academic search engines and bibliographic databases. *Scientometrics, 118*, 177–214.

Guy, C., Cassano, C. R., Cazarre, L., et al. (2016). Evaluating landscape suitability for Golden-headed lion tamarins (*Leontopithecus chrysomelas*) and Wied's black tufted-ear marmosets (*Callithrix kuhlii*) in the Bahian Atlantic Forest. *Tropical Conservation Science, 9*, 735–757.

Hasui, É., Silva, V. X., Cunha, R. G. T., et al. (2017). Additions of landscape metrics improve predictions of occurrence of species distribution models. *Journal of Forest Research, 28*, 963–974.

Hawes, J., Calouro, A. M., & Peres, C. A. (2013). Sampling effort in Neotropical primate diet studies: Collective gains and underlying geographic and taxonomic biases. *International Journal of Primatology, 34*(6), 1081–1104.

Helenbrook, W. D., & Valdez, J. W. (2020). Species distribution modeling and conservation assessment of the black-headed night monkey (*Aotus nigriceps*) – A species of least concern that faces widespread anthropogenic threats. *bioRxiv* 2020.05.20.107383.

Herrera, J. P., Borgerson, C., Tongasoa, L., et al. (2018). Estimating the population size of lemurs based on their mutualistic food trees. *Journal of Biogeography, 45*, 2546–2563.

Hill, S. E., & Winder, I. C. (2019). Predicting the impacts of climate change on *Papio* baboon biogeography: Are widespread, generalist primates 'safe'? *Journal of Biogeography, 46*, 1380–1405.

Holzmann, I., Agostini, I., DeMatteo, K., et al. (2015). Using species distribution modeling to assess factors that determine the distribution of two parapatric howlers (*Alouatta* spp.) in South America. *International Journal of Primatology, 36*, 18–32.

Howard, A. M., Nibbelink, N., Bernardes, S., et al. (2015). Remote sensing and habitat mapping for bearded capuchin monkeys (*Sapajus libidinosus*): Landscapes for the use of stone tools. *Journal of Applied Remote Sensing, 9*, 096020.

IUCN. (2020). *IUCN Red List of Threatened Species. Version 2020-3*. Available at http://www.iucnredlist.org (downloaded 10.01.2021).

Jaman, M. F., & Huffman, M. A. (2013). The effect of urban and rural habitats and resource type on activity budgets of commensal rhesus macaques (*Macaca mulatta*) in Bangladesh. *Primates, 54*(1), 49–59.

Kamilar, J. M., & Tecot, S. R. (2016). Anthropogenic and climatic effects on the distribution of *Eulemur* species: An ecological niche modeling approach. *International Journal of Primatology, 37*, 47–68.

Khanal, L., Chalise, M. K., He, K., et al. (2018a). Mitochondrial DNA analyses and ecological niche modeling reveal post-LGM expansion of the Assam macaque (*Macaca assamensis*) in the foothills of Nepal Himalaya. *American Journal of Primatology, 80*, 1–13.

Khanal, L., Chalise, M. K., & Jiang, X. (2018b). Ecological niche modelling of Himalayan langur (*Semnopithecus entellus*) in the southern flank of the Himalaya. *Journal of Institute of Science and Technology, 23*, 1–9.

Khanal, L., Chalise, M. K., Wan, T., & Jiang, X. (2018c). Riverine barrier effects on population genetic structure of the Hanuman langur (*Semnopithecus entellus*) in the Nepal Himalaya. *BMC Evolutionary Biology, 18*, 1–16.

Korstjens, A. H. (2019). The effect of climate change on the distribution of the genera *Colobus* and *Cercopithecus*. In A. M. Behie, J. A. Teichroeb, & N. Malone (Eds.), *Primate research and conservation in the Anthropocene*. Cambridge University Press.

Korstjens, A. H., & Hillyer, A. P. (2016). Primates and climate change: A review of current knowledge. In S. A. Wich & A. J. Marshall (Eds.), *An introduction to primate conservation*. Oxford University Press.

Korstjens, A. H., Lehmann, J., & Dunbar, R. I. M. (2018). Time constraints do not limit group size in arboreal guenons but do explain community size and distribution patterns. *International Journal of Primatology, 39*, 511–531.

Kumara, H. N., Santhanakrishnan, B., Nitte, M., & Karunakaran, P. (2021). Conservation status and potential distribution of the Bengal slow loris *Nycticebus bengalensis* in Northeast India. *Primate Conservation, 35*, 1–10.

Linero, D., Cuervo-Robayo, A. P., & Etter, A. (2020). Assessing the future conservation potential of the Amazon and Andes Protected Areas: Using the woolly monkey (*Lagothrix lagothricha*) as an umbrella species. *Journal for Nature Conservation, 58*, 125926.

Luo, Z., Zhou, S., Yu, W., et al. (2015). Impacts of climate change on the distribution of Sichuan snub-nosed monkeys (*Rhinopithecus roxellana*) in Shennongjia area, China. *American Journal of Primatology, 77*, 135–151.

Maldonado, C., Molina, C. I., Zizka, A., et al. (2015). Estimating species diversity and distribution in the era of Big Data: To what extent can we trust public databases? *Global Ecology and Biogeography, 24*(8), 973–984.

McDonald, M. M., Johnson, S. M., Henry, E. R., & Cunneyworth, P. M. K. (2019). Differences between ecological niches in northern and southern populations of Angolan black and white colobus monkeys (*Colobus angolensis palliatus* and *Colobus angolensis sharpei*) throughout Kenya and Tanzania. *American Journal of Primatology, 81*, 1–15.

Mercado Malabet, F., Peacock, H., Razafitsalama, J., et al. (2020). Realized distribution patterns of crowned lemurs (*Eulemur coronatus*) within a human-dominated forest fragment in northern Madagascar. *American Journal of Primatology, 82*, e23125.

Moody, J. (2007). *Population genetics, biogeography and conservation of the Indochinese silvered Langur, Trachypithecus germaini, in Cambodia: Is the Mekong River a taxonomic boundary?* (PhD thesis). City University of New York – Hunter College.

Moraes, A. M., Vancine, M. H., Moraes, A. M., et al. (2019). Predicting the potential hybridization zones between native and invasive marmosets within Neotropical biodiversity hotspots. *Global Ecology and Conservation, 20*, e00706.

Moraes, B., Razgour, O., Souza-Alves, J. P., et al. (2020). Habitat suitability for primate conservation in north-east Brazil. *Oryx, 54*, 803–813.

Moyes, C. L., Shearer, F. M., Huang, Z., et al. (2016). Predicting the geographical distributions of the macaque hosts and mosquito vectors of *Plasmodium knowlesi* malaria in forested and non-forested areas. *Parasites and Vectors, 9*, 1–12.

Mwambo, F. M. (2010). *Human and climatic change impact modelling on the habitat suitability for the Chimpanzee (Pan troglodytes ellioti)* (MS thesis). Universidade Nova.

Nekaris, K. A. I., & Stengel, C. J. (2013). Where are they? Quantification, distribution and microhabitat use of fragments by the red slender loris (*Loris tardigradus tardigradus*) in Sri Lanka. In L. K. Marsh & C. A. Chapman (Eds.), *Primates in fragments: Complexity and resilience*. Springer.

Norberg, A., Abrego, N., Blanchet, F. G., et al. (2019). A comprehensive evaluation of predictive performance of 33 species distribution models at species and community levels. *Ecological Monographs, 89*(3), e01370.

Ochoa-Quintero, J., Chang, C., Gardner, T. A., et al. (2017). Habitat loss on Rondon's marmoset potential distribution. *Land, 6*, 8.

Ormsby, L. (2019). *Distribution, abundance and habitat use of the mongoose lemur, Eulemur mongoz, on Anjouan, Comoros* (PhD thesis). University of Bristol.

Ortega Huerta, M. A. (2007). Fragmentation patterns and implications for biodiversity conservation in three biosphere reserves and surrounding regional environments, northeastern Mexico. *Biological Conservation, 134*, 83–95.

Pacifici, M., Visconti, P., Butchart, S. H., et al. (2017). Species' traits influenced their response to recent climate change. *Nature Climate Change, 7*, 205–208.

Peacock, H. (2011). *Assessment of the protected areas network in Madagascar for lemur conservation* (MSc thesis). The University of Western Ontario.

Phillips, S. J., Anderson, R. P., & Schapire, R. E. (2006). Maximum entropy modeling of species geographic distributions. *Ecological Modelling, 190*, 231–259.

Pozo-Montuy, G., Serio-Silva, J. C., & Bonilla-Sánchez, Y. M. (2011). Influence of the landscape matrix on the abundance of arboreal primates in fragmented landscapes. *Primates, 52*(2), 139–147.

Rahman, D. A., Rinaldi, D., Kuswanda, W., & Putro, H. R. (2019). Determining the landscape priority and their threats for the critically endangered *Pongo tapanuliensis* population in Indonesia. *Biodiversitas, 20*, 3584–3592.

Ren, G. P., Yang, Y., He, X. D., et al. (2017). Habitat evaluation and conservation framework of the newly discovered and critically endangered black snub-nosed monkey. *Biological Conservation, 209*, 273–279.

Rezende, G., Sobral-Souza, T., & Culot, L. (2020). Integrating climate and landscape models to prioritize areas and conservation strategies for an endangered arboreal primate. *American Journal of Primatology, 82*, e23202.

Sales, L. P., Neves, O. V., De Marco, P., & Loyola, R. (2017). Model uncertainties do not affect observed patterns of species richness in the Amazon. *PLoS One, 12*, e0183785.

Sales, L. P., Ribeiro, B. R., Pires, M. M., et al. (2019). Recalculating route: Dispersal constraints will drive the redistribution of Amazon primates in the Anthropocene. *Ecography, 42*, 1789–1801.

Sales, L., Ribeiro, B. R., Chapman, C. A., & Loyola, R. (2020). Multiple dimensions of climate change on the distribution of Amazon primates. *Perspectives in Ecology and Conservation, 18*(2), 83–90.

Santini, L., Benítez-López, A., Maiorano, L., Čengić, M., & Huijbregts, M. A. J. (2021). Assessing the reliability of species distribution projections in climate change research. *Diversity and Distributions, 27*(6), 1035–1050.

Schloss, C. A., Nunez, T. A., & Lawler, J. J. (2012). Dispersal will limit ability of mammals to track climate change in the Western hemisphere. *PNAS, 109*(22), 8606–8611.

Shanee, S. (2016). Predicting future effects of multiple drivers of extinction risk in Peru's endemic primate fauna. In M. T. Waller (Ed.), *Ethnoprimatology*. Springer International Publishing.

Shanee, S., Allgas, N., Shanee, N., & Campbell, N. (2015). Distribution, ecological niche modelling and conservation assessment of the Peruvian Night Monkey (Mammalia: Primates: Aotidae: *Aotus miconax* Thomas, 1927) in northeastern Peru, with notes on the distributions of Aotus spp. *Journal of Threatened Taxa, 7*, 6947–6964.

Singh, M., Cheyne, S. M., & Ehlers Smith, D. A. (2018). How conspecific primates use their habitats: Surviving in an anthropogenically-disturbed forest in Central Kalimantan, Indonesia. *Ecological Indicators, 87*, 167–177.

Stalenberg, E. M. (2019). *Biophysical ecology of the white-footed sportive lemur (Lepilemur leucopus) of Southern Madagascar* (PhD thesis). Australian National University.

Struebig, M. J., Fischer, M., Gaveau, D. L., et al. (2015). Anticipated climate and land-cover changes reveal refuge areas for Borneo's orang-utans. *Global Change Biology, 15*(8), 2891–2904.

Thorn, J. S., Nijman, V., Smith, D., & Nekaris, K. A. I. (2009). Ecological niche modelling as a technique for assessing threats and setting conservation priorities for Asian slow lorises (Primates: Nycticebus). *Diversity and Distributions, 15*, 289–298.

Thorne, J. H., Seo, C., Basabose, A., et al. (2013). Alternative biological assumptions strongly influence models of climate change effects on mountain gorillas. *Ecosphere, 4*(art108).

Titeux, N., Henle, K., Mihoub, J. B., et al. (2017). Global scenarios for biodiversity need to better integrate climate and land use change. *Diversity and Distributions, 23*, 1231–1234.

Tran, D. V., & Vu, T. T. (2020). Combining species distribution modeling and distance sampling to assess wildlife population size: A case study with the northern yellow-cheeked gibbon (*Nomascus annamensis*). *American Journal of Primatology, 82*, e23169.

Tran, D. V., Vu, T. T., Tran, Q. B., et al. (2018). Predicting suitable distribution for an endemic, rare and threatened species (Grey-shanked douc langur, *Pygathrix cinerea nadler*, 1997) using MaxEnt model. *Applied Ecology and Environmental Research, 16*, 1275–1291.

Tran, D. V., Vu, T. T., Tran, B. Q., et al. (2020). Modelling the change in the distribution of the black-shanked douc, *Pygathrix nigripes* (Milne-Edwards) in the context of climate change: Implications for conservation. *The Raffles Bulletin of Zoology, 68*, 769–778.

Trisurat, Y. (2018). Planning Thailand's protected areas in response to future land use and climate change. *International Journal of Conservation Science, 9*, 805–820.

Vidal-García, F., & Serio-Silva, J. C. (2011). Potential distribution of Mexican primates: Modeling the ecological niche with the maximum entropy algorithm. *Primates, 52*, 261–270.

Voskamp, A., Rode, E., Coudrat, C. N. Z., et al. (2014). Modelling the habitat use and distribution of the threatened Javan slow loris *Nycticebus javanicus*. *Endangered Species Research, 23*, 277–286.

Vu, T. T., Tran, D. V., Tran, H. T. P., et al. (2020). An assessment of the impact of climate change on the distribution of the grey-shanked douc *Pygathrix cinerea* using an ecological niche model. *Primates, 61*, 267–275.

Wich, S. A., Gaveau, D., Abram, N., et al. (2012). Understanding the impacts of land-use policies on a threatened species: Is there a future for the Bornean orang-utan? *PLoS One, 7*, e49142.

Wich, S. A., Singleton, I., Nowak, M. G., et al. (2016). Land-cover changes predict steep declines for the Sumatran orangutan (*Pongo abelii*). *Science Advances, 2*, e1500789.

Willems, E. P., & Hill, R. A. (2009). Predator-specific landscapes of fear and resource distribution: Effects on spatial range use. *Ecology, 90*, 546–555.

Williams, K. A., Slater, H. D., Gillingham, P., & Korstjens, A. H. (2021). Environmental factors are stronger predictors of primate species' distributions than basic biological traits. *International Journal of Primatology, 42*(3), 404–425.

Windyoningrum, A. (2013). *Estimating habitat loss and identifying refuge area for Javan langur (Trachypithecus auratus) as impact of Merapi eruption 2010* (MSc thesis). Gadjah Mada University/University of Twente.

Wisz, M. S., Hijmans, R. J., Li, J., et al. (2008). Effects of sample size on the performance of species distribution models. *Diversity and Distributions, 14*, 763–773.

Yuh, Y. G., N'Goran, P. K., Dongmo, Z. N., et al. (2020). Mapping suitable great ape habitat in and around the Lobéké National Park, South-East Cameroon. *Ecology and Evolution, 10*, 14282–14299.

Zhang, L., Ameca, E. I., Cowlishaw, G., et al. (2019a). Global assessment of primate vulnerability to extreme climatic events. *Nature Climate Change, 9*, 554–561.

Zhang, Y., Clauzel, C., Li, J., et al. (2019b). Identifying refugia and corridors under climate change conditions for the Sichuan snub-nosed monkey (*Rhinopithecus roxellana*) in Hubei Province, China. *Ecology and Evolution, 9*, 1680–1690.

Zhao, X., Ren, B., Li, D., et al. (2019). Climate change, grazing, and collecting accelerate habitat contraction in an endangered primate. *Biological Conservation, 231*, 88–97.

Zurell, D., Franklin, J., König, C., et al. (2020). A standard protocol for reporting species distribution models. *Ecography, 43*, 1261–1277.

Part II
Primates in Human-Dominated Landscapes

Chapter 7
Community-Based Strategies to Promote Primate Conservation in Agricultural Landscapes: Lessons Learned from Case Studies in South America

Laura A. Abondano, Amanda D. Webber, Lina M. Valencia, Carolina Gómez-Posada, Daniel Hending, Felipe Alfonso Cortes, and Nathalia Fuentes

Contents

L. A. Abondano (✉)
Department of Anthropology, University of Texas at Austin, Austin, TX, USA

Project Dragonfly, Department of Biology, Miami University, Oxford, Ohio, USA
e-mail: abondala@miamioh.edu

A. D. Webber
Bristol Zoological Society, Bristol, UK

Centre for Water, Communities and Resilience, University of the West of England, Bristol, UK

L. M. Valencia
Re:wild (formerly Global Wildlife Conservation), Austin, TX, USA

C. Gómez-Posada
Instituto de Investigación de Recursos Biológicos Alexander von Humboldt, Bogotá, Colombia

D. Hending
School of Biological Sciences, University of Bristol, Bristol, UK

F. A. Cortes · N. Fuentes
Fundación Naturaleza y Arte/Proyecto Washu, Esmeraldas, Ecuador

Abstract The increasing demand for natural resources has led to continued changes in land use, affecting the survival of many wild species, including non-human primates. One of the major challenges for primate conservation in landscapes dominated by agriculture is to find environmentally friendly alternatives that provide economic benefits to local communities while improving the health of the ecosystems that primates and humans rely on. Community-based conservation is an approach whereby researchers and conservationists work in collaboration with local people to plan, implement, and assess conservation projects. This ensures effective and sustainable management of their natural resources based on the specific needs and cultural traditions of each community. In this chapter, we present an overview of primates living in agricultural landscapes and provide some guidelines for developing community-based conservation projects based on experiences of three case studies from Colombia and Ecuador. It is important to create participatory spaces for local communities to become involved in the co-planning and co-design of conservation actions and provide training that strengthens people's capacities to acquire the necessary skills for implementing sustainable practices that bring revenue to the communities while protecting wildlife. Due to the social nature of community-based approaches, these conservation projects must also consider the socioeconomic and political contexts that influence the relationships between people and wildlife at each intervention site.

Keywords Participatory conservation · Sustainable agriculture · Economic alternatives · Capacity building · Local traditions · Multidisciplinary conservation · Sustainable cacao · Sustainable livestock · Empowering communities · Natural resource management

7.1 Introduction

The continued expansion of human populations around the world has resulted in increased demands for food and other natural resources. This leads to intensified deforestation and land-use shifts towards small-scale subsistence farming, large-scale industrial agriculture, cattle ranching, and extractivist activities (Godfray et al., 2010). Due to these changes in land use, many animal species, including non-human primates (hereafter primates), have been subject to habitat loss, fragmentation, and degradation, leading to significant contractions of their home ranges and an increase in their spatial and ecological overlap with humans (Fuentes & Hockings, 2010). Primates, however, are known for their behavioural and ecological flexibility, including a high degree of **vagility**, and, for some species, the ability to survive in degraded landscapes (e.g. Hending, 2021). This results in a shared space with humans and an increased likelihood of direct human-primate interactions. As habitat alteration continues, it is imperative to understand how wild primates and humans use ecological and social spaces to effectively develop strategies that allow the survival of both taxa (Fuentes, 2012; Fuentes et al., 2016).

Primates and humans share coevolutionary histories due to human modification of landscapes occupied by primates, shaping their movement patterns, ranging, habitat use, and genetic variation. However, primates have also played a significant role in shaping the culture, economy, and everyday lives of people living within close proximity to them (reviewed in Fuentes, 2012; Fuentes et al., 2016). Therefore, understanding the human-primate interface is crucial to ensure the continued survival of the species occupying today's urban **ecotones** and complex **heterogeneous human-modified landscapes** (Lee, 2010). To enable **human-primate coexistence**, it is necessary to integrate the interests and concerns of both taxa and undertake conservation actions that consider an ecological framework accounting for both human and primate requirements (Riley, 2006; Wolfe & Fuentes, 2007).

Community-based conservation is focused on including the participation and buy-in of local communities in the planning, implementation, and assessment of conservation projects (Baldauf, 2020). This bottom-up conservation approach ensures that communities take an active role in the management of local species and their habitats. It also allows local people to gain autonomy in the management and control of their territories in a sustainable manner. This approach not only guarantees that the community's interests are taken into account but also promotes a relationship based on trust and collaboration between local communities, conservation managers, and researchers. Rather than relying on interventions from outsiders who may not understand the nuances of the human-wildlife dynamics at each site, may not have appropriate permissions, and/or are focused solely on the protection of wildlife, community-based interventions are centred around the needs of the people living in the area (Adams & Hulme, 2001; Shaffer, 2015). For human and primate coexistence to occur within agricultural landscapes, the ever-growing demand for agricultural products, the needs of local communities, and the successful conservation of primate populations need to occur simultaneously (Hill, 2002). However, successful coexistence depends on the specific needs and opportunities of each community, their relationship with wildlife, and the geographic, historic, and socioeconomic context of each shared landscape. Therefore, it is important to develop inclusive strategies that are environmentally and financially sustainable and meet both primate and human needs.

In this chapter, we present an overview of primates living in agricultural landscapes and provide some guidelines on how to engage with local communities to increase the success of participatory conservation projects. We also present a set of case studies from South America that demonstrate first-hand strategies to co-develop community-based projects to ensure the conservation of primate species in human-modified landscapes. Our goal is that by presenting examples of applied community-based approaches for the conservation of wild primates, we can inform and guide current and future primate conservationists about the different strategies and their varying levels of success in various contexts and scenarios. With this, we want to stress the importance of involving and working with local communities and governments in conservation project decision-making, as well as providing suggestions of what we think are the best practices to implement primate threat mitigation strategies that meet both primate ecological demands and human interests and wellbeing.

7.2   Primates in Agricultural Landscapes

The ongoing expansion of tropical agriculture is particularly catastrophic for primates, as almost all extant primate taxa live within tropical regions, and most of
these primates depend exclusively on tropical forest for their survival (Chapman
et al., 2003). In the last 30 years alone, agricultural land has expanded by over 1.5
million km² into primate habitat, and this expansion and its associated deforestation
are now regarded as one of the biggest threats to global primate populations (Estrada
et al., 2017; Fernández et al., 2021). Croplands now occupy 17.1 million km²
(35.3%) of the 50.2 million km² global primate distribution (Fig. 7.1), much of
which is directly adjacent to remaining tropical forests, resulting in intense anthropogenic encroachment and disturbance of primate habitat (Estrada et al., 2017).

Many tropical landscapes are now dominated by matrices of isolated forest
patches surrounded by croplands and plantations (Perfecto & Vandermeer, 2008).
Although simplified agroecosystems such as pastures and **monocultures** generally
exclude most native biodiversity and restrict the dispersal of many primates, more
complex systems (e.g. **successional agroforestry** and **silvopastoral systems**) can
support more biodiversity and serve as suitable habitat or travel corridors while
providing food resources and dispersal opportunities (Estrada et al., 2012; Guzmán
et al., 2016). Over 60 primate species have been observed foraging within croplands
and plantations or travelling through them to reach adjacent forest fragments

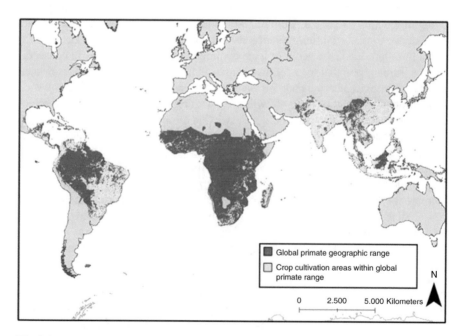

Fig. 7.1 The global distribution of the Primate order (orange) and the land used for crop cultivation within this geographic range (yellow). Map created with a scale of 1:100,000,000. Global crop
geoTIFF data downloaded from Earthstat.org

(Estrada et al., 2012). For example, several arboreal primates in Central and South America have been observed moving within matrices of cacao, coffee, or bamboo *agroecosystems* and through cattle ranching landscapes (e.g. Estrada et al., 2006; Guzmán et al., 2016; Loría & Méndez-Carvajal, 2017; Gómez-Posada, 2014). In Madagascar, several species of lemur have been recorded in vanilla and cacao plantations, and these areas may represent extensions of suitable habitat for these endemic primates (Hending et al., 2018; Webber et al., 2019). Chimpanzees (*Pan troglodytes*) and a multitude of monkey species use banana, palm oil, and mango plantations as temporary habitats, while they travel between forest fragments in sub-Saharan Africa (Estrada et al., 2012). In Asia, tarsiers (*Tarsius* sp.) have adapted to live in cacao, ylang-ylang, and coffee plantations (Merker & Muhlenberg, 2000), and several cercopithecids use palm oil plantations as **travel corridors** (Campbell-Smith et al., 2010). These examples provide evidence that agricultural practices can provide both economic benefits to local communities while maintaining landscapes with enough **ecological integrity** (Grantham et al., 2020) for primate species to survive, if properly managed. However, more long-term data is needed to ensure that primates can indeed survive, reproduce, and maintain stable and healthy populations in these anthropogenically impacted landscapes.

7.3 Community-Based Conservation Approaches

Community-based conservation takes place when researchers and conservationists collaborate with local communities to ensure sustainable and responsible management of their natural resources based on the specific needs and cultural traditions of each community (Horwich & Lyon, 2007). This bottom-up approach deviates from more traditional top-down models of biodiversity conservation that are often based on the creation of protected areas or interventions from large conservation organisations or international researchers (Baldauf, 2020; Berkes, 2004). Top-down approaches to conservation have been criticised for not including local people and local context as part of the decision-making processes (Baldauf, 2020; Berkes, 2004). Local communities are more likely to participate in conservation projects when they are previously consulted, when their traditions and beliefs are taken into account, and especially when they directly benefit from the interventions to be made (Horwich & Lyon, 2007). These projects may then become attractive to neighbouring communities and among multiple stakeholders, becoming a driver of positive changes towards biodiversity conservation at a regional and national level (Horwich & Lyon, 2007; Savage et al., 2010; Shanee et al., 2020).

To develop and implement potential interventions for reducing or mitigating actions that threaten biodiversity, it is important to have participatory spaces (e.g. Fig. 7.2) where local people can express how they interact and coexist with local wildlife through their beliefs, perceptions, and everyday activities (Jacobson, 2010; Savage et al., 2010). Building trust with communities is crucial to fully understanding the dynamics between people and their environment, including any negative interactions between humans and wildlife (Estrada et al., 2017; Savage et al., 2010;

Fig. 7.2 Examples of participatory methods employed by Proyecto Washu in Ecuador (see Box 7.1) to understand the interactions between local communities and wildlife in shared landscapes. (Photo credits: Proyecto Washu)

Shanee, 2012; Shanee & Shanee, 2015; Waylen et al., 2010), which can become major obstacles for primate conservation. These conversations can help conservationists to design specific strategies and messages tailored towards people's attitudes regarding wildlife and their needs and, by doing so, increase the likelihood of people engaging with conservation projects (Jacobson, 2010; Waters et al., 2019). For example, areas where communities perceive primates as pests (e.g. Hockings & McLennan, 2016; Tweheyo et al., 2005; Warren, 2009; Regmi et al., 2013; Saraswat et al., 2015) or as vectors of disease (e.g. Bicca-Marques & de Freitas, 2010) will require different strategies to those sites where primates are revered (e.g. Lutgendorf, 2007) or seen as key for forest **regeneration** (e.g. Stevenson et al., 2002; González-Zamora et al., 2012; Arroyo-Rodríguez et al., 2015; Franquesa-Soler & Serio-Silva, 2017; de Luna et al., 2016). Therefore, participatory spaces are important for local people to communicate how they interact with their surrounding environment and for conservationists to learn about the communities' relationships with nature, to subsequently co-design the most effective strategies that ensure the protection of natural resources.

Local community members have traditionally managed and utilised natural resources for many generations. They have relied on small-scale and subsistence agriculture, as well as forest-based economies, in accordance with their cultural traditions, to supplement their economic activities (Hill, 2002). However, commercial-scale agriculture, extensive cattle production, and large-scale natural resource extraction represent a major income source for large-scale farm owners, national and international companies, and development agencies who may, or may not, be local to the region or represent the local interests (Estrada et al., 2020). Therefore, one of the major solutions to promote primate conservation in landscapes dominated by agriculture is to find alternative income sources and economic incentives that do not rely on conventional unsustainable practices. Although this is not always possible when addressing large-scale commercial operations, it has proven to be a very effective strategy with local communities who have transitioned to more environmentally friendly approaches (e.g. Boxes 7.1 and 7.2). Alternative sustainable farming methods, such as successional agroforestry, **silvopastoral** systems,

Box 7.1 Protecting Brown-Headed Spider Monkeys in Ecuador Through the Creation of a Sustainable Matrix Model

The Chocó region in western Ecuador is a biodiversity hotspot (Myers et al., 2000) that requires immediate conservation action given that it has lost over 95% of its original vegetation cover (Mittermeier et al., 1999; Myers et al., 2000; CEPF, 2005). This forest loss has led to population decreases of several species in the region, including the Critically Endangered brown-headed spider monkey (*Ateles fusciceps fusciceps*) (Moscoso et al., 2021). The protection of the few remaining forest patches has been promoted by establishing private and state reserves. However, the success of this strategy, based on a protected area model, relies on the connectivity and permanence of unprotected forests located in the buffer zones (Checa et al., 2012). Buffer zones in this region have mainly been used for commercial and illegal logging, intensive agriculture, and cattle ranching by private enterprises, *Mestizo* immigrants, and Afro-Ecuadorians and Indigenous (Chachis/Awa) communities (Sierra & Stallings, 1998) that live in extreme poverty (Unidad de Información Socio Ambiental, 2021).

Proyecto Washu was established as an NGO with the goal of promoting the conservation of brown-headed spider monkeys and their habitats using primarily participatory methods with local communities (Fig. 7.2). Since 2013, *Proyecto Washu* has been working with farmers living in the buffer zones of the protected areas in north-western Ecuador, to create a Sustainable Matrix Model (SMM), which integrates concepts of **agroecological matrices**, sustainable development, and land sharing (Perfecto & Vandermeer, 2010; Butsic & Kuemmerle, 2015). Under this model, *Proyecto Washu* and members of the local communities have established socio-environmental agreements to hold themselves accountable for the protection of the forests and wellbeing of their communities. Additionally, *Proyecto Washu* has played an important role in facilitating local capacity building in biodiversity conservation, sustainability, and leadership, to promote autonomy and ownership of their territories, and, in the process, instilling a sense of community stewardship in conservation projects.

Through these efforts with local communities, *Proyecto Washu* has facilitated the inclusion of more than 500 hectares of land within the SMM. More than 300 hectares of forest, owned by 17 families that currently maintain socio-environmental agreements, are protected until 2025. With these community agreements, the project seeks to have both low-intensity farming and areas dedicated to biodiversity conservation within the same territory, promoting a "high-quality matrix" which allows for the migration of species and for preventing regional extinction trends. Local communities involved in the SMM are committed to protecting their forests while strengthening their capacities to produce high-quality cacao. These activities also help in increasing the communities' economic opportunities by establishing direct commercial relationships with buyers and farmers, who acknowledge the added value of the high-quality cacao produced within the SMM. This added value has

(continued)

Box 7.1 (continued)

resulted in a twofold increment in price per kilogram compared to average cacao prices in the country.

With the SMM framework, *Proyecto Washu* has helped in strengthening the capacities and principal economic activities of local communities in this **biodiversity hotspot** and, by doing so, improved both the people's livelihoods and protecting biodiversity, including the Critically Endangered brown-headed spider monkeys that live in the region.

Box 7.2 The Colombian Sustainable Cattle Ranching Project

The transformation of natural forests for livestock production is the third leading cause of habitat loss worldwide, affecting 31% of primate species (Estrada et al., 2018). In Colombia, cattle ranching occupies over 30% of the national territory and is primarily managed by small family businesses (FEDEGAN, 2021; Giraldo et al., 2018). In 2010, the Colombian Sustainable Cattle Ranching Project was launched as an economic programme to improve the income and the quality of life of ranchers and their families (Uribe et al., 2011; Giraldo et al., 2018). The project promotes the adoption of environmentally friendly cattle ranching practices that improve the management of natural resources and expand the provision of **ecosystem services** while increasing the farms' productivity (Giraldo et al., 2018; Uribe et al., 2011).

The project focuses on three main sustainable practices: (1) creating **living fences** to improve connectivity, to act as a windbreak barrier, and to conserve biodiversity; (2) planting scattered trees in pastures to provide shade for livestock, as windbreaks, and to promote biodiversity and soil improvement; and (3) using intensive silvopastoral systems (Giraldo et al., 2018). It also includes the management of pastures with activities focused on avoiding **soil compaction**, protection of watersheds by the reforestation of river banks, development and application of organic compost, and use of forage species generated by the silvopastoral system to feed cattle and to reduce the costs of supplementary feed to maintain cows (Giraldo et al., 2018; TNC, 2020).

By 2019, a total of 4100 families practising cattle ranching had benefited from this project (Calle, 2021). Between 2010 and 2019, participating farms recorded a 3% increase in secondary forest cover (while the mature forest cover remained the same), a 54% increase in the amount of land converted to silvopastoral systems, and a 151% increase in area covered by living fences, with a significant reduction of pastures and degraded soil (TNC, 2020). In total, 38,390 ha of silvopastoral systems with cattle production have been implemented through the project (TNC, 2020; Calle, 2021). These changes have also brought economic benefits to communities, with a reported average increase of 32% in animal load and 29% in dairy productivity (Calle, 2021). Furthermore, changes in land use and biodiversity (including birds, bats, dung beetles, edaphic microfauna, aquatic macrofauna, and plants) are being

(continued)

> **Box 7.2** (continued)
>
> monitored in collaboration with local communities (Calle, 2021). Results suggest that, with the expansion of silvopastoral systems (instead of pastures), there has been an increase in biodiversity and the mobility of wildlife between forest patches has improved due to an increase in living fences and scattered trees (Calle, 2021; TNC, 2020).
>
> This project has been implemented in farms located in lowland and mountain areas in the Caribbean, Andes, plains foothills, and the Orinoquia savannas in Colombia. Because of its geographic extension, this project potentially benefits at least 21 of the Colombian primate species that occupy these regions, including the Critically Endangered cotton-top tamarin (*S. oedipus*) and brown spider monkey (*A. hybridus*), and the Endangered white-bellied spider monkey (*A. belzebuth*) and the white-fronted (*Cebus versicolor*) and Colombian white-faced capuchins (*C. capucinus capucinus*). Primate communities inhabiting these regions may benefit from the increase in forest cover and the connectivity provided by living fences, the reforestation of **riparian areas**, and the increase in mobility between forest patches using scattered trees and silvopastoral systems (e.g. Torres et al., 2021). These sustainable livestock initiatives are a potential tool for the conservation of primates in rural landscapes while also benefiting local communities that depend on productive systems in those landscapes.

productive reforestation, as well as **climate smart** and ecosystem-based adaptation approaches (Colls et al., 2009; Lipper et al., 2014; Wezel et al., 2014), can become crucial to provide economic benefits to communities while improving the health of the ecosystems that primates and humans rely on (Jacobson, 2010; Estrada et al., 2012). In parallel to these sustainable practices, it is important to increase capacity and provide the necessary infrastructure that would allow for the integration of local communities into **productive market chains** at the local, regional, national, and international scale. This is important so that local producers have a demand for their sustainably sourced products and for those products to become marketable at competitive prices (Smith, 2008). Additionally, environmental and **fair-trade certifications** for sustainability (i.e. Rainforest Alliance, Wildlife Friendly Enterprises) can provide economic incentives for farmers by providing platforms so that they can sell their products at a higher price than regular market prices. This is due to the ecosystem conservation, wildlife protection, and fair treatment and good working conditions for workers that these certifications promote and that make the product unique and special (Makita, 2016; Box 7.1). By increasing product price, farmers are more willing to switch to more environmentally friendly practices that protect primates and their habitats and promote social, economic, and environmental standards for agriculture. These market integration strategies can be strengthened by creating local **cooperatives** and associations between individual community members or partnering with NGOs and businesses to collaborate towards shared environmental goals (Jacobson, 2010; Smith, 2008). The creation of these partnerships among local community members, which may be facilitated by NGOs, researchers, or other external organisations, strengthens their participation in sustainable practices. It also empowers communities to become responsible for the ownership and long-term management of the natural resources in their territories (Savage et al., 2010; Shanee et al., 2020).

Conservation initiatives tend to be more successful when implemented alongside a strategy to increase awareness of the threats to target species or ecosystems. Also important are training and **capacity building** programmes that help local communities to gain or strengthen their skills in implementing sustainable practices (that bring revenue to the communities while protecting wildlife) (Horwich & Lyon, 2007; Box 7.3). If conservation projects are not accompanied by awareness and capacity building activities, communities may become dependent on external organisations to adequately manage and maintain sustainable practices within their agro-ecosystems (Horwich & Lyon, 2007). Including financial literacy within training programmes is also important for key participants to acquire the necessary skills to run financially sustainable projects that empower and ensure the autonomy of local communities (Baldauf, 2020). Conservationists and researchers must also keep in mind that several years of consultation, planning, training, and implementation may be required for a community-based conservation project to become successful (e.g.

Box 7.3 Silvery-Brown Tamarin Conservation in Cattle Ranching Farms in Colombia

Silvery-brown tamarins (*Saguinus leucopus*) are endemic to the Andean region of Colombia and currently threatened by increased habitat fragmentation caused by cattle ranching, agriculture, mining, and dam building (Link et al., 2021; Henao-Díaz et al., 2020). Almost 80% of the tamarin's geographic distribution now consists of cattle ranching pastures (Etter, 1997). Many tamarin populations overlap with human settlements, resulting in a close relationship between local people and these primates, which shapes communities' livelihoods and cultural identity both positively and negatively (Valencia, 2018). For example, *S. leucopus* are in high demand in illegal pet trade markets (Henao-Díaz et al., 2020) representing an important source of income for some local people. Tamarins can also be a nuisance for communities, as they forage for food inside houses and move across roads, electric poles, and fences. Nevertheless, silvery-brown tamarins have become a symbol in the region.

Conservación Tití Gris (CTG) is a community-based conservation and research programme that promotes *S. leucopus* population recovery and long-term survival in cattle ranching farms in Norcasia, Caldas. Through multi-stakeholder coalitions with local government, national and international NGOs, universities, and most importantly the local community, *CTG* has built scientific knowledge about the viability of tamarin populations in highly degraded habitats and has raised awareness of the species' importance

---

(continued)

Box 7.2 (continued)

(Valencia, 2018). The project recognises that farmers are key allies for the long-term survival of tamarins in this region. At the project's inception, the goal was to understand the perceptions, behaviours, and attitudes of small holders and large-scale cattle ranchers towards tamarins and their habitat. This was achieved through continuous engagement and open communication with the local community, which included informal and formal interviews (Fig. 7.3), conversations over a cup of coffee, and educational workshops. Initially, the local community was not aware of the conservation status, threats, and endemicity of the species, nor did they understand how their everyday activities impacted tamarin survival. Using a variety of outreach materials and communication strategies specifically targeted to the local community (e.g. ponchos for cattle ranchers, mugs for cattle farm owners, colouring books for students), the project instilled a sense of pride and stewardship in the community (Fig. 7.3), using taglines like "Let's protect it, it's unique and ours", "The tamarin is as Colombian as myself", and "Tamarins are on my farm, and I protect them".

The project is currently addressing drivers of forest loss and identifying and implementing strategies that could increase the tamarins' chance of survival while improving local communities' livelihoods. Tourism is one of the main economic activities in the area, and the project has highlighted the importance of tamarins to tourists by using the primates as a flagship species. Billboards with the message "Welcome to the land of the Titi Gris" have been installed throughout the region. The project also worked with cattle ranchers to understand beef production practices and landscape management, with a view to designing feasible strategies for the implementation of silvopastoral systems on these farms. *CTG* will now focus on empowering the local community to develop long-term and self-sustainable ecotourism plans, as well as silvopastoral and reforestation activities.

Fig. 7.3 Informal conversations between conservationists and cattle ranchers to understand zoning and management mechanisms of cattle farms (left). Instilling stewardship and pride within local communities (right), both strategies for the conservation of silvery-brown tamarins in Norcasia, Caldas (Colombia). (Photo credits: Lina M. Valencia)

Savage et al., 2010; Horwich et al., 2010), so they should plan a long-term strategy with local communities when developing these projects.

Although some of the socioeconomic factors that affect local communities in primate habitat countries can be alleviated by incorporating sustainable practices, it is also important to consider other social, political, and historical realities to fully understand the dynamics between people and the environment in each area (Jacobson, 2010; Estrada et al., 2020). These socioeconomic and political factors can become major hurdles for conservation initiatives, and they should be considered carefully in the design of community-based conservation programmes. If ignored, interventions can cause more harm than good for local communities (Waters et al., 2021). For example, within the last few decades, there has been a rise in targeted violence, threats, and the assassination of activists, Indigenous people, lawyers, and journalists affiliated with environmental and social justice organisations, as well as agrarian communities who were engaged in the defence of environmental rights and their territories (Butt et al., 2019). This wave of violence has been marked by conflicts over natural resources and is more predominant in megadiverse countries in Latin America, the Caribbean, and South-East Asia (Butt et al., 2019), which host a large percentage of all primate species (Estrada et al., 2012, 2017). Although there is pressure from international organisations calling for governments to advocate for social and environmental justice (e.g. Escazu Agreement, CEPAL, 2018), weak and corrupt institutions and governments continue to leave environmental defenders in a very vulnerable, potentially life-threatening position (Butt et al., 2019; López-Cubillos et al., 2021). Therefore, conservationists must consider these socio-political aspects before planning conservation strategies in order to ensure the safety of local communities.

Finally, in an effort to improve the livelihoods of local communities, it is also important to include an approach that reduces gender inequalities, especially in rural areas where the *gender gap* in job opportunities and land ownership is even more pronounced (FAO, 2011). Cultural traditions and conventional gender roles in the division of labour have often left women in a more vulnerable situation. Men tend to join the workforce and have paid jobs, while women focus on unpaid domestic work and taking care of family or other community members (including children and elders) (Elson, 2017). Additionally, women endure more barriers to access educational programmes, including those offered in community-based conservation initiatives. As a consequence, women are less likely to receive the information and training on sustainable practices that they could implement in their households and transmit to their children (Agarwal, 2009; Gutierrez-Montes et al., 2012). Receiving information and training on sustainable activities is especially important for the development of rural and agricultural communities given that women have a dominant role in obtaining firewood, procuring water, and gathering and cooking food (Gutierrez-Montes et al., 2012). Since women often carry out these activities, it is likely they interact more closely with the environment than men. Women's decisions can, therefore, have significant effects on the use and management of natural resources (Gutierrez-Montes et al., 2012). When designing community-based conservation projects, it is important to find livelihood alternatives that alleviate poverty and provide leadership to women and other vulnerable populations, and to do so, women should be included in all aspects of the project, including the planning and implementation stages.

In conclusion, community-based conservation approaches are more effective when local community members participate in all the phases of the project and individuals outside the community take a facilitator role rather than a paternalistic and hands-on approach (Appleton et al., 2021). Researchers and conservationists must include local communities in the design and planning of strategies to protect natural resources, support them in obtaining the skills required to put into action the planned interventions, and provide the foundation to ensure their sustainability. Additionally, it is important to mention that despite potential commonalities across sites, each project will vary depending on the specific interests and needs of each local community and on the historical and socio-political context that shape the relationships between people and wildlife in each site.

7.4 Conclusions

Local communities in many primate habitat countries rely on activities related to crop cultivation, livestock farming, and the extraction of natural resources for their subsistence. It is, therefore, vital that conservation projects taking place in agricultural landscapes consider the local traditions and economic interests of all stakeholders involved (Baldauf, 2020). Participatory spaces allow conservationists to collaborate with stakeholders at different scales (i.e. local, regional, national, and international) in working together towards shared environmental goals that benefit both humans and wildlife. Given the social nature of participatory and inclusive conservation programmes, it is important to have a socioeconomic, socio-political, and socioecological approach for understanding the root causes beyond the immediate drivers of biodiversity loss (Baldauf, 2020).

Primate conservationists often have a background in biology, zoology, or biological anthropology that allow them to understand the biological and ecological factors impacting primate populations. However, primatologists tend to lack proper training to understand the social dynamics that may threaten primates directly or indirectly and the socio-political and historical background influencing land use in each region or to resolve social conflicts between different stakeholders (Jacobson, 2010; Horwich & Lyon, 2007; Estrada et al., 2020; Baldauf, 2020). Because of this, many projects result in a trial-and-error approach leading to several unsuccessful strategies that might discourage communities from wanting to participate in future conservation efforts. It is, therefore, crucial for primatologists to collaborate with people with a background in sociology, sociocultural anthropology, history, conflict management, pedagogy, law and policy, accounting, entrepreneurship, marketing, fundraising, and agricultural sciences, as well as local stakeholders, to ensure the long-term success of community-based projects for primate conservation.

Acknowledgements All co-authors are thankful to the editors for their invitation to participate in this volume. Thanks to Siân Waters and an anonymous reviewer whose edits and suggestions greatly improved this chapter. We also thank Gabriela Rezende, Sedera Solofondranohatra, Catherine Hill, and Sam Cotton for their contributions to earlier chapter drafts. The Colombian Sustainable Cattle Ranching Project is an alliance among the Global Environment Facility (GEF), the Government of the United Kingdom, the Federación Colombiana de Ganaderos (FEDEGAN), The Nature Conservancy (TNC), Fundación Centro para la Investigación en Sistemas Sostenibles de Producción Agropecuaria (CIPAV), and the Fondo para la Acción Ambiental (FA), under the supervision of the World Bank. LMV is first and foremost grateful to the local community of Rio La Miel and la Habana, to the Jaramillo family, and to the workers and staff from La Playita and La Reserva farms. LMV is also grateful to all her students and field assistants, local government authorities and other NGO partners, as well as the National Science Foundation (Award No. BSC-1540270), the Margot Marsh Biodiversity Foundation, Primate Conservation, Inc., the Rufford Foundation, the Primate Society of Great Britain, the Conservation Leadership Programme, and the Schlumberger Foundation – Faculty for the Future programme. FAC and NF from Proyecto Washu would like to thank the farmers' families of ASOPROTESCO and ASOCONCANANDE associations for this joint effort and work; the communities of Tesoro Escondido and Cristobal Colon; the work of Sylvana Urbina, Francesca Angiolani, Sofía Trujillo, and the students and volunteers of ISTOM. Proyecto Washu is also thankful for the support of Conservation Leadership Programme, Mohammed Bin Zayed Species Conservation Fund, University of Sussex, The Network For Social Change Charitable Trust, The Newman's Family, WCCN, Bouga Cacao, Samuel Von Rutte and Ana Basurto, Van Tienhoven Foundation, Bingo - Environmental Foundation of Lower Saxony, Germany, Bioparc de Doué-la-Fontaine, WNEF - Wildlands Nature Education Fund, and Wildlands Adventure Zoo Emmen.

References

Adams, W. M., & Hulme, D. (2001). If community conservation is the answer in Africa, what is the question? *Oryx, 35*, 193–200.

Agarwal, B. (2009). Gender and forest conservation: The impact of women's participation in community forest governance. *Ecological Economics, 68*, 2785–2799.

Appleton, M. R., Barborak, J. R., Daltry, J. C., et al. (2021). How should conservation be professionalized? *Oryx*. https://doi.org/10.1017/S0030605321000594

Arroyo-Rodríguez V, Andresen E, Bravo SP, Stevenson, PR (2015) Seed dispersal by howler monkeys: Current knowledge, conservation implications, and future directions. In: Kowalewski M, Garber PA, Cortés-Ortiz L, Urban B, Youlatos D (eds), Howler monkeys Springer, : pp.111–139.

Baldauf, C. (2020). *Participatory biodiversity conservation: Concepts, experiences, and perspectives*. Springer Nature.

Berkes, F. (2004). Rethinking community-based conservation. *Conservation Biology, 18*, 621–630.

Bicca-Marques, J. C., & de Freitas, D. S. (2010). The role of monkeys, mosquitoes, and humans in the occurrence of a yellow fever outbreak in a fragmented landscape in south Brazil: Protecting howler monkeys is a matter of public health. *Tropical Conservation Science, 3*, 78–89.

Butsic, V., & Kuemmerle, T. (2015). Using optimization methods to align food production and biodiversity conservation beyond land sharing and land sparing. *Ecological Applications, 25*, 589–595.

Butt, N., Lambrick, F., Menton, M., & Renwick, A. (2019). The supply chain of violence. *Nature Sustainability, 2*, 742–747.

Calle, Z. (2021). Transitioning to tree-based grazing systems in Colombia. In J. Ghazoul & D. Schweizer (Eds.), *Forests for the future: Restoration success at landscape scale - what will it take and what have we learned?* (pp. 70–75). WWF-Netherlands, Zeist and Utrecht University.

Campbell-Smith, G., Simanjorang, H. V., Leader-Williams, N., & Linkie, M. (2010). Local attitudes and perceptions toward crop-raiding by orangutans (*Pongo abelii*) and other nonhuman primates in northern Sumatra, Indonesia. *American Journal of Primatology, 72*(10), 866–876.

CEPAL. (2018). *Acuerdo Regional sobre el Acceso a la Información, la Participación Pública y el Acceso a la Justicia en Asuntos Ambientales en América Latina y el Caribe.* Naciones Unidas (UN).

CEPF (Critical Ecosystem Partnership Fund). (2005). *Chocó-Darién-Western Ecuador: Chocó-Manabí conservation corridor briefing book.* https://www.cepf.net/sites/default/files/final. chocodarienwesternecuador.chocomanabi.briefingbook.pdf. Accessed 20 July 2021.

Chapman, C. A., Lawes, M. J., Naughton-Treves, L., & Gillespie, T. (2003). Primate survival in community-owned forest fragments: Are metapopulation models useful amidst intensive use? In L. K. Marsh & C. A. Chapman (Eds.), *Primates in fragments* (pp. 63–78). Springer.

Checa, F., Córdoba, M., Freile, J. F., et al. (2012). *Llamado de acción para la protección del área de Canande.* Informe presentado al Ministerio del Ambiente, MAE Quito.

Colls, A., Ash, N., & Ikkala, N. (2009). *Ecosystem-based adaptation: a natural response to climate change.* IUCN. 16 pp.

de Luna, A. G., García-Morera, Y., & Link, A. (2016). Behavior and ecology of the white-footed tamarin (*Saguinus leucopus*) in a fragmented landscape of Colombia: Small bodied primates and seed dispersal in Neotropical forests. *Tropical Conservation Science, 9*, 788–808.

Elson, D. (2017). Recognize, reduce, and redistribute unpaid care work: how to close the gender gap. *New Labor Forum, 26*(2), 52–61.

Estrada, A., Saenz, J., Harvey, C., et al. (2006). Primates in agroecosystems: Conservation value of some agricultural practices in Mesoamerican landscapes. In A. Estrada, P. A. Garber, M. S. Pavelka, & L. Luecke (Eds.), (pp. 437–470). New perspectives in the study of Mesoamerican primates, Springer.

Estrada, A., Raboy, B. E., & Oliveira, L. C. (2012). Agroecosystems and primate conservation in the tropics: A review. *American Journal of Primatology, 74*, 696–711.

Estrada, A., Garber, P. A., Rylands, A. B., et al. (2017). Impending extinction crisis of the world's primates: Why primates matter. *Science Advances, 3*(1), e1600946.

Estrada, A., Garber, P. A., Mittermeier, R. A., et al. (2018). Primates in peril: The significance of Brazil, Madagascar, Indonesia and the Democratic Republic of the Congo for global primate conservation. *PeerJ, 6*, e4869.

Estrada, A., Garber, P. A., & Chaudhary, A. (2020). Current and future trends in socio-economic, demographic and governance factors affecting global primate conservation. *PeerJ, 8*, e9816.

Etter, A. (1997). Clasificación general de los ecosistemas de Colombia. In *Instituto de recursos biológicos Alexander von Humboldt* (pp. 176–185). Informe nacional sobre el estado de la biodiversidad.

FEDEGAN (Federación Colombiana de Ganaderos). (2021). *Cifras de referencia del sector ganadero colombiano.* https://www.fedegan.org.co/estadisticas/documentos-de-estadistica. Accessed 2 Dec 2021.

Fernández, D., Kerhoas, D., Dempsey, A., et al. (2021). The current status of the world's primates: Mapping threats to understand priorities for primate conservation. *International Journal of Primatology.* https://doi.org/10.1007/s10764-021-00242-2

Food and Agriculture Organisation (FAO). (2011). *The state of food and agriculture 2010–2011: Women in Agriculture* https://www.fao.org/3/i2050e/i2050e.pdf. Accessed 2 Dec 2021.

Franquesa-Soler, M., & Serio-Silva, J. C. (2017). Through the eyes of children: Drawings as an evaluation tool for children's understanding about endangered Mexican primates. *American Journal of Primatology, 79*, 12. https://doi.org/10.1002/ajp.22723

Fuentes, A. (2012). Ethnoprimatology and the anthropology of the human-primate interface. *Annual Review of Anthropology, 41*, 101–117.

Fuentes, A., & Hockings, K. J. (2010). The ethnoprimatological approach in primatology. *American Journal of Primatology, 72*, 841–847.

Fuentes, A., Cortez, A. D., & Peterson, J. V. (2016). Ethnoprimatology and conservation: Applying insights and developing practice. In M. T. Waller (Ed.), *Ethnoprimatology* (pp. 1–19). Springer.

Giraldo, C., Chará, J. D., Uribe, F., et al. (2018). Ganadería colombiana sostenible: Entre la productividad y la conservación de la biodiversidad. In G. Halffter, M. Cruz, & C. Huerta (Eds.), *Ganadería sustentable en el Golfo de México* (pp. 35–61). Instituto de Ecología.

Godfray, H. C. J., Beddington, J. R., Crute, I. R., et al. (2010). Food security: The challenge of feeding 9 billion people. *Science, 327*(5967), 812–818.

Gómez-Posada, C. (2014). Conserving primates in Colombian bamboo forest fragments: logging and landscape impacts on Red Howler Monkeys. Doctoral Dissertation. University of Washington. Seattle, WA.).

González-Zamora, A., Arroyo-Rodríguez, V., Oyama, K., et al. (2012). Sleeping sites and latrines of spider monkeys in continuous and fragmented rainforests: implications for seed dispersal and forest regeneration. *PLoS One, 7*(10), e46852.

Grantham, H. S., Duncan, A., Evans, T. D., et al. (2020). Anthropogenic modification of forests means only 40% of remaining forests have high ecosystem integrity. *Nature Communications, 11*, 1–10.

Gutierrez-Montes, I., Emery, M., & Fernandez-Baca, E. (2012). Why gender matters to ecological management and poverty reduction. In J. C. Ingram, F. DeClerck, & C. Rumbaitis del Rio (Eds.), *Integrating ecology and poverty reduction* (pp. 39–59). Springer.

Guzmán, A., Link, A., Castillo, J. A., & Botero, J. E. (2016). Agroecosystems and primate conservation: Shade coffee as potential habitat for the conservation of Andean night monkeys in the northern Andes. *Agriculture, Ecosystems & Environment, 215*, 57–67.

Henao-Díaz, F., Olaya-Rodríguez, M. E., Noguera-Urbano, E. A., & Gutiérrez, C. (2020). *Atlas de la biodiversidad de Colombia: Primates*. Instituto de Investigación de Recursos Biológicos Alexander von Humboldt. http://hdl.handle.net/20.500.11761/35544. Accessed 9 Aug 2021.

Hending, D. (2021). Environmental drivers of Cheirogaleidae population density: Remarkable resilience of Madagascar's smallest lemurs to habitat degradation. *Ecology and Evolution, 11*, 5874–5891.

Hending, D., Andrianiaina, A., Rakotomalala, Z., & Cotton, S. (2018). The use of vanilla plantations by lemurs: encouraging findings for both lemur conservation and sustainable agroforestry in the Sava Region, Northeast Madagascar. *International Journal of Primatology, 39*, 141–153.

Hill, C. M. (2002). Primate conservation and local communities - ethical issues and debates. *American Anthropologist, 104*, 1184–1194.

Hockings, K. J., & McLennan, M. R. (2016). Problematic primate behaviour in agricultural landscapes: Chimpanzees as 'pests' and 'predators. In M. T. Waller (Ed.), *Ethnoprimatology* (pp. 137–156). Springer.

Horwich, R. H., & Lyon, J. (2007). Community conservation: practitioners' answer to critics. *Oryx, 41*, 376–385.

Horwich, R. H., Islari, R., Bose, A., et al. (2010). Community protection of the Manas Biosphere Reserve in Assam, India, and the Endangered golden langur *Trachypithecus geei. Oryx, 44*(2), 252–260.

Jacobson, S. K. (2010). Effective primate conservation education: gaps and opportunities. *American Journal of Primatology, 72*, 414–419.

Lee, P. C. (2010). Sharing space: Can ethnoprimatology contribute to the survival of nonhuman primates in human-dominated globalized landscapes? *American Journal of Primatology, 72*, 925–931.

Link, A., Guzmán-Caro, D. C., Roncancio, N., et al. (2021). *Saguinus leucopus* (amended version of 2020 assessment). The IUCN Red List of Threatened Species 2021: e.T19819A192550769. https://doi.org/10.2305/IUCN.UK.2021-1.RLTS.T19819A192550769.en. Accessed 9 Aug 2021.

Lipper, L., Thornton, P., Campbell, B. M., et al. (2014). Climate-smart agriculture for food security. *Nature Climate Change, 4*(12), 1068–1072.

López-Cubillos, S., Muñoz-Ávila, L., & Roberson, L. A. (2021). The landmark Escazú Agreement: An opportunity to integrate democracy, human rights, and transboundary conservation. *Conservation Letters*. https://doi.org/10.1111/conl.12838

Loría, L. I., & Méndez-Carvajal, P. G. (2017). Uso de hábitat y patrón de actividad del mono cariblanco (*Cebus imitator*) en un agroecosistema cafetalero en la Provincia de Chiriquí, Panamá. *Tecnociencia, 19*, 61–78.

Lutgendorf, P. (2007). *Hanuman's tale: The messages of a divine monkey*. Oxford University Press.

Makita, R. (2016). A role of fair-trade certification for environmental sustainability. *Journal of Agricultural and Environmental Ethics, 29*(2), 185–201.

Merker, S., & Muhlenberg, M. (2000). Traditional land use and tarsiers – human influences on population densities of *Tarsius dianae*. *Folia Primatologica, 71*, 426–428.

Mittermeier, R. A., Myers, N., Mittermeier, C. G., & Robles Gil, P. (1999). *Hotspots: Earth's biologically richest and most endangered terrestrial ecoregions*. CEMEX/Agrupación Sierra Madre.

Moscoso, P., Link, A., Defler, T. R., et al. (2021). *Ateles fusciceps* (amended version of 2020 assessment). The IUCN Red List of Threatened Species 2021. https://doi.org/10.2305/IUCN. UK.2021-1.RLTS.T135446A191687087.en. Accessed 20 July 2021.

Myers, N., Mittermeier, R. A., Mittermeier, C. G., et al. (2000). Biodiversity hotspots for conservation priorities. *Nature, 403*, 853–858.

Perfecto, I., & Vandermeer, J. (2008). Biodiversity conservation in tropical agroecosystems: a new conservation paradigm. *Annals of the New York Academy of Sciences, 1134*(1), 173–200.

Perfecto, I., & Vandermeer, J. (2010). The agroecological matrix as alternative to the land-sparing/ agriculture intensification model. *Proceedings of the National Academy of Sciences, 107*(13), 5786–5791.

Regmi, G. R., Nekaris, K. A. I., Kandel, K., & Nijman, V. (2013). Crop-raiding macaques: Predictions, patterns and perceptions from Langtang National Park, Nepal. *Endangered Species Research, 20*(3), 217–226.

Riley, E. (2006). Ethnoprimatology: Toward reconciliation of biological and cultural anthropology. *Ecology and Environmental Anthropology, 2*, 75–86.

Saraswat, R., Sinha, A., & Radhakrishna, S. (2015). A god becomes a pest? Human-rhesus macaque interactions in Himachal Pradesh, northern India. *European Journal of Wildlife Research, 61*(3), 435–443.

Savage, A., Guillen, R., Lamilla, I., & Soto, L. (2010). Developing an effective community conservation program for cotton-top tamarins (*Saguinus oedipus*) in Colombia. *American Journal of Primatology, 72*, 379–390.

Shaffer, C. L. (2015). Cautionary thoughts on IUCN protected area management categories V– VI. *Global Ecology and Conservation, 3*, 331–348.

Shanee, N. (2012). Trends in local wildlife hunting, trade and control in the Tropical Andes Biodiversity Hotspot, northeastern Peru. *Endangered Species Research, 19*, 177–186.

Shanee, S., & Shanee, N. (2015). Measuring success in a community conservation project: local population increase in a critically endangered primate, the yellow-tailed woolly monkey (*Lagothrix flavicauda*) at la Esperanza, northeastern Peru. *Tropical Conservation Science, 8*, 169–186.

Shanee, S., Shanee, N., Lock, W., & Espejo-Uribe, M. J. (2020). The development and growth of non-governmental Conservation in Peru: Privately and communally protected areas. *Human Ecology, 48*, 681–693.

Sierra, R., & Stallings, J. (1998). The dynamics and social organization of tropical deforestation in northwest Ecuador, 1983–1995. *Human Ecology, 26*, 135–161.

Smith, B. G. (2008). Developing sustainable food supply chains. *Philosophical Transactions of the Royal Society B: Biological Sciences, 363*(1492), 849–861.

Stevenson, P. R., Castellanos, M. C., Pizarro, J. C., & Garavito, M. (2002). Effects of seed dispersal by three ateline monkey species on seed germination at Tinigua National Park, Colombia. *International Journal of Primatology, 23*, 1187–1204.

TNC (The Nature Conservancy). (2020). *Resultados del Proyecto Ganadería Colombiana Sostenible.* https://www.nature.org/content/dam/tnc/nature/en/documents/TNC_COL_GCS_PRESENTACION.pdf. Accessed 24 July 2021.

Torres, S., Valenzuela, L., Patarroyo, C., et al. (2021). Corridors in heavily fragmented landscapes: Reconnecting populations of critically endangered brown spider monkeys (*Ateles hybridus*) and sympatric terrestrial vertebrates in the lowland rainforests of Central Colombia. *Restoration Ecology.* https://doi.org/10.1111/rec.13556

Tweheyo, M., Hill, C. M., & Obua, J. (2005). Patterns of crop raiding by primates around the Budongo Forest Reserve, Uganda. *Wildlife Biology, 11*, 237–247.

Unidad de Información Socio Ambiental. (2021). *Censo 2010.* https://www.uasb.edu.ec/unidad-de-informacion-socio-ambiental/censo2010/. Accessed 20 July 2021

Uribe, F., Zuluaga, A., Valencia, L., Murgueitio, E., Zapata, A., & Solarte, L. (2011). *Establecimiento y manejo de sistemas silvopastoriles. Manual 1. Proyecto Ganadería Colombiana Sostenible.* GEF, Banco Mundial, Fedegan, Cipav, Fondo Acción, TNC.

Valencia, L. M. (2018). *Effects of anthropogenic habitat fragmentation on silvery brown tamarin (Saguinus leucopus) dispersal and movement patterns: Landscape genetics, habitat connectivity and conservation implications* (Doctoral dissertation). University of Texas at Austin, Austin, TX.

Warren, Y. (2009). Crop-raiding baboons (*Papio anubis*) and defensive farmers: a West African perspective. *West African Journal of Applied Ecology, 14*(1), 1–11.

Waters, S., El Harrad, A., Bell, S., & Setchell, J. M. (2019). Interpreting people's behaviour towards primates using qualitative data: a case study from North Morocco. *International Journal of Primatology, 40*, 316–333.

Waters, S., El Harrad, A., Bell, S., & Setchell, J. M. (2021). Decolonising conservation in practice: A case study from North Morocco. *International Journal of Primatology.* https://doi.org/10.1007/s10764-021-00228-0

Waylen, K. A., Fischer, A., McGowan, P. J., et al. (2010). Effect of local cultural context on the success of community-based conservation interventions. *Conservation Biology, 24*, 1119–1129.

Webber, A. D., Solofondranohatra, J. S., Razafindramoana, S., et al. (2019). Lemurs in cacao: Presence and abundance within the shade plantations of northern Madagascar. *Folia Primatologica, 91*, 96–107.

Wezel, A., Casagrande, M., Celette, F., et al. (2014). Agroecological practices for sustainable agriculture: A review. *Agronomy for Sustainable Development, 34*(1), 1–20.

Wolfe, L. D., & Fuentes, A. (2007). Ethnoprimatology. In C. Campbell, A. Fuentes, K. MacKinnon, S. Bearder, & R. Stumpf (Eds.), *Primates in perspective* (pp. 691–705). Oxford University Press.

Chapter 8
Primates in the Urban Mosaic: Terminology, Flexibility, and Management

Harriet R. Thatcher, Colleen T. Downs, and Nicola F. Koyama

Contents

Abstract Continuous human expansion is affecting landscape composition, in particular through urbanisation. Wildlife persistence in the urban mosaic is generally negatively affected; however, many primate species show behavioural plasticity and thrive in the urban mosaic. Urban primates often show selective behaviours in the urban mosaic, e.g. responses to anthropogenic food resources. However, as the urban mosaic becomes more prominent and important for biodiversity, conservation, and management, clearer definitions and terminology used to describe the urban mosaic are needed. Therefore, we use this chapter to review current definitions and suggest using the term 'mosaic' to discuss urban landscape ecology moving forward. Throughout our chapter, we consider the complexity of the urban mosaic and emphasise the value of considering quantified anthropogenic disturbance and species-specific knowledge in urban primate ecology. We suggest that management focus on the multiple facets of the urban mosaic, both human and

H. R. Thatcher (✉)
Department of Biomedical Sciences, University of Edinburgh, Edinburgh, UK
e-mail: Harriet.Thatcher@ed.ac.uk

C. T. Downs
Centre for Functional Biodiversity, School of Life Sciences, University of KwaZulu-Natal, Pietermaritzburg, South Africa

N. F. Koyama
Research Centre in Evolutionary Anthropology and Palaeoecology, School of Biological and Environmental Sciences, Liverpool John Moores University, Merseyside, UK

primate derived, and discuss the benefits for biodiversity, conservation, and human-primate coexistence.

Keywords Fitness · Human-primate coexistence · Management · Matrix · Mosaic · Urbanisation

8.1 Introduction

Almost all wildlife lives in an environment that is subject to some level of **anthropogenic** disturbance (Soulsbury & White, 2015). The effects of this disturbance vary dramatically with the nature of the environmental change (McKinney, 2008). Research on wildlife living in the **urban mosaic** is increasing (Perry et al., 2020), yet research on non-human primates (hereafter primates) in the urban mosaic is complex because of the multi-dimensional and heterogeneous nature of urban areas. The viability of biodiversity in an urban environment is influenced by multiple aspects such as the environment's ecological structure (Mackenstedt et al., 2015), species-specific physiological and behavioural adaptations (Humle & Hill, 2016), and human-primate relationships (Naughton-Treves et al., 1998). In this chapter, we will first consider terminology used to discuss the urban mosaic and then review research on primate ecological and **behavioural flexibility** in an urban environment. Finally, we will consider the application of this knowledge for management and conservation.

8.2 The Urban Mosaic

Global environmental change, caused by human land use requirements, often has detrimental impacts on ecosystems (e.g. Lambin et al., 2000). The growth of human populations, resulting in anthropogenic changes to landscapes and urban sprawl, is now considered a key driver of environmental change (Grimm et al., 2008). Urbanisation creates a unique landscape ecology through increasing human populations, anthropogenic topography, and habitat fragmentation (Werner, 2011). These urban landscapes vary dramatically from large cities to small settlements; therefore, the ecological effects are difficult to measure quantitatively (Bennett & Gratton, 2012).

Increases in human populations have resulted in changes to the natural ecosystem's function and biodiversity (Bonier et al., 2007). Although effects are species-specific, certain primate species have shown behavioural flexibility to ecological changes and thrive in these conditions (McLennan et al., 2017). Desirable characteristics linked to the urban environment, such as increased resources and a reduction in predation, provide an attractive habitat (Bateman & Fleming, 2012; de Andrade et al., 2020); hence, some species often favour the urban environment to its rural counterpart (Kaplan & Rogers, 2013). Generally, urban primates persist in these areas because of their omnivorous foraging behaviour (Lowry et al., 2013). By optimising their foraging strategies, exploiting human resources (Thatcher et al., 2020), altering ranging patterns for food access or changing foraging activity to avoid increased aggression from humans (Thatcher et al., 2019a, b), and using the city as a refuge (Waite et al., 2007), many primates are able to thrive in the urban mosaic.

Although the urban mosaic has many positive aspects, there are also many negative consequences (Bicca-Marques, 2016; Perry et al., 2020). Primates in the urban mosaic face challenges of habitat destruction and fragmentation resulting in poor habitat quality and connectivity (Bicca-Marques, 2016), as well as challenges associated with anthropogenic topography, human-wildlife interactions, pollution, and food restrictions (Gordo et al., 2013). These challenges can bring an increased risk of stress (Giraudeau et al., 2014) and increased chances of death and/or injury related to the human-primate interface (O'Riain & Hoffman, 2012), for example due to power lines (Lokschin et al., 2007; Lindshield, 2016; Pereira et al., 2020), and dog predation (Chap. 5, this volume). Furthermore, studies on endangered primate populations have shown the genetic risks of increased fragmentation (e.g. banded leaf monkey, *Presbytis femoralis femoralis*, Ang et al., 2012; pied tamarin, *Saguinus bicolor*, Farias et al., 2015), this research stresses the importance of considering the urban sprawl for genetic conservation of endangered species.

8.2.1   Defining the Urban Mosaic

It is commonly acknowledged that landscapes are spatially heterogeneous areas comprised of a mosaic of patches that differ in spatial patterns and ecological processes (Forman & Gordon, 1986; Wu, 2013). Urban patch mosaics form key attributes for wildlife providing anthropogenic resources (Johnson & Munshi-South, 2017) and green space (de Andrade et al., 2020; Downs et al., 2021). Although the term urban mosaic is becoming more widely used within urban ecology (e.g. Corrêa et al., 2018), it is still not clearly defined, likely because of global variation in these landscapes (see Werner, 2011). Here, we will consider the two most commonly accepted definitions of Werner (2011) and Marzluff et al. (2001) and define the urban mosaic as 'a habitat made up of areas of building density, residential human-density, anthropogenic disturbance, green areas and linear anthropogenic structures'.

In conducting this chapter, we found multiple discrete phrases for urban landscapes within the primate literature, including 'urban', 'peri-urban', 'semi-urban', 'urban-city/forest/farm/rural/semi-rural/tourist', 'human disturbance', 'anthropogenic', and 'tourist'. All these studies used the word 'urban' at some point to describe their study within either the abstract, introduction, or methods, yet their study sites varied dramatically. Most studies only described the study site, with few studies defining landscapes and anthropogenic terminology (e.g. Scheun et al., 2019; Chowdhury et al., 2020). We, therefore, suggest that studies should more clearly describe the matrix within their study site, providing a clear ecological description of the habitat composition (Werner, 2011) and considering landscape scales (see Marzluff et al., 2001, pp. 11, Table 1) to clearly define the habitat type. Understanding this matrix of connected habitats and species requirements in the urban mosaic is important for ecosystem services and biodiversity conservation (Downs et al., 2021; Zungu et al., 2020a, b). We acknowledge the value of these diverse and discrete terms within the developing urban mosaic; nevertheless, within this chapter, we will focus primarily on research conducted in 'urban' habitats as clearly defined in the study's methodology and/or following our definition. More detailed analysis of the above subcategories can be seen in the respective chapters throughout this book.

8.2.2 Quantifying Anthropogenic Disturbance

With increasing urban expansion, interest in the human-primate interface is growing, evidenced by recent ethnoprimatology studies (McKinney & Dore, 2018). Currently, most research that focuses on the anthropogenic interface classifies disturbance by habitat type (see Sect. 9.2.1), and data are often compared interchangeably without consideration of the varying ecological pressures within these landscapes (McKinney, 2015). Additionally, as the urban mosaic varies globally, so does the nature of the human-primate interface (Beisner et al., 2015). For example, economic loss to communities (Dickman, 2012) differs from the economic and cultural benefits of **primate tourism** (Zhao, 2005). In response, McKinney (2015) suggested a generalised classification system allowing researchers to clearly report four major variables within their study site including landscape, non-human primate interface, diet, and predation risk. Although McKinney's descriptive system is a valuable initiative, it is not necessarily applicable to accelerating urban landscapes, for example, with respect to the value of opportunistic foraging for urban primates. McKinney's classification has only been used in one urban study so far (Thatcher et al., 2018; Table 1), although with modifications the premise of this system could be used more widely across the urban mosaic.

Nevertheless, some studies do provide a quantified estimate of field site variables (Table 8.1). For example, research has attempted to quantify anthropogenic topography by calculating the density of key urban mosaic features (e.g. buildings and green space) (Santos et al., 2014; Thatcher et al., 2018; de Andrade et al., 2020). Additionally, some studies have considered the effect of noise pollution in the urban environment, measuring noise amplitude (de Andrade et al., 2020; Duarte et al., 2011).

Measuring interactions and associations within the human-primate interface is one of the most common measures of anthropogenic disturbance (Table 8.1). Multiple studies have used behaviour sampling to record all **human-primate interactions**, the initiator and context, providing a detailed account of interactions (e.g. Beisner et al., 2015; Suzin et al., 2017). Additionally, studies have highlighted the

Table 8.1 Different measurements of quantified anthropogenic disturbance for primates in the urban mosaic. Each shows a brief description of the method and the associated study

Measure of urbanisation	Definition	Study
McKinney's classification	Code landscape variables including diet, human-nonhuman primate interface, and predation level (see McKinney, 2015)	Thatcher et al. (2018)
Land cover	Anthropogenic topography	de Andrade et al. (2020), Santos et al. (2014), and Thatcher et al. (2018)
Human presence	Tourism rate	Ilham et al. (2017)
	Human population (per km²)	Thatcher et al. (2018)
	Human traffic scans (humans, bikes, motorcycle, bus, truck, cars)	Beisner et al. (2015)
Noise disturbance	Noise amplitude	de Andrade et al. (2020) and Duarte et al. (2011)
Human-primate interface	Human-primate conflict (injury or death)	O'Riain and Hoffman (2012) and Pragatheesh (2011)
	Human-primate interaction from local human community perspective (questionnaire)	Rodrigues and Martinez (2014), Teixeira et al. (2015), Beisner et al. (2015), Chauhan and Pirta (2010), Kaburu et al. (2018, 2019), Marty et al. (2019a, b), Olaleru and Ogunfuwa (2020), and Patterson et al. (2017)
	Human-primate interactions (behavioural observations)	Suzin et al. (2017), Thatcher (2019), and Thatcher et al. (2019a, b, 2020)
	Human monitoring (behavioural observations)	Kaburu et al. (2018, 2019) and Marty et al. (2019a, b)
	Provisioning, rate, and food type	Ilham et al. (2018), Kaplan et al. (2011), Suzin et al. (2017), and Thatcher et al. (2020)
Human impact index	Human activity score	Fourie et al. (2015)

importance of considering the nature of human-primate interactions in the urban mosaic, both positive and agonistic/negative, providing an understanding of the drivers of urban primate behaviour (Suzin et al., 2017; Thatcher et al., 2019a, b; Thatcher et al., 2020). What is noteworthy is that although studies have considered both human- and primate-orientated interactions (Table 8.1), to our knowledge, no study has simultaneously considered both positive and negative interactions from both the human and primate perspectives, an important consideration to fully understand the multiple facets of the human-primate interface.

As the importance of studying the urban mosaic becomes a more prominent issue (Perry et al., 2020), clearer definitions and understanding within this landscape are needed to allow comparisons of research and support management plans for biodiversity and conservation. Although Table 8.1 highlights the current array of quantified anthropogenic pressures measured in urban studies, the methods and techniques within these studies are still variable, and as the human-primate interface varies with both anthropogenic and ecological pressures, these quantified measures need to be supported with an ecological description of urban mosaic characteristics (Table 8.1).

8.3 Behavioural Flexibility

To successfully thrive in an urban mosaic, animals must display behavioural flexibility to adapt to changing environmental pressures (Wright et al., 2010). Species that display a high degree of behavioural flexibility can adjust to a range of conditions and thrive in the urban mosaic (Healy & Nijman, 2014). Therefore, research has focused on this plasticity in the urban environment to understand fitness implications (Sol et al., 2013) and how this knowledge can be used for management plans (McLennan et al., 2017; Sol et al., 2002).

As time is a bounded resource, its allocation and use reflect ecological pressures (Dunbar et al., 2009). Time budgets have been applied to primates in the urban mosaic to demonstrate trade-offs in behaviour (McLennan et al., 2017). For example, urban properties such as high-value food generally decrease foraging time (Back et al., 2019; Hoffman & O'Riain, 2011; Jaman & Huffman, 2013), often corresponding with reduced movement (Jaman & Huffman, 2013) and associated with an increase in social interactions and resting (Ilham et al., 2018; Jaman & Huffman, 2013; Scheun et al., 2019). Additionally, studies on primates in the urban mosaic have shown that urban primates can flexibly adjust their activity seasonally (Jaman & Huffman, 2013; Thatcher et al., 2019a).

8.3.1 Foraging

Generalist species who display foraging flexibility and dietary plasticity can typically adjust more readily to anthropogenic changes than specialist species; therefore, foraging/dietary plasticity is highlighted as a key attribute to thrive in the urban mosaic (Lowry et al., 2013). Research on primates in the urban mosaic highlights a preference for high-calorie anthropogenic food resources (Hoffman & O'Riain, 2012a; Dasgupta et al., 2020). Foraging patterns in urban primates show that both natural and human foods are important to their diet, but dependence on either resource can differ between species and even within species (Ilham et al., 2017; Thatcher et al., 2020). More so, urban primates have been shown to modify the proportion of anthropogenic and natural food dependent upon food availability, largely influenced by **provisioning** and natural food availability (long-tailed macaques, *Macaca fascicularis*, Ilham et al., 2017; grey langurs *Semnopithecus entellus*, Dasgupta et al., 2020; vervet monkeys, *Chlorocebus pygerythrus*, Thatcher et al., 2020).

Research highlights that human food within a primate's diet can have varied fitness effects across anthropogenic landscapes; studies show this high-value food can increase individual fitness and reproductive success in **agroecosystems** because of higher nutritional content (Warren et al., 2011) and increase intergroup competition (Sinha & Mukhopadhyay, 2013) and subsequently increase anxiety and social tension (Maréchal et al., 2011) in areas of high tourism. Although these examples come from studies across anthropogenic landscapes, the consequences most likely hold true within the urban landscape as urban primates show a high degree of foraging flexibility and increased human-primate proximity (Thatcher et al., 2019a, b, 2020). Additionally, urban research has previously suggested that these increased foraging opportunities in the urban mosaic can lead to increased group size (Patterson et al., 2018). Therefore, an enhanced understanding of urban primates' dependence on **anthropogenic** food, and the potential fitness implications, is necessary to implement and sustain management plans (Thatcher et al., 2020).

8.3.2 Ranging

Primate studies in the **urban mosaic** generally show that increased anthropogenic effects reduce home range size, primarily because of increased urban resources (Hoffman & O'Riain, 2011, 2012b; Klegarth et al., 2017). Research in the urban mosaic has also shown that urban primates express shorter daily path lengths (Corrêa et al., 2018; Thatcher et al., 2019b). Generally, this is highlighted to be an adaptive strategy, with a preference for habitats with greater food resources associated with reducing the energy and need to travel for food (Hoffman & O'Riain, 2012a; Patterson et al., 2019). Further research has shown that such habitat selection can be an adaptive strategy to avoid areas of high noise pollution and human disturbance (Duarte et al., 2011).

Conversely, research on urban vervet monkeys that quantified human-primate interactions reported the opposite, that home ranges increased with anthropogenic disturbance (Thatcher et al., 2019b). This differing result is likely due to the quantified measures of the human-primate interface in Thatcher et al.'s research (Thatcher et al., 2019b) suggesting that this study population show an avoidance strategy to avoid human-directed aggression or that they ranged further to forage at more predictable sources of high-value human food. Granting clear interpretations of the costs and benefits of this strategy cannot be derived from the findings, it nonetheless suggests a complex attraction-avoidance scale within the urban mosaic. Although current research on primates in the urban mosaic is limited and does not show consistent patterns, it does imply that primates show spatial feeding strategies dependent upon anthropogenic pressures, highlighting fitness consequences and the value of species-specific studies.

8.3.3 Sociality

Social flexibility of wildlife has been shown to be plastic to change (review: Smil, 1993) and is an important behavioural trait that persists in the urban mosaic (Skandrani et al., 2017). Increased anthropogenic food availability can allow more time for socialising (Jaman & Huffman, 2013; Thatcher et al., 2019a, b). Kaburu et al. (2019) found that rhesus macaques (*M. mulatta*) who interacted more frequently with humans spent significantly less time resting and grooming, suggesting unpredictable human behaviour is a time constraint. Furthermore, Thatcher et al. (2019a) categorised human interactions as either positive (food) or negative (conflict) and found that human interactions influenced time budget behaviour, suggesting a complex relationship between the costs and benefits of urban living. These studies therefore highlight the benefits of urban living, with more time for socialising (Jaman & Huffman, 2013; Kaburu et al., 2019; Thatcher et al., 2019a) likely because of provisioning and increased dispersed feeding opportunities (Back et al., 2019; Scheun et al., 2019).

There are of course costs to social living in the urban mosaic, for example, greater anthropogenic food availability has been shown to increase competition and aggressive behaviours (Sinha & Mukhopadhyay, 2013). Furthermore, research on the human-primate interface shows the complexity of social dynamics between humans and primates; trends generally highlight that human actions and resources cause primate reactions (e.g. aggression) (Beisner et al., 2015; Chauhan & Pirta, 2010) and that human interactions are exacerbated by primate reactions (Chauhan & Pirta, 2010) and economic loss/damage (Beisner et al., 2015). A wealth of research on inter-individual differences in primate social behaviour has been conducted on macaques in the urban mosaic. Research has shown that bonnet macaques (*M. radiata*) (Balasubramaniam et al., 2020), rhesus macaques (Kaburu et al., 2019), and long-tailed macaques (Marty et al., 2019b) that spend more time monitoring human activity reduce their grooming effort. Furthermore, research has shown that long-tailed macaques spend less time grooming when human presence increases (Ilham et al., 2018). This research therefore highlights the importance of studying the more complex individual social dynamics within the urban mosaic, an area of research that is continually developing.

8.4 Urban Health

Urbanisation is commonly linked to more complex and disturbed habitats increasing human-wildlife interactions, all of which have been suggested to increase the intensity and diversity of parasites (Soto-Calderón et al., 2016; Thatcher et al., 2018). However, the effects of urbanisation on parasite load are not always consistent, and other studies have found that a more anthropogenic environment can lead to a reduction in the intensity and diversity of parasites (Lane et al., 2011) or no difference in parasite prevalence (Adrus et al., 2019). Although parasite diversity trends are mixed throughout the anthropogenic landscape, studies on parasite diversity and intensity are important for human health and wellbeing (Díaz et al., 2006). Primates in the urban mosaic are often found in proximity to humans, and this increases opportunities for **zoonosis** (Singh & Gajadhar, 2014; Sapkota et al., 2020). Although studies have suggested the potential for transmission between urban primates and humans, data are still preliminary and currently inconclusive (Aitken et al., 2016; Debenham et al., 2017).

Parasite load and urbanisation have been linked to further primate health concerns. For example, white-footed tamarins, *Saguinus leucopus*, living in the urban mosaic were found to be overweight and have a higher body mass and cholesterol level than rural tamarins (Soto-Calderón et al., 2016). Additionally, urban populations of the African lesser bushbaby (*Galago moholi*) have a greater body mass index, and females have higher faecal glucocorticoid than their rural counterparts (Scheun et al., 2015). Overall, these studies highlight the risks of increased time and dependence on anthropogenic resources and potential negative health impacts of the urban mosaic, suggesting species may show flexibility and habituate to humans' presence, but not necessarily to the conflict and stress associated with the urban mosaic.

8.5 Managing the Urban Mosaic

Understanding an animal's phenotypic and behavioural flexibility in response to urban challenges provides an educated rationale to form species-specific management techniques for human-wildlife coexistence and conservation management (Lowry et al., 2013). Acquiring further knowledge on the impact of urbanisation on wildlife populations is a priority to be able to implement appropriate management (Redpath et al., 2013). Due to their intelligence and sociality, primates pose a complex challenge to execute effective management plans (Woodroffe et al., 2005).

Knowledge of behavioural flexibility can be beneficial for species management. Research has shown black-tufted marmosets (*Callithrix penicillata*) in the urban mosaic avoid high noise areas, even if the area has high food availability, showing noise has potential benefits as aversive management (Duarte et al., 2011). Additionally, research on chacma baboon (*Papio ursinus*) has been applied for the benefit of species management in the urban mosaic (Hoffman & O'Riain, 2011). For example, Kaplan et al. (2011) studied the effectiveness of a food station in deterring chacma baboons away from urban spaces, showing the need of a combined approach. Additionally, O'Riain and Hoffman (2012) modelled characteristics of chacma baboon spatial ecology to predict potential human-baboon conflict and show the benefit of applying this knowledge to make informed management suggestions. Overall, this research across the urban mosaic highlights the value of behavioural studies for management and the consideration of urban mosaic features.

Research generally highlights that foraging flexibility and anthropogenic food play a key role in wildlife success and should be the focus of management (Bicca-Marques, 2016; Thatcher et al., 2019b, 2020). In Box 8.1, we present a case study

Box 8.1 A Case Study Showing that Positive and Negative Human-Interactions Create a Complex Attraction-Avoidance Scale for Urban Vervet Monkeys that Should Be Considered for Management

Urban ecosystems present complex challenges for the human-primate interface. Thatcher and colleagues analyses consider time budget behaviours (Thatcher et al., 2019a), ranging behaviours (Thatcher et al., 2019b), and foraging flexibility (Thatcher et al., 2020). In these studies, Thatcher et al. measured rates of human-wildlife interaction, considering both positive (e.g. food) and negative (e.g. aggression) urban drivers from a vervet monkey perspective. Results highlighted that vervet monkey behaviour is influenced by human-wildlife interactions, suggesting that urban vervet monkeys express behavioural flexibility. In this case study, we summarise key findings, highlighting the application of this approach for managing the human-primate interface.

Vervet monkeys in an urban mosaic take shorter (Thatcher et al., 2019b) and more direct journeys (Thatcher et al., 2019a) if they have increased access to human food. However, if the rate of positive human-interactions decreases, and negative human-interactions increase, these routes become longer and less direct. Results further highlight that vervet monkeys' movement is highly dependent on the value of available food resources, as vervet monkeys are less likely to move in response to human aggression when anthropogenic food is available. Therefore, managing access to this anthropogenic food can directly affect vervet monkey movement patterns.

The interaction of positive and negative human interactions was also significant for foraging, indicating that if vervet monkeys have access to high-value anthropogenic food, then, despite human aggression, their time spent foraging would increase (Thatcher et al., 2019a). Again, this result has important consequences for management showing the key role of human food, but the ineffective deterrent of human aggression. However, more recent in-depth analysis showed that foraging depended upon availability of resources (human-derived food and horticultural plants) (Thatcher et al., 2020). Thus, vervet monkeys show strong seasonal foraging, but that negative human interactions can reduce foraging rate of specific food resources. These results (Thatcher et al., 2019a, 2020) highlight the foraging flexibility of urban vervet monkeys, but also highlight some conflicting results that could possibly be because Thatcher et al.'s (2019a) study population depends solely on human-derived resources (horticultural garden plants and human-derived food), emphasising the need for refined foraging terminology within the urban landscape.

Overall, Thatcher and colleagues' studies highlight the key role of human food and that increased human aggression does not necessarily reduce the 'unwanted' behaviour of vervet monkeys. Therefore, management strategies should aim to reduce opportunities for human food consumption that may

(continued)

Box 8.1 (continued)

support human-wildlife conflict through education and local management programmes.

Overall, findings emphasise vervet monkey behavioural flexibility, demonstrating how vervet monkeys respond to the urban landscape by altering their behaviour under periods of positive human interactions to benefit from the potentially high calorific food. The interplay between positive and negative aspects of the urban environment creates a complex attraction-avoidance scale, and both aspects must be considered to fully understand behavioural adaptations under anthropogenic pressures for species management.

that highlights the role of anthropogenic food and human-wildlife interactions to create a complex attraction-avoidance scale that should be considered for human-wildlife management. Acknowledgement and incorporation of the human-primate interface in research and the positive and negative consequences of this interface, for both primates and humans, is beneficial to make informed management strategies for primate welfare and biodiversity conservation (Dore et al., 2018; Setchell et al., 2017). Understanding the dynamics and frequencies of human-wildlife interactions is necessary to feed forward into appropriate management (Beisner et al., 2015). Although primatology is moving away from the term human-wildlife conflict and focusing on coexistence (McKinney & Dore, 2018), it is important that we consider both positive (e.g. anthropogenic resources) and negative (e.g. human-directed aggression) aspects of the **human-primate interface** (Thatcher et al., 2019a, b).

Additionally, it is just as important to consider human perceptions and roles within urban species management. In particular, an ethnographic perspective is essential when designing strategies to ensure they are truly inclusive, advocating a decolonial approach to research in the urban mosaic (Ehlers Smith et al., 2021) and wider anthropogenic landscapes (Setchell et al., 2017; Waters et al., 2019). Diverse multi-cultural beliefs, views, and philosophies may be embedded within local human-wildlife relationships, and these need to be fully considered in order to develop effective management strategies, for example, shepherds' perceptions of Barbary macaques (*M. sylvanus*) in the anthropogenic landscape (Waters et al., 2018). Therefore, consultation with local communities and Indigenous populations must be a priority during the conceptualisation of any plan in the urban mosaic.

Community science studies have highlighted human concern for urban primates, emphasising the value of incorporating a human dimension within urban ecology and management (Patterson et al., 2017; Suzin et al., 2017). Although human-focused behaviour change can be more challenging to implement in the urban mosaic (Bicca-Marques, 2016), education is considered a key action in managing the human-primate relationship, and research has shown a public willingness to engage in these measures (Sha et al., 2009). Furthermore, the presence of urban wildlife is also beneficial at a public level because often people in the urban mosaic only encounter urban wildlife (Lunney & Burgin, 2004), and it has been shown that exposure to wildlife at an early age can encourage support for conservation (Soga & Gaston, 2016), even making humans more tolerant of exploitive wildlife (Hosaka et al., 2017). Recent research has also highlighted the mental health benefits of the human-primate interface, suggesting positive consequences for human wellbeing (Barua et al., 2021).

As urbanisation is only predicted to increase with the growing human population, a developed understanding of species-specific primate reactions to urban drivers is needed (Lowry et al., 2013). Species that can thrive and tolerate anthropogenic drivers are currently 'winners' in this developing mosaic (Perry et al., 2020). However, as the urban mosaic becomes more dominant, knowledge of flexible behavioural ecology will be necessary to predict and manage species adaptations to the changing landscape to benefit human-wildlife coexistence and biodiversity conservation. We suggest the best way to facilitate this moving forward is clearer measurements and definitions of urbanisation where possible.

Acknowledgements We thank Tracie McKinney, Siân Waters, and Michelle Rodrigues for their efforts in forming this book. We are also grateful to the above and an anonymous reviewer for their constructive comments that greatly improved this chapter.

References

Adrus, M., Zainudin, R., Ahamad, M., Jayasilan, M. A., & Abdullah, M. T. (2019). Gastrointestinal parasites of zoonotic importance observed in the wild, urban, and captive populations of non-human primates in Malaysia. *Journal of Medical Primatology, 48*, 22–31.

Aitken, E. H., Bueno, M. G., Dos Santos Ortolan, L., et al. (2016). Survey of plasmodium in the golden-headed lion tamarin (*Leontopithecus chrysomelas*) living in urban Atlantic Forest in Rio de Janeiro, Brazil. *Malaria Journal, 15*, 1–6. https://doi.org/10.1186/s12936-016-1155-3

Ang, A., Srivasthan, A., Md-Zain, B. M., et al. (2012). Low genetic variability in the recovering urban banded leaf monkey population of Singapore. *The Raffles Bulletin of Zoology, 60*, 589–594.

Back, J. P., Suzin, A., & Aguiar, L. M. (2019). Activity budget and social behavior of urban capuchin monkeys, *Sapajus* sp. (Primates: Cebidae). *Zoologia (Curitiba), 36*, e30845.

Balasubramaniam, K. N., Marty, P. R., Arlet, M. E., et al. (2020). Impact of anthropogenic factors on affiliative behaviors among bonnet macaques. *American Journal of Physical Anthropology, 171*, 704–717.

Barua, M., Jadhav, S., Kumar, G., Gupta, U., Justa, P., & Sinha, A. (2021). Mental health ecologies and urban wellbeing. *Health & Place, 69*, 102577.

Bateman, P. W., & Fleming, P. A. (2012). Big city life: Carnivores in urban environments. *Journal of Zoology, 287*, 1–23.

Beisner, B. A., Heagerty, A., Seil, S. K., et al. (2015). Human–wildlife conflict: Proximate predictors of aggression between humans and rhesus macaques in India. *American Journal of Physical Anthropology, 156*, 286–294.

Bennett, A. B., & Gratton, C. (2012). Local and landscape scale variables impact parasitoid assemblages across an urbanization gradient. *Landscape and Urban Planning, 104*, 26–33.

Bicca-Marques, J. C. (2016). Urbanization (and primate conservation). In A. Fuentes, M. Bezanson, C. J. Campbell, A. F. Di Fiore, S. Elton, A. Estrada, & J. Yamagiwa (Eds.),

The international encyclopedia of primatology (pp. 1–5). Wiley-Blackwell. https://doi. org/10.1002/9781119179313.wbprim0153

Bonier, F., Martin, P. R., Sheldon, K. S., et al. (2007). Sex-specific consequences of life in the city. *Behavioral Ecology, 18*, 121–129.

Chauhan, A., & Pirta, R. S. (2010). Agonistic interactions between humans and two species of monkeys (rhesus monkey *Macaca mulatta* and hanuman langur *Semnopithecus entellus*) in Shimla, Himachal Pradesh. *The Journal of Psychology, 1*, 9–14.

Chowdhury, S., Brown, J., & Swedell, L. (2020). Anthropogenic effects on the physiology and behaviour of chacma baboons in the Cape Peninsula of South Africa. *Conservation Physiology, 8*, coaa066.

Corrêa, F. M., Chaves, Ó. M., Printes, R. C., & Romanowski, H. P. (2018). Surviving in the urban–rural interface: Feeding and ranging behavior of brown howlers (*Alouatta guariba clamitans*) in an urban fragment in southern Brazil. *American Journal of Primatology, 80*, e22865.

Dasgupta, D., Banerjee, A., Karar, R., et al. (2020). Altered food habits? Understanding the feeding preference of free-ranging Grey langur (*Semnopithecus entellus*) within an urban settlement. *bioRxiv*. https://doi.org/10.1101/2020.10.22.350926

de Andrade, A. C., Medeiros, S., & Chiarello, A. G. (2020). City sloths and marmosets in Atlantic Forest fragments with contrasting levels of anthropogenic disturbance. *Mammal Research*. https://doi.org/10.1007/s13364-020-00492-0

Debenham, J. J., Tysnes, K., Khunger, S., & Robertson, L. J. (2017). Occurrence of Giardia, Cryptosporidium, and Entamoeba in wild rhesus macaques (*Macaca mulatta*) living in urban and semi-rural North-West India. *International Journal for Parasitology: Parasites and Wildlife, 6*, 29–34.

Díaz, S., Fargione, J., Iii, F. S. C., & Tilman, D. (2006). Biodiversity loss threatens human well-being. *PLoS Biology, 4*, e277. https://doi.org/10.1371/journal.pbio.0040277

Dickman, A. J. (2012). From cheetahs to chimpanzees: A comparative review of the drivers of human-carnivore conflict and human-primate conflict. *Folia Primatologica, 83*, 377–387.

Dore, K. M., Radford, L., Alexander, S., & Waters, S. (2018). Ethnographic approaches in primatology. *Folia Primatologica, 89*, 5–12.

Downs, C. T., Alexander, J., Brown, M., et al. (2021). Modification of the third phase in the framework for vertebrate species persistence in urban mosaic environments. *Ambio, 50*, 1866–1878.

Duarte, M. H. L., Vecci, M. A., Hirsch, A., & Young, R. J. (2011). Noisy human neighbours affect where urban monkeys live. *Biology Letters, 7*, 840–842.

Dunbar, R. I. M., Korstjens, A. H., & Lehmann, J. (2009). Time as an ecological constraint. *Biological Reviews, 84*, 413–429.

Ehlers Smith, Y. C., Maseko, M. S. T., Sosibo, M., et al. (2021). Indigenous knowledge of South African bird and rangeland ecology is effective for informing conservation science. *Journal of Environmental Management, 284*, 112041.

Farias, I. P., Santos, W. G., Gordo, M., & Hrbek, T. (2015). Effects of forest fragmentation on genetic diversity of the critically endangered primate, the pied tamarin (*Saguinus bicolor*): Implications for conservation. *The Journal of Heredity, 106*, 512–521.

Forman, R. T. T., & Gordon, M. (1986). *Landscape ecology* (p. 619). Wiley.

Fourie, N. H., Turner, T. R., Brown, J. L., et al. (2015). Variation in vervet (*Chlorocebus aethiops*) hair cortisol concentrations reflects ecological disturbance by humans. *Primates, 56*, 365–373.

Giraudeau, M., Mousel, M., Earl, S., & McGraw, K. (2014). Parasites in the city: Degree of urbanization predicts poxvirus and coccidian infections in house finches (*Haemorhous mexicanus*). *PLoS One, 9*, e86747. https://doi.org/10.1371/journal.pone.0086747

Gordo, M., Calleia, F. O., Vasconcelos, S. A., et al. (2013). The challenges of survival in a concrete jungle: Conservation of the pied tamarin (*Saguinus bicolor*) in the urban landscape of Manaus, Brazil. In L. K. Marsh & C. A. Chapman (Eds.), (pp. 357–370). Springer.

Grimm, N. B., Faeth, S. H., Golubiewski, N. E., et al. (2008). Global change and the ecology of cities. *Science, 319*, 756–760.

Healy, A., & Nijman, V. (2014). Pets and pests: Vervet monkey intake at a specialist South African rehabilitation centre. *Animal Welfare, 23*, 353–360.

Hoffman, T. S., & O'Riain, M. J. (2011). The spatial ecology of chacma baboons (*Papio ursinus*) in a human-modified environment. *International Journal of Primatology, 32*, 308–328.

Hoffman, T. S., & O'Riain, M. J. (2012a). Landscape requirements of a primate population in a human-dominated environment. *Frontiers in Zoology, 9*, 1–17.

Hoffman, T. S., & O'Riain, M. J. (2012b). Troop size and human-modified habitat affect the ranging patterns of a chacma baboon population in the Cape Peninsula, South Africa. *American Journal of Primatology, 74*, 853–863.

Hosaka, T., Sugimoto, K., & Numata, S. (2017). Childhood experience of nature influences the willingness to coexist with biodiversity in cities. *Palgrave Communications, 3*, 1–8.

Humle, T., & Hill, C. (2016). People–primate interactions: Implications for primate conservation. In S. A. Wich & A. J. Marshall (Eds.), *Introduction to primate conservation* (pp. 219–240). Oxford University Press.

Ilham, K., Nurdin, J., & Tsuji, Y. (2017). Status of urban populations of the long-tailed macaque (*Macaca fascicularis*) in West Sumatra, Indonesia. *Primates, 58*, 295–305.

Ilham, K., Nurdin, J., & Tsuji, Y. (2018). Effect of provisioning on the temporal variation in the activity budget of urban long-tailed macaques (*Macaca fascicularis*) in West Sumatra, Indonesia. *Folia Primatologica, 89*, 347–356.

Jaman, M. F., & Huffman, M. A. (2013). The effect of urban and rural habitats and resource type on activity budgets of commensal rhesus macaques (*Macaca mulatta*) in Bangladesh. *Primates, 54*, 49–59.

Johnson, M. T. J., & Munshi-South, J. (2017). Evolution of life in urban environments. *Science, 358*. https://doi.org/10.1126/science.aam8327

Kaburu, S. S. K., Marty, P. R., Beisner, B., et al. (2018). Rates of human–macaque interactions affect grooming behavior among urban-dwelling rhesus macaques (*Macaca mulatta*). *American Journal of Physical Anthropology, 168*, 92–103.

Kaburu, S. S. K., Beisner, B., Balasubramaniam, K. N., et al. (2019). Interactions with humans impose time constraints on urban-dwelling rhesus macaques (*Macaca mulatta*). *Behaviour, 156*, 1255–1282.

Kaplan, G., & Rogers, L. J. (2013). Stability of referential signalling across time and locations: Testing alarm calls of Australian magpies (*Gymnorhina tibicen*) in urban and rural Australia and in Fiji. *PeerJ, 1*, e112.

Kaplan, B. S., O'Riain, M. J., van Eeden, R., & King, A. J. (2011). A low-cost manipulation of food resources reduces spatial overlap between baboons (*Papio ursinus*) and humans in conflict. *International Journal of Primatology, 32*, 1397–1412.

Klegarth, A. R., Hollocher, H., Jones-Engel, L., et al. (2017). Urban primate ranging patterns: GPS-collar deployments for *Macaca fascicularis and M. sylvanus*. *American Journal of Primatology, 79*, 1–17.

Lambin, E. F., Rounsevell, M., & Geist, H. (2000). Are current agricultural land use models able to predict changes in land use intensity? *Agriculture, Ecosystems and Environment, 82*, 321–331.

Lane, K. E., Holley, C., Hollocher, H., & Fuentes, A. (2011). The anthropogenic environment lessens the intensity and prevalence of gastrointestinal parasites in Balinese long-tailed macaques (*Macaca fascicularis*). *Primates, 52*, 117–128.

Lindshield, S. M. (2016). Protecting nonhuman primates in peri-urban environments: A case study of Neotropical monkeys, corridor ecology, and coastal economy in the Caribe Sur of Costa Rica. In M. Waller (Ed.), *Ethnoprimatology*. Springer.

Lokschin, L. X., Rodrigo, C. P., Hallal Cabral, J. N., & Buss, G. (2007). Power lines and howler monkey conservation in Porto Alegre, Rio Grande do Sul, Brazil. *Neotropical Primates, 14*, 76–80.

Lowry, H., Lill, A., & Wong, B. (2013). Behavioural responses of wildlife to urban environments. *Biological Reviews, 88*, 537–549.

Lunney, D., & Burgin, S. (2004). Urban wildlife management: An emerging discipline. In *Proceedings of a forum held by the Royal Zoological Society of New South Wales at Taronga Zoo on 20 October 2001.*

Mackenstedt, U., Jenkins, D., & Romig, T. (2015). The role of wildlife in the transmission of parasitic zoonoses in peri-urban and urban areas. *International Journal for Parasitology: Parasites and Wildlife, 4,* 71–79.

Maréchal, L., Semple, S., Majolo, B., et al. (2011). Impacts of tourism on anxiety and physiological stress levels in wild male Barbary macaques. *Biological Conservation, 144,* 2188–2193.

Marty, P., Beisner, B., Kaburu, S. S. K., et al. (2019a). Time constraints and stress imposed by human presence alter social behaviour in urban long-tailed macaques. *Animal Behaviour, 150,* 157–165.

Marty, P. R., Balasubramaniam, K. N., Kaburu, S. S. K., et al. (2019b). Individuals in urban dwelling primate species face unequal benefits associated with living in an anthropogenic environment. *Primates, 61,* 249–255.

Marzluff, J. M., Bowman, R., & Donnelly, R. (2001). A historical perspective on urban bird research: Trends, terms, and approaches. In J. M. Marzluff, R. Bowman, & R. Donnelly (Eds.), *Avian ecology and conservation in an urbanizing world* (pp. 1–17). Springer.

McKinney, M. L. (2008). Effects of urbanization on species richness: A review of plants and animals. *Urban Ecosystem, 11,* 161–176.

McKinney, T. (2015). A classification system for describing anthropogenic influence on nonhuman primate populations. *American Journal of Primatology, 77,* 715–726.

McKinney, T., & Dore, K. M. (2018). The state of ethnoprimatology: Its use and potential in today's primate research. *International Journal of Primatology, 39,* 730–748.

McLennan, M. R., Spagnoletti, N., & Hockings, K. J. (2017). The implications of primate behavioral flexibility for sustainable human–primate coexistence in anthropogenic habitats. *International Journal of Primatology, 38,* 105–121.

Naughton-Treves, L., Treves, A., Chapman, C., & Wrangham, R. (1998). Temporal patterns of crop-raiding by primates: Linking food availability in croplands and adjacent forest. *Journal of Applied Ecology, 35,* 596–606.

O'Riain, M. J., & Hoffman, T. S. (2012). Monkey management: Using spatial ecology to understand the extent and severity of human-baboon conflict in the Cape Peninsula, South Africa. *Ecology and Society, 17,* 13–29.

Olaleru, F., & Ogunfuwa, A. A. (2020). An assessment of human-monkey conflict in urban communities in Lagos State, Nigeria. *UNILAG Journal of Medicine, Science and Technology, 8,* 160–175.

Patterson, L., Kalle, R., & Downs, C. (2017). A citizen science survey: Perceptions and attitudes of urban residents towards vervet monkeys. *Urban Ecosystem, 20,* 617–628.

Patterson, L., Kalle, R., & Downs, C. (2018). Factors affecting presence of vervet monkey troops in a suburban matrix in KwaZulu-Natal, South Africa. *Landscape and Urban Planning, 169,* 220–228.

Patterson, L., Kalle, R., & Downs, C. T. (2019). Living in the suburbs: Space use by vervet monkeys (*Chlorocebus pygerythrus*) in an eco-estate, South Africa. *African Journal of Ecology, 57,* 539–551.

Pereira, A. A. B. G., Dias, B., Castro, S. I., et al. (2020). Electrocutions in free-living black-tufted marmosets (*Callithrix penicillata*) in anthropogenic environments in the Federal District and surrounding areas, Brazil. *Primates, 61,* 321–329.

Perry, G., Boal, C., Verble, R., & Wallace, M. (2020). "Good" and "bad" urban wildlife. In F. Angelici & L. Rossi (Eds.), *Problematic wildlife II* (pp. 141–170). Springer.

Pragatheesh, A. (2011). Effect of human feeding on the road mortality of Rhesus Macaques on National Highway-7 routed along Pench Tiger Reserve, Madhya Pradesh, India. *Journal of Threatened Taxa, 3,* 1656–1662.

Redpath, S. M., Young, J., Evely, A., et al. (2013). Understanding and managing conservation conflicts. *Trends in Ecology & Evolution, 28,* 100–109.

Rodrigues, N. N., & Martinez, R. A. (2014). Wildlife in our backyard: Interactions between Wied's marmoset *Callithrix kuhlii* (Primates: *Callitrichidae*) and residents of Ilhéus, Bahia, Brazil. *Wildlife Biology, 20,* 91–96.

Santos, M., Duarte, M., & Young, R. J. (2014). Behavioural and ecological aspects of black tufted-ear marmosets, *Callithrix penicillata* (Geoffroy, 1812) (Primates: *Callitrichidae*) in a semi-urban environment. *Revista de Etologia = Journal of Ethology, 13,* 37–46.

Sapkota, B., Adhikari, R. B., Regmi, G. R., Bhattarai, B. P., & Ghimire, T. R. (2020). Diversity and prevalence of gut parasites in urban macaques. *Applied Science and Technology Annals, 30,* 34–41.

Scheun, J., Bennett, N. C., Ganswindt, A., & Nowack, J. (2015). The hustle and bustle of city life: Monitoring the effects of urbanisation in the African lesser bushbaby. *The Science of Nature, 102,* 57.

Scheun, J., Greeff, D., & Nowack, J. (2019). Urbanisation as an important driver of nocturnal primate sociality. *Primates, 60,* 375–381.

Setchell, J. M., Fairet, E., Shutt, K., et al. (2017). Biosocial conservation: Integrating biological and ethnographic methods to study human–primate interactions. *International Journal of Primatology, 38,* 401–426.

Sha, J. C. M., Gumert, M. D., Lee, B. P., et al. (2009). Macaque–human interactions and the societal perceptions of macaques in Singapore. *American Journal of Primatology, 71,* 825–839.

Singh, B. B., & Gajadhar, A. A. (2014). Role of India's wildlife in the emergence and re-emergence of zoonotic pathogens, risk factors and public health implications. *Acta Tropica, 138,* 67–77.

Sinha, A., & Mukhopadhyay, K. (2013). The monkey in the town's commons, revisited: An anthropogenic history of the Indian bonnet macaque. In S. Radhakrishna, M. Huffman, & A. Sinha (Eds.), *The Macaque Connection* (pp. 187–208). Springer.

Skandrani, Z., Bovet, D., Gasparini, J., et al. (2017). Sociality enhances birds' capacity to deal with anthropogenic ecosystems. *Urban Ecosystem, 20,* 609–615.

Smil, V. (1993). *Global ecology: Environmental change and social flexibility.* Routledge.

Soga, M., & Gaston, K. J. (2016). Extinction of experience: The loss of human–nature interactions. *Frontiers in Ecology and the Environment, 14,* 94–101.

Sol, D., Timmermans, S., & Lefebvre, L. (2002). Behavioural flexibility and invasion success in birds. *Animal Behaviour, 63,* 495–502.

Sol, D., Lapiedra, O., & González-Lagos, C. (2013). Behavioural adjustments for a life in the city. *Animal Behaviour, 85,* 1101–1112.

Soto-Calderón, I. D., Acevedo-Garcés, Y. A., Hernández-Castro, J. Á.-C. C., & García-Montoya, G. M. (2016). Physiological and parasitological implications of living in a city: The case of the white-footed tamarin (*Saguinus leucopus*). *American Journal of Primatology, 78,* 1272–1281.

Soulsbury, C. D., & White, P. C. L. (2015). Human-wildlife interactions in urban areas: A review of conflicts, benefits and opportunities. *Wildlife Research, 42,* 541–553.

Suzin, A., Back, J. P., Garey, M. V., & Aguiar, L. M. (2017). The relationship between humans and capuchins (*Sapajus* sp.) in an urban green area in Brazil. *International Journal of Primatology, 38,* 1058–1071.

Teixeira, B., Hirsch, A., Goulart, V. D. L. R., et al. (2015). Good neighbours: Distribution of black-tufted marmoset (*Callithrix penicillata*) in an urban environment. *Wildlife Research, 42,* 579–589.

Thatcher, H. (2019). Anthropogenic influences on the behavioural ecology of urban vervet monkeys (Doctoral dissertation, Liverpool John Moores University).

Thatcher, H. R., Downs, C. T., & Koyama, N. F. (2018). Using parasitic load to measure the effect of anthropogenic disturbance on vervet monkeys. *EcoHealth, 15,* 676–681.

Thatcher, H. R., Downs, C. T., & Koyama, N. F. (2019a). Anthropogenic influences on the time budgets of urban vervet monkeys. *Landscape and Urban Planning, 181,* 38–44.

Thatcher, H. R., Downs, C. T., & Koyama, N. F. (2019b). Positive and negative interactions with humans concurrently affect vervet monkey (*Chlorocebus pygerythrus*) ranging behavior. *International Journal of Primatology, 40,* 496–510.

Thatcher, H. R., Downs, C. T., & Koyama, N. F. (2020). Understanding foraging flexibility in urban vervet monkeys, *Chlorocebus pygerythrus*, for the benefit of human-wildlife coexistence. *Urban Ecosystem, 23*, 1349–1357.

Waite, T. A., Chhangani, A. K., Campbell, L. G., et al. (2007). Sanctuary in the city: Urban monkeys buffered against catastrophic die-off during ENSO-related drought. *EcoHealth, 4*, 278–286.

Warren, Y., Higham, J. P., Maclarnon, A. M., & Ross, C. (2011). Crop-raiding and commensalism in olive baboons: The costs and benefits of living with humans. In V. Sommer & C. Ross (Eds.), *Primates of Gashaka* (pp. 359–384). Springer.

Waters, S., Bell, S., & Setchell, J. M. (2018). Understanding human-animal relations in the context of primate conservation: A multispecies ethnographic approach in North Morocco. *Folia Primatologica, 89*, 13–29.

Waters, S., El Harrad, A., Bell, S., & Setchell, J. M. (2019). Interpreting people's behaviour towards primates using qualitative data: A case study from North Morocco. *International Journal of Primatology, 40*, 316–333.

Werner, P. (2011). The ecology of urban areas and their functions for species diversity. *Landscape and Ecological Engineering, 7*, 231–240.

Woodroffe, R., Thirgood, S., & Rabinowitz, A. (2005). *People and wildlife, conflict or coexistence?* Cambridge University Press.

Wright, T. F., Eberhard, J. R., Hobson, E. A., et al. (2010). Behavioral flexibility and species invasions: The adaptive flexibility hypothesis. *Ethology Ecology and Evolution, 22*, 393–404.

Wu, J. (2013). Key concepts and research topics in landscape ecology revisited: 30 years after the Allerton Park workshop. *Landscape Ecology, 28*, 1–11.

Zhao, Q.-K. (2005). Tibetan macaques, visitors, and local people at Mt. Emei: Problems and countermeasures. In J. D. Paterson & J. Wallis (Eds.), *Commensalism and conflict: The human-primate interface* (pp. 376–399). American Society of Primatologists.

Zungu, M. M., Maseko, M. S. T., Kalle, R., et al. (2020a). Effects of landscape context on mammal richness in the urban forest mosaic of EThekwini Municipality, Durban, South Africa. *Global Ecology and Conservation, 21*, e00878.

Zungu, M. M., Maseko, M. S. T., Kalle, R., et al. (2020b). Factors affecting the occupancy of forest mammals in an urban-forest mosaic in EThekwini Municipality, Durban, South Africa. *Urban Forestry & Urban Greening, 48*, 126562.

Chapter 9
Infectious Diseases in Primates in Human-Impacted Landscapes

Marina Ramon, Matthew R. McLennan, Carlos R. Ruiz-Miranda, Gladys Kalema-Zikusoka, Joana Bessa, Elena Bersacola, Américo Sanhá, Maimuna Jaló, Aissa Regalla de Barros, Fabian H. Leendertz, and Kimberley J. Hockings

Contents

Abstract The close phylogenetic relationship between humans and nonhuman primates (hereafter primates), coupled with mounting anthropogenic impacts, such as habitat change, wildmeat hunting, pet-keeping, and tourism, increases disease risks for primates and humans by facilitating zoonotic pathogen exchange and altering host-pathogen interactions. Infectious diseases, and particularly emerging infectious diseases, have the potential to threaten the long-term survival of many primate populations and pose a risk to public health. Adoption of holistic approaches such

M. Ramon (✉)
Centre for Ecology and Conservation, Faculty of Environment, Science and Economy, University of Exeter, Penryn, UK

School of Biosciences, Cardiff University, Cardiff, United Kingdom
e-mail: mr637@exeter.ac.uk

M. R. McLennan
Bulindi Chimpanzee and Community Project, Hoima, Uganda

Faculty of Humanities and Social Sciences, Oxford Brookes University, Oxford, UK

C. R. Ruiz-Miranda
Laboratório de Ciências Ambientais, Universidade Estadual do Norte Fluminense, Campos dos Goytacazes, Rio de Janeiro, Brazil

Associação Mico Leão Dourado, Silva Jardim, Rio de Janeiro, Brazil

© The Author(s), under exclusive license to Springer Nature Switzerland AG 2023
T. McKinney et al. (eds.), *Primates in Anthropogenic Landscapes*, Developments in Primatology: Progress and Prospects, https://doi.org/10.1007/978-3-031-11736-7_9

as One Health is critical to provide a framework for protecting the health of wild primates and humans and the ecosystems we share. Here, we describe multiple practices of this transdisciplinary strategy, including disease risk analysis, disease prevention, health monitoring and disease surveillance, clinical interventions, and community engagement.

Keywords Pathogen · Parasite · Disease · Zoonosis · EID · Health · Surveillance · Vaccination · One Health

9.1   Introduction

An infectious disease is a disorder caused by an infectious agent, a parasite, or a pathogen, which lives in or on another organism (the host), at some cost to the host (Nunn & Gillespie, 2016). Parasites are categorised into microparasites and macroparasites. Microparasites are microbes, including viruses, bacteria, protozoa, and fungi, while macroparasites are multi-cellular organisms, such as helminths and arthropods. Due to differences in generation times, microparasites are often associated with acute disease (appears rapidly and lasts a relatively short period of time), in contrast to macroparasites, which tend to produce infections of longer duration (Delahay et al., 2009). However, some chronic diseases in primates are caused by bacteria, such as tuberculosis, or viruses, such as simian immunodeficiency virus (SIV).

Transmission of parasites follows three main routes: direct contact, involving physical contact with body fluids of an infected individual; indirect contact, occurring through exposure to an environment contaminated with pathogens; and vectors, agents – mainly arthropods but also gastropods and helminths – which transmit parasites to a secondary host (Cormier & Jolly, 2018).

Zoonoses are infectious diseases that are naturally transmissible from nonhuman animals to humans (WHO, 2020), whereas **anthroponoses** (also known as anthrozoonoses, anthropozoonoses, zooanthroponoses, or reverse zoonoses) are human infectious diseases that can be naturally transmitted to other animals (Muehlenbein, 2017). These terms are often used interchangeably, and experts recommend the use of 'zoonoses' to refer to infectious diseases naturally transmitted between

G. Kalema-Zikusoka
Conservation Through Public Health, Entebbe, Uganda

J. Bessa
Department of Zoology, University of Oxford, Oxford, UK

E. Bersacola · K. J. Hockings
Centre for Ecology and Conservation, Faculty of Environment, Science and Economy, University of Exeter, Penryn, UK

A. Sanhá · M. Jaló · A. Regalla de Barros
Instituto da Biodiversidade e das Áreas Protegidas (IBAP), Bissau, Guinea-Bissau

F. H. Leendertz
Helmholtz Institute for One Health, Greifswald, Germany

Fig. 9.1 Infectious diseases are the result of factors acting at the host, pathogen, and environment level. Anthropogenic disturbance in the forms of habitat change, hunting or tourism, can affect host-pathogen interactions

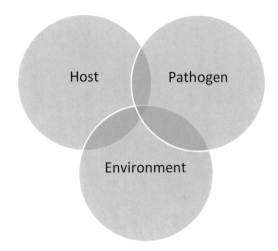

nonhuman animals and humans (Hubálek, 2003). Zoonotic infections are triggered by a **spillover** event, involving **pathogen transmission** from one reservoir population to a novel host population (Daszak et al., 2000).

The probability of acquiring and transmitting an infectious disease (disease risk) is influenced by the interaction of three factors, known as the epidemiologic triad: the animal host, the pathogen, and the environment (Fig. 9.1). A change in one of these factors can alter the entire system and modify disease patterns. Human activities can strongly influence these interactions (Chapman et al., 2005). The clearest example of this is **emerging infectious diseases (EIDs)**, which are rising due to expanding human populations and increasing exploitation of natural environments and resources (Travis et al., 2006).

EIDs are identified as affecting a host population for the first time or as previously existing but rapidly increasing, either in number of new cases in a population or by spreading to new geographical areas, such as the Ebola virus disease (EVD) (WHO, 2014). Taylor et al. (2001) calculated that 61% of identified human pathogens, and about 75% of EIDs affecting people, had a zoonotic origin. The specific **pathogen** and its mode of transmission are key factors in its potential to become an emerging infection in humans, but ecological and environmental factors, often linked to human-induced changes, also play an essential role in disease emergence.

9.2   Anthropogenic Disturbance and Infectious Disease Risk

The close phylogenetic relationship between humans and primates combined with unprecedented interspecies contacts derived from anthropogenic disturbance increases the potential for bidirectional pathogen exchange (Gillespie et al., 2008). With ongoing conversion of natural habitats, human pathogens are being introduced into immunologically naïve wild primate populations and vice versa, threatening the persistence of primates and human health (Nunn & Altizer, 2006) (Table 9.1).

Table 9.1 Examples of pathogens exchanged between humans and primates

Pathogen	Species affected	Route of exchange	Anthropogenic risk factor	Direction of exchange	Impact	References
Ebola virus disease (EVD)	Gorilla, chimpanzee	Body fluids	Hunting and butchering	Primate to human	Acute disease, causing sudden die-offs	Bermejo et al. (2006); Leroy et al. (2004)
Respiratory diseases	Great apes	Respiratory droplets and/or aerosols	Tourism and research	Human to primate	High morbidity; considerable mortality	Grützmacher et al. (2016, 2018b); Köndgen et al. (2008); Negrey et al. (2019); Palacios et al. (2011); Patrono et al. (2018)
Yellow fever	Neotropical primates	Vector	Habitat change, human encroachment	Both directions	High mortality	Bicca-Marques et al. (2017); Dietz et al. (2019); Holzmann et al. (2010)
Gastrointestinal parasites	All primates	Faecal-oral	Human-primate spatial overlap	Both directions	From asymptomatic to fatal; can cause high morbidity	Chapman et al. (2006a); Johnston et al. (2010); Kalousová et al. (2016); Mbora and McPeek (2009); McLennan et al. (2018); Nizeyi et al. (1999)
Herpesvirus B	Macaques	Bite	Tourism	Primate to human	Mild symptoms in macaques; high mortality in humans	Engel et al. (2002)

9.2.1   Habitat Change

Habitat change – i.e. habitat loss, fragmentation, and degradation – can be associated with zoonotic disease outbreaks (Wilkinson et al., 2018). For example, vector-borne diseases such as yellow fever and zoonotic malaria, caused by a spillover from a primate sylvatic cycle to humans and other animals, are associated with expansion of human settlements into forests (Wilcox & Ellis, 2006). EVD outbreaks in West and Central Africa are linked to recent deforestation, mostly in hotspots of forest fragmentation (Olivero et al., 2017; Rulli et al., 2017).

Habitat loss and fragmentation might increase gastrointestinal parasite prevalence and richness in primates, due to higher host densities and edge effects. High host density can lead to repeated use of the same area, increasing contact with substrates containing parasites at infective stages (McCallum & Dobson, 2002), and contribute to social and nutritional stress in primates, compromising their immune response (Chapman et al., 2006b). Moreover, higher host densities can increase among conspecifics, exposing individuals to directly transmitted pathogens (Anderson & May, 1992). Mbora and McPeek (2009) found that Tana River red colobus (*Piliocolobus rufomitratus*) and Tana River mangabey (*Cercocebus galeritus galeritus*) living in highly fragmented habitats in Kenya had higher prevalence of parasites than many primates elsewhere, which was associated with primate density.

The edge effect concept predicts increased biodiversity at ecosystem edges, where species from adjacent ecosystems and edge-specific species overlap, potentially increasing cross-species contact and pathogen transmission (Borremans et al., 2019). In line with this, ashy red colobus (*Piliocolobus tephrosceles*) and eastern black-and-white colobus (*Colobus guereza*) from forest edge groups in Kibale National Park, Uganda, had a higher proportion of multiple infections and prevalence of specific gastrointestinal parasites than core forest groups (Chapman et al., 2006a). Notably, red colobus were infected with *Giardia duodenalis* assemblages BIV and E, which characteristically infect humans and livestock, respectively (Johnston et al., 2010). These results suggest that in areas of human-livestock-wildlife overlap, pathogen sharing may be especially frequent.

Primates living in fragmented and degraded habitats frequently feed on agricultural crops (Hill, 2017). Dietary changes associated with eating human foods might modify primate-parasite dynamics. Weyher (2009) showed that crop feeding olive baboons (*Papio anubis*) in Nigeria had a lower intensity of helminth parasites (*Trichuris* sp. and *Physaloptera* sp.) compared to wild-foraging troops. Conversely, crop-foraging troops had a higher *Balantidium coli* cyst output, because this protozoan parasite benefited from increased starch intake.

Box 9.1 Self-Medication in 'Village Chimpanzees'
Matthew R. McLennan

When primates are sick, they can employ behavioural strategies to help control the disease or alleviate symptoms – behaviour referred to as self-medication (Huffman, 1997, 2001). Although primates living in diverse habitats self-medicate, anthropogenic factors including habitat encroachment may influence these behaviours.

In Bulindi, Uganda, wild chimpanzees inhabit a human-modified environment comprising small forest fragments amidst farmland and villages. The chimpanzees eat crops and encounter humans and domestic animals daily (McLennan et al., 2017). An early study reported an unusually high frequency of 'leaf swallowing' among these chimpanzees. This behaviour involves ingesting leaves with a rough or hairy surface without chewing (Huffman, 1997). Huffman et al. (1996) found that whole undigested leaves in chimpanzee faeces at Mahale, Tanzania, tended to co-occur with adult *Oesophagostomum stephanostomum* worms – a pathogenic nematode that commonly parasitises mammals including primates. Swallowing bristly leaves increases gut motility causing expulsion of adult worms, helping to alleviate the infection (Huffman & Caton, 2001).

Over 13 months, 10.4% of chimpanzee faecal samples at Bulindi contained whole bristly leaves, compared to frequencies of 0.4–4.0% at sites where chimpanzees inhabit less-disturbed environments (McLennan & Huffman, 2012). Elevated levels of self-medication suggested these chimpanzees were especially vulnerable to certain gastrointestinal parasites. Similar to findings from Mahale, leaf swallowing at Bulindi was linked to expulsion of adult *Oesophagostomum stephanostomum* worms in chimpanzee faeces.

A subsequent study at Bulindi used microscopic analysis to evaluate the relationship between leaf swallowing and parasite infections with known or likely pathogenicity. Consistent with earlier findings, un-chewed leaves occurred in chimpanzee faeces at a high frequency (11.8%), and, as before, the leaves were often associated with egested *Oesophagostomum* worms. However, microscopic analysis also revealed independent associations between seasonal increases in leaf swallowing and prevalence of two other pathogenic nematodes (*Strongyloides* sp. and *Necator* sp.) in addition to *Oesophagostomum* (McLennan et al., 2017).

Chimpanzees in Bulindi are exposed to various pathogens with zoonotic potential, including some possibly acquired directly or indirectly from people or livestock (e.g. enterobacteria such as *Salmonella* sp., *Escherichia coli*, and *Shigella* sp.; McLennan et al., 2018). Frequent leaf swallowing by these 'village chimpanzees' (Fig. 9.2) might be a generalised response to gastrointestinal discomfort caused by multiple infections (McLennan et al., 2017) while still functioning to reduce *Oesophagostomum* infections specifically by causing adult worms to be expelled, thereby disrupting this pathogenic nematode's lifecycle and reducing the intensity of infection (Huffman & Caton, 2001).

Fig. 9.2 Subadult male chimpanzee at Bulindi swallowing rough leaves of *Pseudarthria hookeri*. He selects a leaf one at a time (top), folds it between his tongue and palette, and swallows the leaf whole without chewing (bottom). (Photo: Bulindi Chimpanzee and Community Project)

9.2.2 Wildmeat and Pet-Keeping

Wildmeat is not only a threat to wildlife conservation but also a means through which humans are exposed to primate pathogens (Kilonzo et al., 2013). *Wildmeat* activities include hunting, handling, transporting, butchering, selling, purchasing, cooking, and consuming wild animals, all of which provide contamination opportunities. Butchering carries the highest risk, since it increases exposure to animal fluids and tissues via cut wounds (Wolfe et al., 2004a; Mossoun et al., 2017). During hunting, individuals exposed to scratches, bites, and injuries from wild primates are at high risk of infection, especially considering that sick animals are more easily captured (LeBreton et al., 2006; Calattini et al., 2007). Hunting and butchering wild primates have been linked to virus introduction to human populations, including HIV (Hahn et al., 2000), simian foamy virus (SFV) (Wolfe et al., 2004b), EVD (Leroy et al., 2004), monkeypox (Jezek et al., 1986), and human T-lymphotropic viruses (HTLV) (Mossoun et al., 2017).

The disease risks of **wildmeat** are not limited to forest-living people, since meat is transported to urban areas and overseas destinations (Chaber et al., 2010). Despite health risks linked to trade being more acute in local **wildmeat** markets, there have been reports of introduction of infectious diseases potentially originating from **wildmeat** smuggling in European cities (Smith et al., 2012).

Pet ownership offers opportunities for pathogen transfer through exposure to body fluids, faecal matter, air-borne pathogens, and shared vectors. Although primate pets may have a lower prevalence of chronic diseases due to their typically young age when captured, owners still risk exposure to zoonotic diseases (Wolfe et al., 2004a; Pourrut et al., 2011). For example, common marmosets (*Callithrix jacchus*) transmitted rabies by biting pet owners in Brazil (Favoretto et al., 2001). A case of humans contracting Marburg virus disease after exposure to primates in a laboratory setting illustrates the potential for transmission of fatal diseases from primate pets to their owners (Martini, 1973).

Contrastingly, pet primates can acquire diseases from their owners (Michel & Huchzermeyer, 1998). If escaped or released, primates can introduce human-adapted pathogens into wild populations, such as drug-resistant human *Staphylococcus aureus* (Schaumburg et al., 2012). Released individuals can even become invasive, as occurred with golden-headed lion tamarins (*Leontopithecus chrysomelas*) that formed a peri-urban population in Niterói city, Brazil. These ex-captive primates get close to homes and pose a disease risk not only to humans and domestic animals but also to the native golden lion tamarins (*L. rosalia*) (Miranda et al., 2019). Another mode of pathogen introduction into wild populations takes place through the interaction of pet primates with wild conspecifics that approach villages (Jones-Engel et al., 2005b).

9.2.3 Tourism and Research

Although research and tourism are important conservation tools, these activities expose wild primates to human pathogens. **Habituation** and proximity to humans can also increase stress levels in wild populations, potentially suppressing their immune system and increasing susceptibility to infectious diseases (Maréchal et al., 2011; Shutt et al., 2014).

The most frequently described diseases of human origin affecting great apes are caused by respiratory pathogens, which often result in high morbidity and potentially high mortality. Human respiratory syncytial virus (HRSV) and human metapneumovirus (HMPV) have been reported in western chimpanzees (*Pan troglodytes verus*) (Köndgen et al., 2008); eastern chimpanzees (*Pan troglodytes schweinfurthii*) (Negrey et al., 2019); mountain gorillas (*Gorilla beringei beringei*) (Palacios et al., 2011); western lowland gorillas (*G. gorilla gorilla*) (Grützmacher et al., 2016); and bonobos (*P. paniscus*) (Grützmacher et al., 2018b).

Although SARS-CoV-2, the coronavirus responsible for **COVID-19**, has not been reported in wild primates to date, apes and African and Asian monkeys, and some lemurs, are likely to be highly susceptible to it (Melin et al., 2020). SARS-CoV-2 caused moderate respiratory disease in rhesus macaques (*Macaca mulatta*) after experimental inoculation (Munster et al., 2020) and mild to moderate respiratory symptoms in captive western lowland gorillas (Gibbons, 2021; Zoo Atlanta, 2021), similar to those caused by human coronavirus OC43 in wild western chimpanzees in Taï National Park, Côte d'Ivoire (Patrono et al., 2018).

Even if measures to reduce risk of transmission of air-borne infections are in place at many great ape tourism sites (MacFie & Williamson, 2010), a widespread lack of rule adherence has been documented (Hanes et al., 2018; Weber et al., 2020; Molyneaux et al., 2021). Moreover, tourists are often unprotected against certain vaccine-preventable infections (Muehlenbein et al., 2008). In other primate tourism settings, tourists get very close to primates, feeding and touching them, as occurs with Barbary macaques (*M. sylvanus*) at Ifrane National Park, Morocco (Maréchal et al., 2011, 2016). Similar interactions occur at Brasilia National Park, Brazil, with bearded capuchin monkeys (*Sapajus libidinosus*) (Sabbatini et al., 2006). These situations put primates at risk of disease acquisition, but also threaten tourists' health. At Hindu and Buddhist temples throughout South and Southeast Asia, humans are sometimes bitten or scratched by free-ranging macaques (Fuentes & Gamerl, 2005), exposing them to viruses, including rhesus cytomegalovirus, simian virus 40, *Cercopithecine herpesvirus* 1, SFV, and herpesvirus B (Engel et al., 2002; Jones-Engel et al., 2005a, 2006). In other places, primates are attracted to tourist lodges by food opportunities. Sapolsky and Share (2004) reported a high mortality of olive baboons at the Masai Mara Reserve, Kenya, after contracting tuberculosis when eating infected meat at a rubbish pit.

Beyond tourism, conservation personnel and researchers are the suspected origin of some diseases in primates. Nizeyi et al. (1999) screened faeces for the protozoan *Cryptosporidium* sp. from non-habituated and habituated gorillas in Bwindi Impenetrable National Park, Uganda, and found that most infections occurred in habituated individuals. The prevalence of *Campylobacter* spp. and *Salmonella* spp. infections among habituated gorillas doubled in 4 years, potentially resulting from habituation efforts (Nizeyi et al., 2001). In Kibale National Park, chimpanzees harboured bacteria genetically more similar to those of humans involved in research than those of local villagers, suggesting transmission between staff and apes (Goldberg et al., 2007). Researchers and conservation personnel also risk exposure to wild primates' pathogens, as occurred with researchers who were infected with gorilla hookworms during their fieldwork in the Central African Republic (Kalousová et al., 2016).

9.3   Promoting Primate and Human Health

9.3.1   Disease Risk Analysis

Disease risk analysis is a multidisciplinary process that estimates the likelihood and consequences of disease occurring in a population to aid in the development of management strategies and inform policies (Jones-Engel & Engel, 2006; Travis et al., 2006). It consists of four interconnected phases: (1) hazard identification aims to identify what can go wrong and which diseases have potential to be harmful; (2) risk assessment evaluates the likelihood and consequences of a disease event (i.e. disease transmission, illness, death) occurring in a population; (3) risk management aims to decrease the likelihood and consequences of an adverse outcome; and (4) risk communication is a continuous process of communication among stakeholders (Deem, 2016). For example, Engel et al. (2006) developed a risk assessment model to predict the likelihood of long-tailed macaques (*M. fascicularis*) infecting visitors with SFV at monkey temples in Asia.

9.3.2 Disease Prevention

Research sites and protected areas should have detailed protocols to minimise risk of disease transmission between humans and wild primates. Standardised regulations for visiting wild great apes were developed by the IUCN, including avoiding trekking when sick, maintaining a minimum distance from the animals (7 m if wearing N95 masks; 10 m if not wearing masks), hygiene measures, a 1-hour time limit per visit, and limited numbers of visits and visitors (MacFie & Williamson, 2010). Nevertheless, these rules are not enforceable when visiting other primate species at highly anthropogenic sites, where tourists get very close to primates and physically interact with them (Carne et al., 2017).

In order to minimise the **COVID-19** threat, the IUCN issued a statement with preventive measures (IUCN, 2021), and disease prevention guidelines have been adapted at many great ape tourism sites (Gilardi et al., 2021). The 'Protect Great Apes From Disease' project (www.protectgreatapesfromdisease.com) provides recommendations to reduce disease transmission, including **COVID-19**, to African apes and offers freely available multimedia education and training materials to tourism sites.

Health risks posed by researchers from overseas can be reduced through vaccination and quarantine. For example, the Taï Chimpanzee Project's restrictive rules for researchers have resulted in a decreased incidence of severe respiratory outbreaks in the chimpanzees (Grützmacher et al., 2018a). In the context of the **COVID-19** pandemic, field research projects are advised to reinforce protective measures to reduce disease transmission risks (Lappan et al., 2020). Moreover, it is essential to evaluate and improve the health of personnel working in primate habitats, especially those in close contact with habituated groups. A formal approach to reducing risk of pathogen transfer between staff and primates is an Employee Health Programme (EHP), such as the Mountain Gorilla Veterinary Project (Ali et al., 2004).

9.3.3 Health Monitoring and Disease Surveillance

As defined by Gilardi et al., (2015), 'monitoring is designed to detect and regularly report on any change in the normal health status of a population, whereas surveillance aims to identify the first cases of a disease in a population in order to minimise impact'. Data from surveillance and monitoring form part of the risk analysis process and make it possible to characterise the 'normal' amount of a particular disease in a population, known as baseline disease levels, and improve outbreak detection (Decision Tree Writing Group, 2006).

Both monitoring and surveillance include systematic and standardised data collection on presence or absence of individuals, observable clinical signs, laboratory testing, and other environmental and/or behavioural factors that could indicate disease (Gilardi et al., 2015). Monitoring potential reservoirs of disease (e.g. bats, rodents) as well as livestock is also crucial (Leendertz et al., 2006). In addition, sampling primate-associated flies can be a safe and useful tool for monitoring pathogens (Gogarten et al., 2019).

When primates are habituated, observational health assessments can be made during routine behavioural data collection (Wolf et al., 2019). When groups are unhabituated, this information can be gathered from distant viewing platforms (Levréro et al., 2007) or using camera traps (Hockings et al., 2021). Moreover, biological material can be collected non-invasively from faeces, hair, saliva, blood, and urine (Gillespie et al., 2008; Gilardi et al., 2015) (Fig. 9.3).

Great apes in their natural habitats can function as **sentinel species** for zoonotic outbreaks (Calvignac-Spencer et al., 2012). Identification of EVD in chimpanzee and gorilla carcasses enabled scientists to alert the health authorities of the Republic

Fig. 9.3 Monitoring infectious disease in wild primates must ensure the safety of primates and researchers, as illustrated in this image of Idrissa Galiza and Marina Ramon wearing protective gloves and masks while collecting western chimpanzee faecal samples in Cantanhez National Park, Guinea-Bissau. (Photo: Canthanez Chimpanzee Project)

of Congo and Gabon of an imminent human outbreak (Rouquet et al., 2005). Other primates, such as howler monkeys (*Alouatta* spp.), are effective *sentinels* of yellow fever. Sudden high mortality of monkeys from this virus prompted early public health responses in Argentina (Holzmann et al., 2010) and Brazil (Fernandes et al., 2017).

9.3.4 Clinical Interventions

The spread of infectious diseases among wild primates can potentially be controlled with clinical veterinary interventions (i.e. medicating, treating, or vaccinating primates). For great apes, interventions should be considered if a group is experiencing an outbreak, the illness is clearly human-induced, and/or the death of an individual is expected to have population-level consequences (Gilardi et al., 2015). Veterinary interventions in habituated groups of mountain gorillas have proven to be a key contributor to population growth rate (Robbins et al., 2011). However, vaccinating wild great apes should only be considered under certain circumstances, due to technical and financial challenges (Ryan & Walsh, 2011; Leendertz et al., 2017).

Similar criteria have been applied to justify vaccinating endangered golden lion tamarins (hereafter GLT) against yellow fever, which is responsible for high population losses (Dietz et al., 2019). Vaccinating GLT would interrupt virus circulation and reduce the risk of spillover to humans (Massad et al., 2018).

Box 9.2 An Intervention to Reverse the Impact of Yellow Fever on Endangered Golden Lion Tamarins

Carlos R. Ruiz-Miranda

Over the last 20 years, Brazil has seen periodic resurgences of yellow fever epidemics (de Oliveira Figueiredo et al., 2020). Prior to the 2016 epidemic, the impacts of the sylvatic cycle of this vector-borne disease had only been documented in howler monkeys (Bicca-Marques et al., 2017). This recent outbreak caused high mortality of several new world monkey species, including howler monkeys, marmosets (*Callithrix* spp.), and tamarins (*Leontopithecus* spp.) (Bicca-Marques et al., 2017; Possas et al., 2018; Silva et al., 2020).

Population effects of yellow fever on GLT posed a risk that could set back 35 years of conservation efforts recovering the species from the brink of extinction to about 3700 individuals (Ruiz-Miranda et al., 2019). The Associação-Mico-Leão-Dourado (AMLD), the main steward of the conservation efforts, documented cases of unexplained mortality attributed to yellow fever. A new population survey estimated a total population reduction of 32%, with two locations showing losses as high as 90% (Dietz et al., 2019). Highest losses occurred in larger fragments with better-quality forests and presence of howler monkeys, potential amplifiers of the epidemic. Forest quality could

(continued)

Box 9.2 (continued)

have favoured the proliferation of the *Haemagogus* mosquito vector (Abreu et al., 2019), and humans that enter these forests could have brought the disease in.

Researchers from the Oswaldo Cruz Immunobiological Technology Institute (FIOCRUZ), producers of the human yellow fever vaccine, and the Rio de Janeiro Primate Centre successfully adapted an attenuated human vaccine against yellow fever for howlers (*A. guariba clamitans*) and tamarins (*L. rosalia* and *L. chrysomelas*). The AMLD and its research partners developed an intervention vaccination plan to safeguard the GLT population. Population extinction risk models were applied to estimate the needed number of vaccinated animals for this species to survive another outbreak. The plan includes vaccinating at least 500 tamarins (i.e. 25% of the population) spread over 8 of the 10 populations (Fig. 9.4) and biological sample collection for disease surveillance. Six vaccinated social groups would then be translocated to the high-loss areas to reinforce the dwindling population. The AMLD, along with the State Health Secretariat, strengthened a campaign to have at least 80% of people vaccinated against yellow fever. Vaccinating both monkeys and humans would contribute to an overall 'population immunisation' that would significantly attenuate future outbreaks in the region (Massad et al., 2018; Possas et al., 2018).

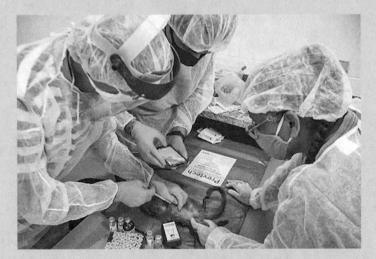

Fig. 9.4 An adult golden lion tamarin being vaccinated against yellow fever at the AMLD field lab in Silva Jardim, Brazil. Veterinarians and biologists are shown wearing protective gear during the COVID-19 pandemic. (Photo: AMLD archives)

9.3.5   Holistic Approaches

One Health is a collaborative and transdisciplinary effort – working locally, nationally, and globally – to achieve optimal health outcomes recognising the interconnection between people, wildlife, domestic animals, plants, and their shared environment (CDC, 2021). With this approach, Conservation Through Public Health (CTPH) was created as a grassroots NGO to promote biodiversity conservation by enabling people to coexist with gorillas and other wildlife at Bwindi Impenetrable National Park and other protected areas in Uganda.

Box 9.3 Community Engagement Using a One Health Approach to Gorilla Conservation

Gladys Kalema-Zikusoka

Two scabies outbreaks in 1996 and 2000/2001 were brought under control among mountain gorillas at Bwindi Impenetrable National Park (BINP). Intervention teams from Uganda Wildlife Authority (UWA) veterinary unit discovered that the affected gorillas ranged outside BINP and foraged on banana plants in community land, where they most likely came into contact with dirty clothing on scarecrows (Fig. 9.5). This contagious skin disease caused mortality of one infant and morbidity in two gorilla groups, which only recovered with ivermectin treatment (Graczyk et al., 2001; Kalema-Zikusoka et al., 2002). This led to the establishment in 2003 of Conservation Through Public Health (CTPH).

CTPH integrated programmes include wildlife conservation, community health, and alternative livelihoods to reduce poaching in gorillas' habitat. CTPH conducts regular comparative pathogen analysis of gorillas, people, and livestock and trains community health workers to improve public health and conservation practices. Village Health and Conservation Teams (VHCT) reach 6000 homes and conduct behaviour change communication promoting good hygiene and sanitation, infectious disease prevention and control, community-based family planning, nutrition, sustainable agriculture, gorilla and forest conservation, as well as reporting homes visited by gorillas (Fig. 9.6).

This *One Health* approach has contributed to growth of the mountain gorilla population from 700 individuals in 2003 to 1063 in 2018, with a reduced incidence of scabies and giardiasis. Among the local communities, the approach has contributed to a threefold increase in voluntary family planning from 22% in 2007 to 67% in 2017; an increase in hand-washing facilities from 10% to 75%; and an 11-fold increase in suspected tuberculosis referrals.

CTPH has mitigated the impact of COVID-19 on gorilla conservation by working with conservation and health partners to prevent the spread of disease among people and from people to gorillas. VHCTs and Gorilla Guardians have been trained to sensitise their communities to wear masks when herding gorillas away from community land. Park staff have been trained to better manage tourists through mandatory face mask use, increasing the minimum

(continued)

Box 9.2 (continued)

viewing distance of gorillas from 7 to 10 m and measuring temperatures of people before entering the forest. This concerted effort has resulted in greater protection for gorillas from COVID-19 and other respiratory diseases and upgraded great ape viewing regulations in Uganda.

Fig. 9.5 Gorilla feeding on banana crops at BINP, Uganda. (Photo: Emmanuel Akampulira, ITFC)

Fig. 9.6 CTPH team with the Batawa VHCTs in BINP, Uganda. (Photo: CTPH)

Other long-term primate research field sites have developed integrated health and conservation projects at the human-primate intersection within the EcoHealth paradigm, an ecosystem approach to health, focused on environmental and socioeconomic issues in the context of biodiversity conservation (Roger et al., 2016). This is the case of the Kibale EcoHealth Project, an ecological study of multi-species infection dynamics, including primates, livestock, and humans around Kibale National Park in Uganda (Goldberg et al., 2012).

Similarly, conservation medicine focuses on the ecological context of health and reversion of ecological health problems (Aguirre et al., 2012). Within this framework, the Mountain Gorilla Veterinary Project in Central Africa incorporates gorilla health and the impact of humans and livestock living in the same area, alongside research and capacity building (Cranfield & Minnis, 2007).

9.4 Conclusion

Unquestionably, infectious diseases are a leading challenge to the long-term survival of wild primates. The risk may be especially great in human-impacted landscapes, where primates face numerous anthropogenic pressures. Further research is needed to identify and study the range of pathogens harboured by coexisting primates, humans, and domestic animals to evaluate the potential for cross-species disease transmission and threats to primate and public health. As the COVID-19 pandemic has shown, local disease outbreaks can quickly become global health problems, causing huge socioeconomic impacts. With the global human population expected to reach 9.8 billion by 2050, and ever-increasing rates of anthropogenic impacts on primate habitats, the zoonotic threat is predicted to become increasingly serious, and we need to prepare by developing holistic and transdisciplinary approaches to reduce health risks at the human-primate intersection.

Acknowledgements MR thanks the NERC GW4+ Doctoral Training Partnership and the College of Life and Environmental Sciences of the University of Exeter, and Re:Wild (Grant SMA-CCO-G0000000059) for funding support and IBAP collaborators for their work in the field. MM thanks the staff of the Bulindi Chimpanzee and Community Project for help in the field and Blair Drummond Safari Park, Born Free Foundation, Friends of Chimps, and Jane Goodall Institut Schweiz for funding support. CR thanks Save the Golden Lion Tamarin Foundation, FIOCRUZ, Universidade Estadual do Norte Fluminense, Dept of Genetics-UFRJ, Zoo Atlanta, Copenhagen Zoo, Disney Conservation Fund, Association of Zoos and Aquariums, The Mohamed bin Zayed Species Conservation Fund, Margot Marsh Biodiversity Foundation, and the International Fund for Animal Welfare. KH thanks the Darwin Initiative (Project 26-018 and CV19RR18) for financial support.

References

Abreu, F. V. S., Ribeiro, I. P., Ferreira-de-Brito, A., et al. (2019). *Haemagogus leucocelaenus* and *Haemagogus janthinomys* are the primary vectors in the major yellow fever outbreak in Brazil, 2016–2018. *Emerging Microbes & Infections, 8*, 218–231.

Aguirre, A. A., Tabor, G. M., & Ostfeld, R. S. (2012). Conservation medicine. Ontogeny of an emerging discipline. In A. A. Aguirre, R. S. Ostfeld, & P. Daszak (Eds.), *New directions in conservation medicine: Applied cases of ecological health* (pp. 3–16). Oxford University Press.

Ali, R., Cranfield, M., Gaffikin, L., et al. (2004). Occupational health and gorilla conservation in Rwanda. *International Journal of Occupational and Environmental Health, 10*, 319–325.

Anderson, R. M., & May, R. M. (1992). *Infectious diseases of humans: Dynamics and control.* Oxford University Press.

Bermejo, M., Rodríguez-Teijeiro, J. D., Illera, G., et al. (2006). Ebola outbreak killed 5000 gorillas. *Science, 314*, 1564.

Bicca-Marques, J., Calegaro-Marques, C., Rylands, A. B., et al. (2017). Yellow fever threatens Atlantic Forest primates. *Science Advances, 3*, e1600946.

Borremans, B., Faust, C., Manlove, K. R., et al. (2019). Cross-species pathogen spillover across ecosystem boundaries: Mechanisms and theory. *Philosophical Transactions of the Royal Society B, 374*, 20180344.

Calattini, S., Betsem, E. B., Froment, A., et al. (2007). Simian foamy virus transmission from apes to humans, rural Cameroon. *Emerging Infectious Diseases, 13*, 1314–1320.

Calvignac-Spencer, S., Leendertz, S. A., Gillespie, T. R., & Leendertz, F. H. (2012). Wild great apes as sentinels and sources of infectious disease. *Clinical Microbiology and Infection, 18*, 521–527.

Carne, C., Semple, S., MacLarnon, A., et al. (2017). Implications of tourist–macaque interactions for disease transmission. *EcoHealth, 14*, 704–717.

CDC. (2021). *One Health.* Accessed January 26, 2021 from https://www.cdc.gov/onehealth/index.html

Chaber, A.-L., Alleborne-Webb, S., Ligneruex, Y., et al. (2010). The scale of illegal meat importation from Africa to Europe via Paris. *Conservation Letters, 3*, 317–321.

Chapman, C. A., Gillespie, T. R., & Goldberg, T. L. (2005). Primates and the ecology of their infectious diseases: How will anthropogenic change affect host-parasite interactions? *Evolutionary Anthropology: Issues, News, and Reviews, 14*, 134–144.

Chapman, C. A., Speirs, M. L., Gillespie, T. R., et al. (2006a). Life on the edge: Gastrointestinal parasites from the forest edge and interior primate groups. *American Journal of Primatology, 68*, 397–409.

Chapman, C. A., Wasserman, M., Gillespie, T. R., et al. (2006b). Do food availability, parasitism, and stress have synergistic effects on red colobus populations living in forest fragments? *American Journal of Physical Anthropology, 131*, 525–534.

Cormier, L. A., & Jolly, P. E. (2018). *The primate zoonoses: Culture change and emerging diseases.* Routledge.

Cranfield, M., & Minnis, R. (2007). An integrated health approach to the conservation of mountain gorillas *Gorilla beringei beringei. International Zoo Yearbook, 41*, 110–121.

Daszak, P., Cunningham, A. A., & Hyatt, A. D. (2000). Emerging infectious diseases of wildlife: Threats to biodiversity and human health. *Science, 287*, 443–449.

de Oliveira Figueiredo, P., Stoffela-Dutra, A. G., Barbosa Costa, G., et al. (2020). Re-emergence of yellow fever in Brazil during 2016–2019: Challenges, lessons learned, and perspectives. *Viruses, 12*, 1233.

Decision Tree Writing Group. (2006). Clinical response decision tree for the mountain gorilla (*Gorilla beringei*) as a model for great apes. *American Journal of Primatology, 68*, 909–927.

Deem, S. L. (2016). Conservation medicine: A solution-based approach for saving nonhuman primates. In M. T. Waller (Ed.), *Ethnoprimatology* (pp. 63–76). Springer.

Delahay, R. J., Smith, G. C., & Hutchings, M. R. (2009). The science of wildlife disease management. In R. J. Delahay, G. C. Smith, & M. R. Hutchings (Eds.), *Management of diseases in wild mammals* (pp. 1–8). Springer.

Dietz, J. M., Hankerson, S. J., Alexandre, B. R., et al. (2019). Yellow fever in Brazil threatens successful recovery of endangered golden lion tamarins. *Scientific Reports, 9*, 12926.

Engel, G. A., Jones-Engel, L., Schillaci, M. A., et al. (2002). Human exposure to herpesvirus B-seropositive macaques, Bali, Indonesia. *Emerging Infectious Diseases, 8*, 789–795.

Engel, G. A., Hungerford, L. L., Jones-Engel, L., et al. (2006). Risk assessment: A model for predicting cross-species transmission of simian foamy virus from macaques (*M. fascicularis*) to humans at a monkey temple in Bali, Indonesia. *American Journal of Primatology, 68*, 934–948.

Favoretto, S. R., De Mattos, C. C., Morais, N. B., et al. (2001). Rabies in marmosets (*Callithrix jacchus*), Ceará, Brazil. *Emerging Infectious Diseases, 7*, 1062–1065.

Fernandes, N. C. C. A., Cunha, M. S., Guerra, J. M., et al. (2017). Outbreak of yellow fever among nonhuman primates, Espirito Santo, Brazil, 2017. *Emerging Infectious Diseases, 23*, 2038–2041.

Fuentes, A., & Gamerl, S. (2005). Disproportionate participation by age/sex classes in aggressive interactions between long-tailed macaques (*Macaca fascicularis*) and human tourists at Padangtegal monkey forest, Bali, Indonesia. *American Journal of Primatology, 66*, 197–204.

Gibbons, A. (2021). *Captive gorillas test positive for coronavirus*. Accessed January 12, 2021 from https://www.sciencemag.org/news/2021/01/captive-gorillas-test-positive-coronavirus

Gilardi, K. V., Gillespie, T. R., Leendertz, F. H., et al. (2015). *Best practice guidelines for health monitoring and disease control in great ape populations*. IUCN SSC Primate Specialist Group.

Gilardi, K. V., Nziza, J., Ssebide, B., et al. (2021). Endangered mountain gorillas and COVID-19: One health lessons for prevention and preparedness during a global pandemic. *American Journal of Primatology, 84*(4–5), e23291.

Gillespie, T. R., Nunn, C. L., & Leendertz, F. H. (2008). Integrative approaches to the study of primate infectious disease: Implications for biodiversity conservation and global health. *American Journal of Physical Anthropology, 47*, 53–69.

Gogarten, J. F., Dux, A., Mubemba, B., et al. (2019). Tropical rainforest flies carrying pathogens form stable associations with social nonhuman primates. *Molecular Ecology, 28*, 4242–4258.

Goldberg, T. L., Gillespie, T. R., Rwego, I. B., et al. (2007). Patterns of gastrointestinal bacterial exchange between chimpanzees and humans involved in research and tourism in western Uganda. *Biological Conservation, 135*, 511–517.

Goldberg, T. L., Paige, S. B., & Chapman, C. A. (2012). The Kibale EcoHealth Project: Exploring connections among human health, animal health, and landscape dynamics in western Uganda. In A. A. Aguirre, R. Ostfeild, & P. Daszak (Eds.), *New directions in conservation medicine. Applied cases of ecological health* (pp. 452–465). Oxford University Press.

Graczyk, T. K., Mudakikwa, A. B., Cranfield, M. R., & Eilenberger, U. (2001). Hyperkeratotic mange caused by *Sarcoptes scabiei* (Acariformes: Sarcoptidae) in juvenile human-habituated mountain gorillas (*Gorilla gorilla beringei*). *Parasitology Research, 87*, 1024–1028.

Grützmacher, T. K., Köndgen, S., Keil, V., et al. (2016). Codetection of respiratory syncytial virus in habituated wild western lowland gorillas and humans during a respiratory disease outbreak. *EcoHealth, 13*, 499–510.

Grützmacher, T. K., Keil, V., Leinert, V., et al. (2018a). Human quarantine: Toward reducing infectious pressure on chimpanzees at the Taï Chimpanzee Project, Côte d'Ivoire. *American Journal of Primatology, 80*, e22619.

Grützmacher, T. K., Keil, V., Metzger, S., et al. (2018b). Human respiratory syncytial virus and *Streptococcus pneumoniae* infection in wild bonobos. *EcoHealth, 15*, 462–466.

Hahn, B. H., Shaw, G. M., De Cock, K. M., & Sharp, P. (2000). AIDS as a zoonosis: Scientific and public health implications. *Science, 287*, 607–614.

Hanes, A. C., Kalema-Zikusoka, G., Svensoon, M. S., & Hill, C. M. (2018). Assessment of health risks posed by tourists visiting mountain gorillas in Bwindi Impenetrable National Park, Uganda. *Primate Conservation, 32*, 123–132.

Hill, C. M. (2017). Primate crop feeding behavior, crop protection, and conservation. *International Journal of Primatology, 38*, 385–400.

Hockings, K. J., Mubemba, B., Avanzi, C., et al. (2021). Leprosy in wild chimpanzees. *Nature, 598*, 652–656.

Holzmann, I., Agostini, I., Areta, J. I., et al. (2010). Impact of yellow fever outbreaks on two howler monkey species (*Alouatta guariba clamitans* and *A. caraya*) in Misiones, Argentina. *American Journal of Primatology, 72*, 475–480.

Hubálek, Z. (2003). Emerging human infectious diseases: Anthroponoses, zoonoses, and saprono-ses. *Emerging Infectious Diseases, 9*, 403–404.

Huffman, M. A. (1997). Current evidence for self-medication in primates: A multidisciplinary perspective. *American Journal of Physical Anthropology, 104*, 171–200.

Huffman, M. A. (2001). Self-medicative behavior in the African great apes: An evolutionary per-spective into the origins of human traditional medicine. *BioScience, 51*(8), 651–661.

Huffman, M. A., & Caton, J. M. (2001). Self-induced increase of gut motility and the control of parasitic infections in wild chimpanzees. *International Journal of Primatology, 22*, 329–346.

Huffman, M. A., Page, J. E., Sukhdeo, M. V. K., Gotoh, S., et al. (1996). Leaf-swallowing by chimpanzees: A behavioral adaptation for the control of strongyle nematode infections. *International Journal of Primatology, 17*, 475–503.

IUCN. (2021). *Great apes, COVID-19 and the SARS-CoV-2*. Section on Great Apes, IUCN SSC Primate Specialist Group. April 2021 revision. IUCN. Accessed September 15, 2022 from https://www.iucngreatapes.org/covid-19

Jezek, Z., Arita, I., Mutombo, M., et al. (1986). Four generations of probable person-to-person transmission of human monkeypox. *American Journal of Epidemiology, 123*, 1004–1012.

Johnston, A. R., Gillespie, T. R., Rwego, I. B., et al. (2010). Molecular epidemiology of cross-spe-cies *Giardia duodenalis* transmission in western Uganda. *PLoS Neglected Tropical Diseases, 4*, e683.

Jones-Engel, L., & Engel, G. A. (2006). Disease risk analysis: A paradigm for using health-based data to inform primate conservation and public health. *American Journal of Primatology, 68*, 851–854.

Jones-Engel, L., Engel, G. A., Schillaci, M. A., et al. (2005a). Primate-to-human retroviral trans-mission in Asia. *Emerging Infectious Diseases, 11*, 1028–1035.

Jones-Engel, L., Schillaci, M. A., Engel, G. A., et al. (2005b). Characterizing primate pet owner-ship in Sulawesi: Implications for disease transmission. In J. D. Paterson & J. Wallis (Eds.), *Commensalism and conflict: The human-primate interface*. American Society of Primatologists.

Jones-Engel, L., Engel, G. A., Heidrich, J., et al. (2006). Temple monkeys and health implications of commensalism, Kathmandu, Nepal. *Emerging Infectious Diseases, 12*, 900–906.

Kalema-Zikusoka, G., Kock, R. A., & MacFie, E. J. (2002). Scabies in free-ranging mountain gorillas (*Gorilla beringei beringei*) in Bwindi Impenetrable National Park, Uganda. *The Veterinary Record, 150*, 12–15.

Kalousová, B., Hasegawa, H., Petrželková, K. J., et al. (2016). Adult hookworms (*Necator* spp.) collected from researchers working with wild western lowland gorillas. *Parasites & Vectors, 9*, 1–6.

Kilonzo, C., Stopka, T. J., & Chomel, B. (2013). Illegal animal and (bush)meat trade associated risk of spread of viral infections. In S. K. Singh (Ed.), *Viral infections and global change* (pp. 179–194). Wiley-Blackwell.

Köndgen, S., Kühl, H., N'Goran, P. K., et al. (2008). Pandemic human viruses cause decline of endangered great apes. *Current Biology, 18*, 260–264.

Lappan, S., Malaivijitnond, S., & Radhakrishna, S. (2020). The human–primate interface in the New Normal: Challenges and opportunities for primatologists in the COVID-19 era and beyond. *American Journal of Primatology, 82*, e23176.

LeBreton, M., Prosser, A. T., Tamoufe, U., et al. (2006). Patterns of bushmeat hunting and percep-tions of disease risk among central African communities. *Animal Conservation, 9*, 357–363.

Leendertz, F. H., Pauli, G., Maetz-Rensing, K., et al. (2006). Pathogens as drivers of population declines: The importance of systematic monitoring in great apes and other threatened mammals. *Biological Conservation, 131*, 325–337.

Leendertz, S. A. J., Wich, S., Ancrenaz, M., et al. (2017). Ebola in great apes – Current knowledge, possibilities for vaccination, and implications for conservation and human health. *Mammal Review, 47*, 98–111.

Leroy, E. M., Rouquet, P., Formenty, P., et al. (2004). Multiple Ebola virus transmission events and rapid decline of central African wildlife. *Science, 303*, 387–390.

Levréro, F., Gatti, S., Gautier-Hion, A., & Ménard, N. (2007). Yaws disease in a wild gorilla population and its impact on the reproductive status of males. *American Journal of Physical Anthropology, 132*, 568–575.

MacFie, E. J., & Williamson, E. A. (2010). *Best practice guidelines for great ape tourism.* IUCN/SSC Primate Specialist Group (PSG).

Maréchal, L., Semple, S., Majolo, B., et al. (2011). Impacts of tourism on anxiety and physiological stress levels in wild male Barbary macaques. *Biological Conservation, 144*, 2188–2193.

Maréchal, L., Semple, S., Majolo, B., & MacLarnon, A. (2016). Assessing the effects of tourist provisioning on the health of wild barbary macaques in Morocco. *PLoS One, 11*, e0155920.

Martini, G. A. (1973). Marburg virus disease. *Postgraduate Medical Journal, 49*, 542–546.

Massad, E., Miguel, M. M., & Coutinho, F. A. B. (2018). Is vaccinating monkeys against yellow fever the ultimate solution for the Brazilian recurrent epizootics? *Epidemiology and Infection, 146*, 1622–1624.

Mbora, D. N., & McPeek, M. A. (2009). Host density and human activities mediate increased parasite prevalence and richness in primates threatened by habitat loss and fragmentation. *The Journal of Animal Ecology, 78*, 210–218.

McCallum, H., & Dobson, A. (2002). Disease, habitat conservation and fragmentation. *Proceedings of the Royal Society of London. Series B, 269*, 2041–2049.

McLennan, M. R., & Huffman, M. A. (2012). High frequency of leaf swallowing and its relationship to intestinal parasite expulsion in 'village' chimpanzees at Bulindi, Uganda. *American Journal of Primatology, 74*, 642–650.

McLennan, M. R., Hasegawa, H., Bardi, M., & Huffman, M. A. (2017). Gastrointestinal parasite infections and self-medication in wild chimpanzees surviving in degraded forest fragments within an agricultural landscape mosaic in Uganda. *PLoS One, 12*, e0180431.

McLennan, M. R., Mori, H., Mahittikorn, A., et al. (2018). Zoonotic enterobacterial pathogens detected in wild chimpanzees. *EcoHealth, 15*, 143–147.

Melin, A. D., Janiak, M. C., Marrone, F. 3rd, et al. (2020). Comparative ACE2 variation and primate COVID-19 risk. *Communications Biology, 3*, 641.

Michel, A. L., & Huchzermeyer, H. F. (1998). The zoonotic importance of *Mycobacterium tuberculosis*: Transmission from human to monkey. *Journal of the South African Veterinary Association, 69*, 64–65.

Miranda, T. S., Muniz, C. P., Moreira, S. B., et al. (2019). Eco-epidemiological profile and molecular characterization of Simian Foamy Virus in a recently-captured invasive population of *Leontopithecus chrysomelas* (Golden-Headed Lion Tamarin) in Rio de Janeiro, Brazil. *Viruses, 11*, 931.

Molyneaux, A., Hankinson, E., Kaban, M., et al. (2021). Primate selfies and anthropozoonotic diseases: Lack of rule compliance and poor risk perception threatens orangutans. *Folia Primatologica, 92*(5–6), 296–305.

Mossoun, A., Calvignac-Spencer, S., Anoh, A. E., et al. (2017). Bushmeat hunting and zoonotic transmission of Simian T-lymphotropic virus 1 in tropical West and Central Africa. *Journal of Virology, 91*, e02479–16.

Muehlenbein, M. P. (2017). Anthroponoses. In A. Fuentes (Ed.), *The international encyclopedia of primatology*. Wiley-Blackwell.

Muehlenbein, M. P., Martinez, L. A., Lemke, A. A., et al. (2008). Perceived vaccination status in ecotourists and risks of anthropozoonoses. *EcoHealth, 5*, 371–378.

Munster, V. J., Feldmann, F., Williamson, B. N., et al. (2020). Respiratory disease in rhesus macaques inoculated with SARS-CoV-2. *Nature, 585*, 268–272.

Negrey, J. D., Reddy, R. B., Scully, E. J., et al. (2019). Simultaneous outbreaks of respiratory disease in wild chimpanzees caused by distinct viruses of human origin. *Emerging Microbes & Infections, 8*, 139–149.

Nizeyi, J. B., Mwebe, R., Nanteza, A., et al. (1999). *Cryptosporidium* sp. and *Giardia* sp. infections in mountain gorillas (*Gorilla gorilla beringei*) of the Bwindi Impenetrable National Park, Uganda. *The Journal of Parasitology, 85*, 1084–1088.

Nizeyi, J. B., Innocent, R. B., Erume, J., et al. (2001). Campylobacteriosis, salmonellosis, and shigellosis in free-ranging human-habituated mountain gorillas of Uganda. *Journal of Wildlife Diseases, 37*, 239–244.

Nunn, C. L., & Altizer, S. (2006). *Infectious diseases in primates: Behavior, ecology and evolution.* Oxford University Press.

Nunn, C. L., & Gillespie, T. R. (2016). Infectious diseases and primate conservation. In S. A. Wich & A. J. Marshall (Eds.), *An introduction to primate conservation* (pp. 157–173). Oxford University Press.

Olivero, J., Fa, J. E., Real, R., et al. (2017). Recent loss of closed forests is associated with Ebola virus disease outbreaks. *Scientific Reports, 7*, 14291.

Palacios, G., Lowenstine, L. J., Cranfield, M. R., et al. (2011). Human metapneumovirus infection in wild mountain gorillas, Rwanda. *Emerging Infectious Diseases, 17*, 711–713.

Patrono, L. V., Samuni, L., Corman, V. M., et al. (2018). Human coronavirus OC43 outbreak in wild chimpanzees, Côte d'Ivoire, 2016. *Emerging Microbes & Infections, 7*, 118.

Possas, C., Lourenço-de-Oliveira, R., Tauil, P. L., et al. (2018). Yellow fever outbreak in Brazil: The puzzle of rapid viral spread and challenges for immunisation. *Memórias do Instituto Oswaldo Cruz, 113*, e180278.

Pourrut, X., Diffo, J. L., Somo, R. M., et al. (2011). Prevalence of gastrointestinal parasites in primate bushmeat and pets in Cameroon. *Veterinary Parasitology, 175*, 187–191.

Robbins, M. M., Gray, M., Fawcett, K. A., et al. (2011). Extreme conservation leads to recovery of the Virunga mountain gorillas. *PLoS One, 6*, e19788.

Roger, F., Caron, A., Morand, S., et al. (2016). One Health and EcoHealth: The same wine in different bottles? *Infection Ecology & Epidemiology, 6*, 30978.

Rouquet, P., Froment, J. M., Bermejo, M., et al. (2005). Wild animal mortality monitoring and human Ebola outbreaks, Gabon and Republic of Congo, 2001-2003. *Emerging Infectious Diseases, 11*, 283–290.

Ruiz-Miranda, C. R., de Morais, M. M. Jr., Dietz, L. A., et al. (2019). Estimating population sizes to evaluate progress in conservation of endangered golden lion tamarins (*Leontopithecus rosalia*). *PLoS One, 14*, e0216664.

Rulli, M. C., Santini, M., Hayman, D. T., & D'Odorico, P. (2017). The nexus between forest fragmentation in Africa and Ebola virus disease outbreaks. *Scientific Reports, 7*, 41613.

Ryan, S. J., & Walsh, P. D. (2011). Consequences of non-intervention for infectious disease in African great apes. *PLoS One, 6*, e29030.

Sabbatini, G., Stammati, M., Travares, M. C. H., et al. (2006). Interactions between humans and capuchin monkeys (*Cebus libidinosus*) in the Parque Nacional de Brasília, Brazil. *Applied Animal Behaviour Science, 97*, 272–283.

Sapolsky, R. M., & Share, L. J. (2004). A pacific culture among wild baboons: Its emergence and transmission. *PLoS Biology, 2*, 0534–0541.

Schaumburg, F., Mugisha, L., Peck, B., et al. (2012). Drug-resistant human *Staphylococcus aureus* in sanctuary apes pose a threat to endangered wild ape populations. *American Journal of Primatology, 74*, 1071–1075.

Shutt, K., Heistermann, M., Kasim, A., et al. (2014). Effects of habituation, research and ecotourism on faecal glucocorticoid metabolites in wild western lowland gorillas: Implications for conservation management. *Biological Conservation, 172*, 72–79.

Silva, N. I. O., Sacchetto, L., de Rezende, I. M., et al. (2020). Recent sylvatic yellow fever virus transmission in Brazil: The news from an old disease. *Virology Journal, 17*, 9.

Smith, K. M., Anthony, S. J., Switzer, W. M., et al. (2012). Zoonotic viruses associated with illegally imported wildlife products. *PLoS One, 7*, e29505.

Taylor, L., Latham, S., & Woolhouse, M. (2001). Risk factors for human disease emergence. *Philosophical Transactions of the Royal Society B: Biological Sciences, 356*, 983–989.

Travis, D. A., Hungerford, L., Engel, G. A., & Jones-Engel, L. (2006). Disease risk analysis: A tool for primate conservation planning and decision making. *American Journal of Primatology, 68*, 855–867.

Weber, A., Kalema-Zikusoka, G., & Stevens, N. J. (2020). Lack of rule-adherence during mountain gorilla tourism encounters in Bwindi Impenetrable National Park, Uganda, places gorillas at risk from human disease. *Frontiers in Public Health, 8*, 1.

Weyher, A. H. (2009). Crop raiding: The influence of behavioral and nutritional changes on primate-parasite relationships. In M. A. Huffman & C. A. Chapman (Eds.), *Primate parasite ecology. The dynamics and study of host-parasite relationships* (pp. 403–422). Cambridge University Press.

WHO. (2014). *A brief guide to emerging infectious diseases and zoonoses.* WHO Regional Office for South-East Asia.

WHO. (2020). *Zoonoses.* Accessed December 30, 2020 from https://www.who.int/news-room/fact-sheets/detail/zoonoses

Wilcox, B. A., & Ellis, B. (2006). Forests and emerging infectious diseases of humans. *Unasylva, 57*, 11–18.

Wilkinson, D. A., Marshall, J. C., French, N. P., & Hayman, D. T. S. (2018). Habitat fragmentation, biodiversity loss and the risk of novel infectious disease emergence. *Journal of the Royal Society Interface, 15*, 20180403.

Wolf, T. M., Wang, W. A., Lonsdorf, E. V., et al. (2019). Optimizing syndromic health surveillance in free ranging great apes: The case of Gombe National Park. *Journal of Applied Ecology, 56*, 509–518.

Wolfe, N. D., Prosser, T. A., Carr, J. K., et al. (2004a). Exposure to nonhuman primates in rural Cameroon. *Emerging Infectious Diseases, 10*, 2094–2099.

Wolfe, N. D., Switzer, W. M., Carr, J. K., et al. (2004b). Naturally acquired simian retrovirus infections in central African hunters. *The Lancet, 363*, 932–937.

Zoo Atlanta. (2021). *Uptade on gorilla population – Sept 17.* Accessed September 15, 2022 from https://zooatlanta.org/update-on-gorilla-population-sept-17/

Chapter 10
Primate Conservation in Shared Landscapes

Elena Bersacola, Kimberley J. Hockings, Mark E. Harrison, Muhammad Ali Imron, Joana Bessa, Marina Ramon, Aissa Regalla de Barros, Maimuna Jaló, Américo Sanhá, Carlos R. Ruiz-Miranda, Luis Paulo Ferraz, Mauricio Talebi, and Matthew R. McLennan

Contents

Abstract The majority of nonhuman primates are found in habitats impacted by humans. Therefore, conservation interventions in anthropogenic landscapes are critical for the long-term survival of primate populations. Due to their intelligence and socioecological flexibility, many primates exhibit behaviours deemed problematic such as crop feeding, property damage, and livestock depredation. Large-bodied primates may also pose a physical risk to people. In this chapter, we first revise the

E. Bersacola (✉) · K. J. Hockings
Centre for Ecology and Conservation, Faculty of Environment, Science and Economy, University of Exeter, Penryn, UK
e-mail: e.bersacola@exeter.ac.uk

M. E. Harrison
Centre for Ecology and Conservation, Faculty of Environment, Science and Economy, University of Exeter, Penryn, UK

School of Geography, Geology and the Environment, University of Leicester, Leicester, UK

M. A. Imron
Faculty of Forestry, Universitas Gadjah Mada, Yogyakarta, Indonesia

J. Bessa
Department of Zoology, University of Oxford, Oxford, UK

common criteria for selecting primate conservation priorities and consider them in the context of shared landscapes. We discuss the importance of inclusive conservation approaches and provide recommendations for addressing negative human-primate interactions based on existing information. Three case studies that illustrate conservation efforts in shared environments are presented: (1) the Bulindi Chimpanzee and Community Project in Uganda, (2) community conservation of orangutans and Javan slow lorises in Indonesia, and (3) inclusive conservation of golden lion tamarins in Brazil's Atlantic Forest. The active participation of a diverse group of stakeholders, including local community groups, in all conservation stages is essential to fully understand the complexities of human-primate interactions in shared landscapes, address negative interactions, mitigate conservation conflicts, advocate for equity, and promote long-term human-primate coexistence.

Keywords Anthropocene · Anthropogenic · Brazil · Coexistence · Community-based conservation · Inclusive conservation · Human-primate interactions · Indonesia · Multi-stakeholder conservation · Uganda

10.1   Introduction

Humans are possibly the dominant force shaping our planet's ecosystems and environment, leading many scientists to define a new geological epoch: the Anthropocene (Lewis & Maslin, 2015). It is now estimated that nearly all of our world's terrestrial

M. Ramon
Centre for Ecology and Conservation, Faculty of Environment, Science and Economy, University of Exeter, Penryn, UK

School of Biosciences, Cardiff University, Cardiff, UK

A. Regalla de Barros · M. Jaló · A. Sanhá
Instituto da Biodiversidade e das Áreas Protegidas (IBAP), Bissau, Guinea-Bissau

C. R. Ruiz-Miranda
Laboratório de Ciências Ambientais, Universidade Estadual do Norte Fluminense, Campos dos Goytacazes, Rio de Janeiro, Brazil

Associação Mico Leão Dourado, Silva Jardim, Rio de Janeiro, Brazil

L. P. Ferraz
Associação Mico Leão Dourado, Silva Jardim, Rio de Janeiro, Brazil

M. Talebi
Universidade Federal de São Paulo (UNIFESP), Departamento de Ciências Ambientais, Campus Diadema & Instituto Pró-Muriqui, São Paulo, Brazil

M. R. McLennan
Bulindi Chimpanzee and Community Project, Hoima, Uganda

Faculty of Humanities and Social Sciences, Oxford Brookes University, Oxford, UK landscapes have been altered by humans (Kennedy et al., 2020). Given that contemporary ecosystem dynamics are rarely disconnected from humans, ecologists must explicitly take their interconnectedness into account in their research frameworks (Ellis & Ramankutty, 2008). At present, 75% of the world's nonhuman primate species (hereafter primates) have declining populations due to human activities, in particular forest conversion into agriculture, logging, and hunting (Estrada et al., 2017). Thus, the ways in which primates respond to and interact with humans are an increasingly important area of research and of growing conservation focus (Humle & Hill, 2016; Kalbitzer & Chapman, 2018; McLennan et al., 2017). **Human-primate interactions** within '**shared landscapes**' may range from exposure to **anthropogenic** noise and infrequent encounters to high spatiotemporal overlap and direct contact, particularly where primates and humans use the same resources or in areas with high hunting pressure (Hockings et al., 2009; McKinney, 2015; McLennan & Hockings, 2016; McLennan et al., 2017; Mormile & Hill, 2017).

The majority of primates inhabit tropical or subtropical forest habitats (Galán-Acedo et al., 2019a) and are susceptible to land use change (Estrada et al., 2017). As human-induced environmental change continues to increase, primates are more often found in modified habitats including forest-agricultural/**urban mosaics** and commercial plantation landscapes (Spehar & Rayadin, 2017). Many primatologists are arguing for increasing recognition of the critical role of human-impacted landscapes for primate conservation in the twenty-first century, particularly for species able to adapt to some level of land use change (Estrada et al., 2012; Galán-Acedo et al., 2019b). With some level of landscape connectivity enabled by remnant forest fragments, an absence or low levels of hunting, the potential use of arboreal and diversified agriculture environments, and sufficient food sources, many primate species can temporarily or permanently persist in **anthropogenic landscapes** (Estrada et al., 2012; Chap. 8, this volume).

In this chapter, we (i) review criteria to select primate conservation priorities and consider them in the context of shared landscapes, (ii) discuss the importance of inclusive conservation approaches, and (iii) provide recommendations for addressing negative human-primate interactions based on 'lessons learned'. We present three case studies across three continents that illustrate conservation efforts in shared environments: chimpanzees (*Pan troglodytes schweinfurthii*) in Uganda, orangutans (*Pongo pygmaeus morio*) and Javan slow lorises (*Nycticebus javanicus*) in Indonesia, and golden lion tamarins (*Leontopithecus rosalia*) in Brazil.

10.2   Primate Conservation Priorities in the Anthropocene

Resources available for conservation, including funds, time, and expertise, are limited. It is therefore necessary to develop resource- and cost-effective conservation priorities. But how can we define priorities in primate conservation? Primate conservation priorities typically focus on 'important' and/or threatened primate species and/or areas with high primate diversity.

Fig. 10.1 (**a**) Infant Bornean orangutan and (**b**) Temminck's red colobus in Cantanhez National Park, Guinea Bissau. Due to their charismatic appearance and threatened status, orangutans and colobines can be considered flagship species. (Photo credits: Andrew Walmsley/Yayasan Borneo Nature Indonesia (**a**) and Elena Bersacola (**b**))

'Important' species may consist of keystone, indicator, flagship, and/or umbrella species (Arponen, 2012) (Fig. 10.1). Keystone species play important roles in ecosystem processes and functioning, e.g. they can be critical seed dispersers (McConkey, 2018) and top-down regulators (predators) or suppress competitors (Bond, 1994). Indicator species are considered to reflect some wider aspect of environmental condition, expressing relatively rapid and consistent responses to environmental change (Lawton & Gaston, 2001). Indicator species richness can also act as surrogate to diversity of other wildlife: across Madagascar, for example, lemurs were found to predict non-primate mammal diversity (Muldoon & Goodman, 2015). Flagship species are high-profile taxa mainly used as tools for leveraging conservation, including to raise public conservation awareness and conservation funds (Simberloff, 1998). They are often considered 'charismatic', a definition that may differ amongst regions, cultures, and/or groups of people. Albert et al. (2018) identified 20 species considered most charismatic by the Western public. The species included chimpanzees (*P. troglodytes*) and gorillas (*Gorilla* spp.), which were considered charismatic for aesthetic reasons or because they are impressive or threatened with extinction (Albert et al., 2018). The iconic status of certain species such as great apes is not universal, however. For example, in Central Kalimantan, indigenous people consider fish more important than orangutans (*P. pygmaeus*) (Thornton et al., 2020). Wide-ranging, flagship species may be categorised as umbrella species, i.e. a species whose home range and/or minimum land requirements are large enough to include a high number of taxa, so that the protection of their habitat will also benefit the conservation of other, sympatric species (Simberloff, 1998). Primates identified as umbrella species include orangutans (Burivalova et al., 2020), guenons (*Cercopithecus* spp.: Lambert, 2011), golden lion tamarins (Ruiz-Miranda et al., 2019), and woolly monkeys (*Lagothrix lagothricha*: Linero et al., 2020).

Prioritisation of species or subspecies that are most vulnerable to extinction is typically based upon assessments by the International Union for Conservation of Nature's (IUCN) Red List of Threatened Species (IUCN, 2020), which is the leading provider of conservation data, assessments, and analysis and includes the IUCN

Primate Specialist Group. It is also worth noting that some species or subspecies may be classified as 'Least Concern' globally (IUCN, 2020), but nevertheless are threatened with extinction in some localities or regions. Additionally, classifications can be revised, meaning populations can suddenly move from being of lower to very high conservation concern, for example, due to taxonomic re-classification into distinct species, such as in the case of Tapanuli orangutan (*P. tapanuliensis*, which was split from *P. abelii*) (Nater et al., 2017). Besides the global conservation status, it is always worth considering the local context (see case studies). Another way to prioritise in primate conservation is to select areas high in (primate) biodiversity, especially where these areas are also experiencing high threat levels. '**Biodiversity hotspots**' are threatened areas with high biodiversity and/or species endemism, where conservation action has the potential to have a large impact (Mittermeier et al., 2011).

Given broad variability in goals, scale, and scope, it is difficult to generalise about conservation priorities, particularly within anthropogenic environments (Hockings et al., 2015; Kalbitzer & Chapman, 2018; McLennan et al., 2017). For example, although many human-influenced ecosystems such as agroforests can retain high biodiversity (Estrada et al., 2012), applying the second criterion alone for selecting primate conservation priority areas may exclude opportunities for conservation interventions in important human-primate shared systems which are essential to ensure large-scale, metapopulation connectivity, given that human-influenced ecosystems may have lower biodiversity compared to pristine forests or remote locations in certain regions (Torres-Romero & Olalla-Tárraga, 2015). In addition, some highly imperilled taxa may not range in areas with high biodiversity (e.g. Barbary macaques *Macaca sylvanus*, Wallis et al. (2020); golden snub-nosed monkey *Rhinopithecus roxellana*, Long and Richardson (2020)). Considering that certain primate species may nonetheless rely on conservation policies that support and integrate these human-primate contexts, we argue that excluding shared landscapes from conservation priorities risks missing opportunities to develop inclusive, new, and effective conservation approaches that may be applicable to a significant portion of the primates' geographical range. In addition, human-primate interactions in shared landscapes can generate considerable attention (good and/or bad), including in the media, especially where interactions are 'negative' such as when primates damage crops or property or pose a threat to human physical safety (e.g. @ NatGeoUK, 2019). These complex interactions require conservation interventions and management for the benefit of local people, for the conservation and welfare of primates, and for the conservation of species – if negative interactions are not addressed, this can weaken public support for conservation (Chua et al., 2020). It is also argued that conservation approaches must now look beyond the one species/ habitat patch interventions and should instead aim at restoring ecosystems at the large scale and integrating multi-stakeholder processes (Norris et al., 2020; Reed, 2008). However, habitat restoration might not be possible in many human-dominated landscapes. With the growing evidence that some taxa can cope with modified habitats, conservation aims may focus more on maintaining sufficient ecosystem function to allow species survival.

10.3   The Importance of Inclusive Primate
Conservation Approaches

A crucial goal of conservation practice is to balance the costs and benefits of conservation interventions to people living in proximity to wildlife to promote long-
term coexistence (Harrison et al., 2020). Importantly, practitioners must avoid
colonial or 'fortress' conservation approaches, i.e. exclusionary and often violent
conservation approaches based on the human-nature dichotomy view, which have
been (and are sometimes still) prevalent across Africa and Asia (Brockington &
Igoe, 2006; Colchester, 2004; Mkumbukwa, 2008). Within the scientific community, it is now widely recognised that to be successful in the long term, conservation
strategies must ultimately improve local people's lives (Adams et al., 2004). Many
primate species in need of conservation occur in areas where human poverty is high.
For example, over 67% of the human population in Guinea-Bissau, where much of
the remaining populations of Critically Endangered Temminck's red colobus
(*Piliocolobus badius temminckii*) are found, live below the poverty line (Bersacola
et al., 2018; World Bank, 2021). Poverty is not only economic but encompasses a
range of diverse issues that are often country and context specific, such as lack of
access to education and healthcare. One way to improve people's wellbeing is
through poverty alleviation and sustainable development programmes (United
Nations, 2021). There are many different ways conservation programmes may be
able to contribute to reduce poverty, for example, through activities that help generate financial income, but also indirectly via safeguarding Indigenous rights, education and capacity building, as well as approaches that aim to increase socioecological
resilience (United Nations, 2021). However, aid-centric approaches that do not
question the economic status quo have some heavy critics (Hickel, 2017; Norris
et al., 2020). We must remember that widespread structural inequalities, poverty,
and lack of equal opportunities not only affect more people in rural or remote areas
but also exist between social groups. Economic and social inequalities such as
power imbalances between conservation stakeholders and gender inequality remain
a problematic issue in contemporary conservation (Chua et al., 2020; Rubis &
Theriault, 2020). Primate conservationists must therefore engage with diverse
stakeholders to attempt to fully understand and acknowledge existing systemic
social issues and develop conservation strategies that also explicitly aim to promote
social equity and human wellbeing. A cross-disciplinary, multi-stakeholder team is
necessary to fully understand the complexities of human-primate interactions.

Conservation plans in human-wildlife systems can be applied at different scales.
At the global/national/regional level, economic policies should reflect the needs of
rural communities, including providing incentives to adopt sustainable approaches,
particularly in food production, to minimise environmental damage and promote
healthy human-influenced ecosystems as well as social equality (Díaz et al., 2019).
Likewise, site-level conservation initiatives should incorporate the needs and

perspectives of local residents, address the challenges these people experience (by living in proximity to primates), and ideally improve (or, at minimum, not worsen) local people's lives and livelihoods. Involving different stakeholders in a co-creation process is therefore crucial when developing large-scale conservation plans, such as those at the national level, as well as small-scale projects at the site level. Multi-stakeholder participation can offer opportunities to explore equity issues, address socio-cultural and environmental complexities, and develop trust between policy makers, scientists, and citizens (Reed, 2008). In the case of primate conservation in human-influenced landscapes, stakeholders may often include local farmers, hunters, women and youth associations, traditional and Indigenous authorities, national and international NGOs, government bodies responsible for the management of wildlife or forest resources (e.g. Agriculture, Forestry, or Biodiversity departments), as well as researchers from multiple disciplines (e.g. social scientists, ecologists, economists, agronomists) and educators (Bersacola et al., 2021; Chazdon et al., 2020; Chesney et al., 2020, Case Study 3). Pre-existing socio-political power imbalances amongst conservation stakeholders can easily preclude equity and fairness in the participation process if left unaddressed (Rubis & Theriault, 2020). Multi-stakeholder approaches must ensure a fair and equal exchange of ideas throughout the conservation process, from research to planning, implementation, and monitoring and evaluation. Besides ensuring an inclusive conservation approach, the role of primatologists may also include bridging conversations between local communities and national or international organisations.

10.4   Addressing Negative Human-Primate Interactions in Anthropogenic Landscapes

Interactions between people and primates in shared landscapes are often complex. Due to their intelligence and socioecological flexibility, many primate species exhibit behaviours deemed problematic by local people. Risks to people living alongside primates include costs to livelihoods due to crop feeding, destruction of stored food, property damage, and livestock depredation (Campbell-Smith et al., 2010; Hill, 2017; Mormile & Hill, 2017). Aggressive interactions between large-bodied primates and humans can also result in human injury and sometimes even death, particularly in young children (Hockings et al., 2010; Hockings & McLennan, 2016; McLennan & Hockings, 2016) (Case Study 1). Primates also pose risks of disease transmission to humans (Jones-Engel et al., 2005; Pedersen & Davies, 2009; Chap. 9, this volume). These risks for humans can be a major cause of primate mortality, for example, when farmers engage in pre-emptive or retaliatory killing to protect crops or livestock (Kibaja, 2014; Kifle & Bekele, 2020; Meijaard et al., 2011; McLennan et al., 2012). It is also important to point out that human-primate interactions are not always negative. Some primates play an important role in human culture and folklore (Cormier, 2006; Fuentes & Gamerl, 2005; Parathian et al.,

2018; Riley & Priston, 2010). In some areas, primates are provisioned with food by people (Fuentes & Gamerl, 2005). The presence of primates may also provide economic opportunities to local people through tourism. However, tourism revenues are often distributed unequally, and/or benefits may be limited to a selected few (Cobbinah et al., 2017; Ezebilo & Mattsson, 2010; Sabuhoro et al., 2017).

The applied field of human-wildlife interactions has traditionally focused on resolving 'problematic[1]' wildlife behaviour such as primate crop feeding (Hill & Wallace, 2012) (Fig. 10.2). Although some technical interventions aimed at reducing crop foraging have been evaluated in a primate context (Webber et al., 2007; Frank et al., 2019), technical measures alone will not be enough in the long term. In some cases, the challenges to finding long-term solutions to primate crop feeding may be compounded by human fear of some primate species (e.g. orangutans: Campbell-Smith et al., 2010). Additionally, primates can adapt to the most sophisticated repellent devices, and traditional fences are mostly ineffective at keeping primates out (Osborn & Hill, 2005). Primates can also learn to navigate electric fences, which are expensive, need high maintenance, and are unaffordable to most farmers in developing countries (Suzuki & Muroyama, 2010; Priston & McLennan, 2013). The use of scarecrows or farmers chasing and shouting at crop feeding primates are common deterrent methods; continuous guarding during the day can sometimes help reduce crop damage (Byamukama & Asuma, 2006; Hockings & Humle, 2009). Alternative crops can be utilised to mitigate crop loss, for example, via planting of unpalatable crop buffer zones at forest edges and by changing the principal crops grown, but such techniques are only effective if these crops are economically profitable and there are existing market chains. For example, chilli is less

Fig. 10.2 Chimpanzees readily incorporate agricultural foods introduced into their habitats. (**a**) Wild chimpanzees in Bulindi, Uganda, sharing a cultivated jackfruit in a village garden; behind them is a stand of exotic eucalyptus trees. (**b**) An adult male chimpanzee at Bulindi eats jackfruit, while a female and her offspring wait for him to finish and leave the fruit for them. (Photo credits: Matthew R. McLennan)

---

[1] Here, we refrain from using the term 'conflict' and choose instead to use 'problematic behaviours', 'negative interactions', or 'risks'; we also refer to animals feeding on cultivated foods as 'crop feeding' or 'crop foraging', rather than 'crop raiding' (Hill, 2015, 2018).

vulnerable to baboon and other wildlife damage in Zimbabwe while also having economic value (Parker & Osborn, 2006). Likewise, tea plantations surrounding the Kibale National Park in Uganda are unappealing to wildlife, including primates, but economically valuable (Hartter & Goldman, 2009). Translocation, as a measure to move problematic primates from a particular area, is extremely invasive and requires significant money and effort. It can be lethal to animals due to stress and injuries caused while trying to capture them, particularly when using darts. Searching for and identifying suitable areas for release takes time and effort, and released animals must be monitored for years (e.g. see Palmer, 2018 for a recent detailed treatment of the ethics of rehabilitation and reintroduction in orangutans). In some cases, it may even be detrimental towards landscape-level primate conservation through removing primates in fragments that help maintain connectivity and gene flow between populations (Ancrenaz et al., 2021). Additionally, removing primates from fragmented landscapes may result in negative changes to ecosystem dynamics due to many primates' role as seed dispersers and 'pest' control (e.g. rodents, snakes).

Negative human-primate interactions affect the conservation of a species in a certain geographic area and may also influence public support for conservation programmes, particularly when coupled with a lack of local community involvement in conservation decision-making processes (Sabuhoro et al., 2017). To develop strategies that aim to promote sustainable coexistence in the long term, besides human-wildlife dynamics, we must also fully understand the socio-political nature of conservation-related issues in shared landscapes (Fuentes, 2012). Social, political, and economic power imbalances between stakeholders (including local persons, researchers, policy makers, industry, and community stakeholder groups) underlie conservation conflicts (Temudo, 2012; Leblan, 2016; Hill et al., 2017). A large branch of ecologists now recognise that conservation and human-wildlife coexistence goals can only be achieved by understanding and addressing the socio-political dimension, as well as monitoring the impacts of human-wildlife interactions on human livelihoods and wildlife persistence (Dore et al., 2017; Hill et al., 2017; Pooley et al., 2020). Strategies may necessitate some unusual shifts in focus, for example, studying fish and fishing livelihoods in the context of peatland fires in Indonesia, as a contribution towards conservation of orangutans and other primates (Chua et al., 2020; Thornton et al., 2020). Although cross-disciplinary research on human-primate interactions (i.e. ethnoprimatology; see Fuentes, 2012; Waters et al., 2018) might not directly prevent negative human-wildlife scenarios, it forms an integral part of conservation, for example, by informing risk mitigation schemes

Box 10.1 The Bulindi Chimpanzee and Community Project: Conservation at the Human-Chimpanzee Interface in Western Uganda
In western Uganda, wild chimpanzees occur outside, as well as within, protected areas. The Budongo and Bugoma Forest Reserves support two of Uganda's largest chimpanzee populations, but are separated by 50 km. The intervening landscape is densely settled and dominated by agriculture, exotic timber plantations, villages, and urban centres. Since the 1990s, riverine

(continued)

Box 10.1 (continued)

forest, which formerly provided habitat for chimpanzees and other wildlife, was converted to farmland by landowners. About 300 chimpanzees survive in this fast-developing landscape, however, using remnant forest fragments on private land in remarkably close contact with villagers (McCarthy et al., 2015; McLennan, 2008) (Fig. 10.3). Besides habitat loss, these chimpanzees are threatened by infrastructure development including road upgrades, construction of an oil pipeline, and advancing urbanisation (McLennan et al., 2021). Given these circumstances, their long-term survival might appear doubtful. Why then, should we conserve them?

First, conserving these chimpanzees is necessary to avoid the large populations in Budongo and Bugoma forests from becoming genetically isolated. Second, increasingly negative interactions between the region's chimpanzees and human residents need addressing and mitigating. Forest clearance caused chimpanzees to feed habitually on agricultural crops, resulting in economic losses for farmers and occasional trapping or killing of chimpanzees (McLennan et al., 2012). Additionally, local people, especially children, have been seriously injured by chimpanzees and several human babies have been killed. These negative interactions have been reported in Uganda's press, potentially reducing public support for conservation of the species.

The Bulindi Chimpanzee and Community Project (BCCP) was established in 2015 to address these problems, initially concentrating on one site (Bulindi) where chimpanzees were the focus of long-term research, but where recent deforestation had shrunk local forests by 80% (McLennan et al., 2020). Informal discussions with landowners helped to understand their priorities

Fig. 10.3 For chimpanzees living in human-dominated landscapes outside protected areas, such as in Bulindi, Uganda, encounters with humans can be unpredictable. (**a**) Chimpanzees looking towards the sound of people approaching in the distance. (**b**) Adult males crossing a village road watched by local residents. (Photo credits: Matthew R. McLennan)

(continued)

Box 10.1 (continued)

and constraints. Residents commonly cited their need to raise cash to fund their children's education as a reason for clearing forest for farming or selling timber. When BCCP offered to contribute to school fees in return for an end to forest cutting, most landowners agreed. This voluntary initiative ended major forest clearance in Bulindi. BCCP helped landowners establish a formal community-based organisation, with a constitution governing conditions of membership that included entrusting members with shared responsibility for conserving local forests. After 6 years of this initiative, forest in Bulindi is regenerating. More recently, the programme was expanded to help landowners conserve unprotected forest used by other chimpanzee groups regionally.

Many landowners no longer have forest on their land, yet are still impacted by chimpanzees. Therefore, a suite of integrated programmes was developed to more widely enhance local capacity to accommodate chimpanzees and engage in conservation. Central to this effort is largescale tree planting. BCCP supplies landowners with tree seedlings as an alternative livelihood to reduce reliance on natural forest. Woodlots of fast-growing species offer an alternative (non-forest) source of wood and income from timber sales, while coffee provides a 'chimp-friendly' alternative to tobacco and rice cash-cropping (both major drivers of deforestation) and sugarcane growing. Unlike sugarcane, chimpanzees and other primates are not reported to eat coffee; thus, coffee farming doesn't generate negative human-primate interactions. Indigenous trees are planted to supplement natural forest regeneration. Other initiatives include energy-efficient stoves that reduce fuelwood consumption; water wells (boreholes) that provide clean water away from forest streams, where children risk encountering chimpanzees; education clubs to promote child safety; savings groups to support alternative livelihoods; and a popular 'chimpanzee football league' that sponsors local teams with kits and tournaments. These community-based programmes are combined with research, yielding long-term data on chimpanzee demography, ranging, and behaviour and identifying site-specific threats to help direct conservation efforts. As of 2021, the project reached over 150 villages.

Nevertheless, these interventions offer no quick fix to the complex challenges inherent in conserving wildlife in human-dominated landscapes outside protected areas. Natural forest regionally is unlikely to ever regenerate such that chimpanzees no longer range and forage around villages. Meanwhile, the human population will continue increasing alongside expanding infrastructural development. Human-dominated environments are characterised by diverse priorities and interests of residents and other stakeholders, which create unanticipated challenges. Patience, understanding, and long-term livelihood support and economic opportunities for local residents, alongside careful management of the chimpanzees, will be required for decades to come.

Box 10.2 Primate Conservation in Shared Landscapes in Indonesia

Indonesia is home to an estimated 48 nonhuman primate species, 45 of which are threatened (Estrada et al., 2018), and the world's fourth largest human population. It possesses diverse habitat types, land uses, and peoples, with 83% of primate ranges outside protected areas (PAs) (Estrada et al., 2018), representing a diversity of human-primate interaction contexts and conservation challenges. We illustrate two contrasting Indonesian primate conservation contexts in non-PA, multi-use landscapes.

The world's most populated island, Java, still harbours relatively large Javan slow loris habitats, largely distributed in high-altitude agroecosystems in West Java (Nekaris et al., 2017) and lower-altitude, secondary forest with coffee agroforestry in Central Java (Sodik et al., 2020). In Central Java particularly, slow loris habitat is relatively small within existing PAs and mostly occurs in production forest (Sodik et al., 2020).

The presence of a small slow loris population (~7–9 individuals) in Kemuning lowland secondary forest (400 ha) in densely populated Central Java provides new conservation hope (Sodik et al., 2019). This forest is managed by the state-owned enterprise, PERHUTANI, and local people have access to the forest through *Pengelolaan Hutan Bersama Masyarakat* (PHBM) or social forestry management, for planting shade coffee. However, slow lorises only use parts of this small fragment (Sodik et al., 2019). Due to their high territoriality (Campera et al., 2020; Nekaris et al., 2020), plus the small population in Kemuning, local extinction may be occurring. Promising initiatives include the successful use of **artificial canopy bridges** to connect loris populations in West Java (Birot et al., 2020) and the promotion of wildlife-friendly coffee production (Campera et al., 2021) by local NGO, JAWI, and Universitas Gadjah Mada, supported by Oxford Brookes University. Local people's involvement in these initiatives helps to reduce poaching, which is also prohibited in village regulations.

In Kalimantan (Indonesian Borneo), orangutan conservation efforts have historically focused on PAs and 'undisturbed' forests, yet >75% of orangutans inhabit areas open for development (Wich et al., 2012). Recent orangutan studies have revealed a high tolerance to forest disturbance in the absence of killing, generating calls to focus on integrated management of multi-use landscapes, including (connecting) orangutan populations in concessions, stakeholder engagement, and killing avoidance (Spehar et al., 2018). One area where this is relevant is Rungan Forest (1500 km^2), Central Kalimantan, which supports around 2220–3275 orangutans, plus five other primate species (Buckley et al., 2018; Husson et al., 2019). The forest is bordered by 20 villages, with 22% of it currently protected, 14% allocated for oil palm, and the remainder as pulp and paper concessions (Husson et al., 2019). Borneo Nature Foundation (BNF) and partners are pursuing a multi-stakeholder conservation plan, aiming to safeguard Rungan's orangutan population while enhancing local community wellbeing. This involves landscape-level orangutan

(continued)

Box 10.2 (continued)

population and habitat surveys, establishing a forest research base, supporting local community forest management rights acquisition, engaging concession managers to conserve **High Conservation Value Forest**, and implementing sustainable livelihood initiatives, including **permaculture** (BNF, 2020) (Fig. 10.4). Initial results are encouraging, though further long-term work is required to achieve desired benefits for both orangutans and people (BNF, 2020).

Fig. 10.4 Permaculture development in Rungan Landscape, Central Kalimantan. (Photo credit: Yayasan Borneo Nature Indonesia)

Box 10.3 Inclusive Conservation in Brazil's Atlantic Forest

Brazil has the highest diversity of primate species (Estrada et al., 2018), and, except for some Amazonian primates, most (including many Endangered species) inhabit landscapes strongly influenced, if not dominated, by human activities including areas of heavy agricultural or urban development. The National Primate Action Plans show concern for dealing with the shared landscape issue. But conservation efforts must move beyond; plans must include the community as actors or stewards.

The Golden Lion Tamarin Conservation Programme, a successful Atlantic Forest flagship species project, is a good example of a conservation strategy with community participation (Kierulff et al., 2012; Ruiz-Miranda et al., 2019). Golden lion tamarins live in a landscape of forest fragments within an agricultural and urban matrix, situated between Rio de Janeiro (80 km away) and major oil and gas production areas, and it is the only water source for a

(continued)

Box 10.3 (continued)

major coastal tourist area in the state. From its foundation in 1984, the programme hired local people to work as research assistants and educators and provide field site logistics. Several of those original employees still work in the project, 40 years later. Environmental education was set up to foster knowledge about the golden lion tamarins and support for forest conservation (Dietz & Nagagata, 1995; Dietz, 1998; Engels & Jacobson, 2007; Pádua et al., 2002). A key strategy, the reintroduction of captive born animals, was only possible through the participation of local landowners, with all the release sites (after the initial experimental release) on private land (Kierulff et al., 2012; Ruiz-Miranda et al., 2010). In 1992, Associação Mico Leão Dourado (Golden Lion Tamarin Association, AMLD) was created (Rambaldi et al., 2002; Ruiz-Miranda et al., 2020) as a community-based NGO with local landowners serving as active and/or board members. Other members include the Chico Mendes Institute for Biodiversity Conservation and local government officials. The AMLD adaptive management strategic plan is organised around monitoring the population and reducing threats to golden lion tamarins and their habitat (Dietz et al., 2010; Ruiz-Miranda et al., 2019). Activities such as reforestation and establishment of forest corridors depend on local landowner participation (Fernandes et al., 2008). For reforestation, the AMLD established a programme to build capacity for six landowners to develop commercial nurseries for native tree species to be used in all reforestation efforts. The AMLD also assists the community to develop economic activities that rely on sustainable land use such as agroforestry and **ecotourism**. The reforestation and forest protection efforts of the AMLD and the Ministry of the Environment have benefited the municipal government; the region receives the largest amount of green tax funds in the State of Rio. The AMLD and local community continue to work together to make the region a multi-use conservation landscape that protects biodiversity and fosters economic activities and quality of life.

and conservation management approaches. At both the research and conservation planning stages, cross-disciplinary research – including psychology, economics, anthropology, political sciences, and ecology – is essential to ensure that pragmatic, effective, and inclusive conservation impacts are achieved (Bartuszevige et al., 2016; Waters et al., 2018).

10.5 Conclusion

Achieving sustainable coexistence between humans and wildlife is one of the greatest challenges we face in the Anthropocene (Lewis & Maslin, 2015; Hockings et al., 2015; McLennan et al., 2017; Frank et al., 2019; Bersacola et al., 2021).

Human-wildlife dynamics are influenced by direct interactions, including competition over space and resources, as well as the socio-political, economic, and environmental contexts. Primate conservation strategies in anthropogenic environments must be based on cross-disciplinary research approaches that are able to resolve these multiple, complex socioecological dimensions. Crucially, as demonstrated in the three case studies presented, conservation practitioners must work directly with local people and ensure equity in decision-making and long-term collaboration amongst stakeholders in all phases including in research and planning and within primate conservation strategies' adaptive frameworks.

Acknowledgements   KJH's work was funded by the Darwin Initiative (project 26-018). MEH thanks the Indonesian Ministry of Research and Technology (RISTEK-BRIN) for research permissions and Universitas Palangka Raya and Universitas Muhammadiyah Palangkaraya for support. BNF's work in Rungan was funded by the Arcus Foundation, Danish Civil Society in Development, Rainforest Trust, The Orangutan Project, US Fish and Wildlife Service Great Apes Conservation Fund, and Prince Bernhard Nature Fund. MRM thanks Born Free Foundation, Friends of Chimps, Jane Goodall Institute-Switzerland, Blair Drummond Safari Park, the Primate Society of Great Britain, and Uganda Biodiversity Fund for supporting the Bulindi Chimpanzee and Community Project.

References

Adams, W. M., Aveling, R., Brockington, D., Dickson, B., Elliott, J., Hutton, J., Roe, D., Vira, B., & Wolmer, W. (2004). Biodiversity conservation and the eradication of poverty. *Science, 306*(5699), 1146–1149.

Albert, C., Luque, G. M., & Courchamp, F. (2018). The twenty most charismatic species. *PLoS One, 13*(7), e0199149. https://doi.org/10.1371/journal.pone.0199149

Ancrenaz, M., Oram, F., Nardiyono, N., et al. (2021). Importance of small forest fragments in agricultural landscapes for maintaining orangutan metapopulations. *Frontiers in Forests and Global Change, 4*, 560944. https://doi.org/10.3389/ffgc.2021.560944

Arponen, A. (2012). Prioritizing species for conservation planning. *Biodiversity and Conservation, 21*(4), 875–893.

Bartuszevige, A. M., Taylor, K., Daniels, A., & Carter, M. F. (2016). Landscape design: Integrating ecological, social, and economic considerations into conservation planning. *Wildlife Society Bulletin, 40*(3), 411–422.

Bersacola, E., Bessa, J., Frazão-Moreira, A., et al. (2018). Primate occurrence across a human-impacted landscape in Guinea-Bissau and neighbouring regions in West Africa: Using a systematic literature review to highlight the next conservation steps. *PeerJ, 6*, e4847. https://doi.org/10.7717/peerj.4847

Bersacola, E., Parathian, H., Frazão-Moreira, A., et al. (2021). Developing an evidence-based coexistence strategy to promote human and wildlife health in a biodiverse agroforest landscape. *Frontiers in Conservation Science, 2*, 735367. https://doi.org/10.3389/fcosc.2021.735367

Birot, H., Campera, M., Imron, M. A., & Nekaris, K. A. I. (2020). Artificial canopy bridges improve connectivity in fragmented landscapes: The case of Javan slow lorises in an agroforest environment. *American Journal of Primatology, 82*(4), e23076. https://doi.org/10.1002/ajp.23076

BNF. (2020). *Initiating biodiversity conservation in the Rungan River Landscape, Central Kalimantan, Indonesia.* Annual report 2019. Borneo Nature Foundation, Palangka Raya, Indonesia.

Bond, W. J. (1994). Keystone species. In E.-D. Schulze & H. A. Mooney (Eds.), *Biodiversity and ecosystem function* (pp. 237–253). Springer.

Brockington, D., & Igoe, J. (2006). Eviction for conservation: A global overview. *Conservation and Society, 4*(3), 424.

Buckley, B. J. W., Ripoll Capilla, B., et al. (2018). *Biodiversity, forest structure & conservation importance of the Mungku Baru Education Forest, Rungan, Central Kalimantan, Indonesia.* Borneo Nature Foundation in collaboration with Universitas Muhammadiyah Palangkaraya and University of Exeter, Palangka Raya, Indonesia.

Burivalova, Z., Game, E. T., Wahyudi, B., et al. (2020). Does biodiversity benefit when the logging stops? An analysis of conservation risks and opportunities in active versus inactive logging concessions in Borneo. *Biological Conservation, 241*, 108369. https://doi.org/10.1016/j.biocon.2019.108369

Byamukama, J., & Asuma, S. (2006). Human-gorilla conflict resolution (HuGo) – The Uganda experience. *Gorilla Journal, 32*, 10–12.

Campbell-Smith, G., Simanjorang, H. V. P., Leader-Williams, N., & Linkie, M. (2010). Local attitudes and perceptions toward crop-raiding by orangutans (Pongo abelii) and other nonhuman primates in northern Sumatra, Indonesia. *American Journal of Primatology, 72*(10), 866–876.

Campera, M., Brown, E., Imron, M. A., & Nekaris, K. A. I. (2020). Unmonitored releases of small animals? The importance of considering natural dispersal, health, and human habituation when releasing a territorial mammal threatened by wildlife trade. *Biological Conservation, 242*, 108404.

Campera, M., Budiadi, B., Adinda, E., et al. (2021). Fostering a wildlife-friendly program for sustainable coffee farming: The case of small-holder farmers in Indonesia. *Land, 10*(2), 1–16.

Chazdon, R. L., Cullen, L., Padua, S. M., & Padua, C. V. (2020). People, primates and predators in the Pontal: From endangered species conservation to forest and landscape restoration in Brazil's Atlantic Forest. *Royal Society Open Science, 7*(12), 200939. https://doi.org/10.1098/rsos.200939

Chesney, C., Bolaños, N. C., Kanneh, B. A., Sillah, E., et al. (2020). Mobonda Community Conservation Project: Chimpanzees, oysters, and community engagement in Sierra Leone. *American Journal of Primatology, 83*(1), e23219. https://doi.org/10.1002/ajp.23219

Chua, L., Harrison, M. E., Fair, H., et al. (2020). Conservation and the social sciences: Beyond critique and co-optation. A case study from orangutan conservation. *People and Nature, 2*(1), 42–60.

Cobbinah, P. B., Amenuvor, D., Black, R., & Peprah, C. (2017). Ecotourism in the Kakum Conservation Area, Ghana: Local politics, practice and outcome. *Journal of Outdoor Recreation and Tourism, 20*, 34–44.

Colchester, M. (2004). Conservation policy and indigenous peoples. *Environmental Science & Policy, 7*(3), 145–153.

Cormier, L. (2006). A preliminary review of neotropical primates in the subsistence and symbolism of indigenous lowland South American peoples. *Ecological and Environmental Anthropology, 2*, 14–32.

Díaz, S., Settele, J., Brondízio, E. S., Ngo, H. T., et al. (2019). Pervasive human-driven decline of life on Earth points to the need for transformative change. *Science, 366*(6471), eaax3100. https://doi.org/10.1126/science.aax3100

Dietz, L. A. (1998). Community conservation education programme for the Golden Lion Tamarin in Brazil: Building support for habitat conservation. In Hoarge, R. J. and Moran, K. (Ed.), *Culture: The missing element in conservation and development* (pp. 85–94). Washington, DC: Kendal/Hunt Publishing Company and National Zoological Park, Smithsonian Institution.

Dietz, L., & Nagagata, E. Y. (1995). Golden lion tamarin conservation program: A community educational effort for forest conservation in Rio de Janeiro State, Brazil. In S. K. Jacobson (Ed.), *Conserving wildlife: International education and communication approaches* (pp. 64–86). Columbia University Press.

Dietz, L. A., Brown, M., & Swaminathan, V. (2010). Increasing the impact of conservation projects. *American Journal of Primatology, 72*(5), 425–440.

Dore, K. M., Riley, E. P., & Fuentes, A. (2017). *Ethnoprimatology*. Cambridge University Press.

Ellis, E. C., & Ramankutty, N. (2008). Putting people in the map: Anthropogenic biomes of the world. *Frontiers in Ecology and the Environment, 6*(8), 439–447.

Engels, C. A., & Jacobson, S. K. (2007). Evaluating long-term effects of the golden lion tamarin environmental education program in Brazil. *The Journal of Environmental Education, 38*(3), 3–14.

Estrada, A., Raboy, B. E., & Oliveira, L. C. (2012). Agroecosystems and primate conservation in the tropics: A review. *American Journal of Primatology, 74*(8), 696–711.

Estrada, A., Garber, P. A., Rylands, A. B., et al. (2017). Impending extinction crisis of the world's primates: Why primates matter. *Science Advances, 3*(1), e1600946. https://doi.org/10.1126/sciadv.1600946

Estrada, A., Garber, P. A., Mittermeier, R. A., et al. (2018). Primates in peril: The significance of Brazil, Madagascar, Indonesia and the Democratic Republic of the Congo for global primate conservation. *PeerJ, 6*, e4869. https://doi.org/10.7717/peerj.4869

Ezebilo, E. E., & Mattsson, L. (2010). Socio-economic benefits of protected areas as perceived by local people around Cross River National Park, Nigeria. *Forest Policy and Economics, 12*(3), 189–193.

Fernandes, R. V., Rambaldi, D. M., & de Godoy Teixeira, A. M. (2008). Restauração e proteção legal da paisagem – Corredores florestais e RPPNs. In P. P. de Oliveira, A. D. Gravitol, & C. R. Miranda (Eds.), *Conservação do mico-leão-dourado* (pp. 160–179). Biblioteca do Centro de Biociências e Biotecnologia da Universidade Estadual do Norte Fluminense.

Frank, B., Glikman, J. A., & Marchini, S. (2019). *Human–wildlife interactions: Turning conflict into coexistence*. Cambridge University Press.

Fuentes, A. (2012). Ethnoprimatology and the anthropology of the human-primate interface. *Annual Review of Anthropology, 41*(1), 101–117.

Fuentes, A., & Gamerl, S. (2005). Disproportionate participation by age/sex classes in aggressive interactions between long-tailed macaques (*Macaca fascicularis*) and human tourists at Padangtegal monkey forest, Bali, Indonesia. *American Journal of Primatology, 66*(2), 197–204.

Galán-Acedo, C., Arroyo-Rodríguez, V., Andresen, E., & Arasa-Gisbert, R. (2019a). Ecological traits of the world's primates. *Scientific Data, 6*(1), 1–5. https://doi.org/10.1038/s41597-019-0059-9

Galán-Acedo, C., Arroyo-Rodríguez, V., Andresen, E., et al. (2019b). The conservation value of human-modified landscapes for the world's primates. *Nature Communications, 10*(1), 1–8. https://doi.org/10.1038/s41467-018-08139-0

Harrison, M. E., Ottay, J. B., D'Arcy, L. J., et al. (2020). Tropical forest and peatland conservation in Indonesia: Challenges and directions. *People and Nature, 2*(1), 4–28.

Hartter, J., & Goldman, A. C. (2009). Life on the edge: Balancing biodiversity, conservation, and sustaining rural livelihoods around Kibale National Park, Uganda. *Focus on Geography, 52*(1), 11–17.

Hickel, J. (2017). *The divide: A brief guide to global inequality and its solutions*. Random House.

Hill, C. M. (2015). Perspectives of 'conflict' at the wildlife–agriculture boundary: 10 years on. *Human Dimensions of Wildlife, 20*(4), 296–301.

Hill, C. M. (2017). Primate crop feeding behavior, crop protection, and conservation. *International Journal of Primatology, 38*(2), 385–400.

Hill, C. M. (2018). Crop foraging, crop losses, and crop raiding. *Annual Review of Anthropology, 47*(1), 377–394.

Hill, C. M., & Wallace, G. E. (2012). Crop protection and conflict mitigation: Reducing the costs of living alongside non-human primates. *Biodiversity and Conservation, 21*(10), 2569–2587.

Hill, C. M., Webber, A. D., & Priston, N. E. C. (2017). *Understanding conflicts about wildlife: A biosocial approach*. Berghahn Books.

Hockings, K. J., & Humle, T. (2009). *Best practice guidelines for the prevention and mitigation of conflict between humans and great apes.* IUCN.

Hockings, K. J., & McLennan, M. R. (2016). Problematic primate behaviour in agricultural landscapes: Chimpanzees as 'pests' and 'predators'. In M. T. Waller (Ed.), *Ethnoprimatology: Primate conservation in the 21st century* (pp. 137–156). Springer International Publishing.

Hockings, K. J., Anderson, J. R., & Matsuzawa, T. (2009). Use of wild and cultivated foods by chimpanzees at Bossou, Republic of Guinea: Feeding dynamics in a human-influenced environment. *American Journal of Primatology, 71*(8), 636–646.

Hockings, K. J., Yamakoshi, G., Kabasawa, A., & Matsuzawa, T. (2010). Attacks on local persons by chimpanzees in Bossou, Republic of Guinea: Long-term perspectives. *American Journal of Primatology, 72*(10), 887–896.

Hockings, K. J., McLennan, M. R., Carvalho, S., et al. (2015). Apes in the Anthropocene: Flexibility and survival. *Trends in Ecology & Evolution, 30*(4), 215–222. https://doi.org/10.1016/j.tree.2015.02.002

Humle, T., & Hill, C. M. (2016). People–primate interactions: Implications for primate conservation. In S. A. Wich & A. J. Marshall (Eds.), *An introduction to primate conservation* (pp. 219–240). Oxford University Press.

Husson, S. J., Ripoll Capilla, B., Ottay, J. B., & Buckley, B. J. W. (2019). *Initiating orangutan conservation in the Rungan Landscape, Central Kalimantan, Indonesia.* 2019 Annual report. Borneo Nature Foundation, Palangka Raya, Indonesia.

IUCN. (2020). *The IUCN red list of threatened species.* IUCN. https://www.iucnredlist.org/en. Accessed 3 Dec 2020.

Jones-Engel, L., Engel, G. A., Schillaci, M. A., et al. (2005). Primate-to-human retroviral transmission in Asia. *Emerging Infectious Diseases, 11*(7), 1028–1035.

Kalbitzer, U., & Chapman, C. A. (2018). Primate responses to changing environments in the Anthropocene. In U. Kalbitzer & K. M. Jack (Eds.), *Primate life histories, sex roles, and adaptability: Essays in honour of Linda M. Fedigan* (pp. 283–310). Springer International Publishing.

Kennedy, C. M., Oakleaf, J. R., Theobald, D. M., et al. (2020). *Global human modification of terrestrial systems.* NASA Socioeconomic Data and Applications Center (SEDAC).

Kibaja, M. (2014). Diet of the ashy red colobus (*Piliocolobus tephrosceles*) and crop-raiding in a forest-farm mosaic, Mbuzi, Rukwa Region, Tanzania. *Primate Conservation, 2014*(28), 109–116.

Kierulff, M. C. M., Ruiz-Miranda, C. R., Oliveira, P. P., et al. (2012). The golden lion tamarin *Leontopithecus rosalia*: A conservation success story. *International Zoo Yearbook, 46*(1), 36–45.

Kifle, Z., & Bekele, A. (2020). Human–gelada conflict and attitude of the local community toward the conservation of the southern gelada (*Theropithecus gelada obscurus*) around Borena Sayint National Park, Ethiopia. *Environmental Management, 65*(3), 399–409.

Lambert, J. E. (2011). Primate seed dispersers as umbrella species: A case study from Kibale National Park, Uganda, with implications for Afrotropical forest conservation. *American Journal of Primatology, 73*(1), 9–24.

Lawton, J. H., & Gaston, K. J. (2001). Indicator species. In S. A. Levin (Ed.), *Encyclopedia of biodiversity* (pp. 437–450). Elsevier.

Leblan, V. (2016). Territorial and land-use rights perspectives on human-chimpanzee-elephant coexistence in West Africa (Guinea, Guinea-Bissau, Senegal, nineteenth to twenty-first centuries). *Primates, 57*(3), 359–366.

Lewis, S. L., & Maslin, M. A. (2015). Defining the Anthropocene. *Nature, 519*(7542), 171–180.

Linero, D., Cuervo-Robayo, A. P., & Etter, A. (2020). Assessing the future conservation potential of the Amazon and Andes Protected Areas: Using the woolly monkey (*Lagothrix lagothricha*) as an umbrella species. *Journal for Nature Conservation, 58*, 125926. https://doi.org/10.1016/j.jnc.2020.125926

Long, Y., & Richardson, M. (2020). Rhinopithecus roxellana. In *IUCN red list threatened species*. IUCN. https://www.iucnredlist.org/en. Accessed 3 Dec 2020.

McCarthy, M. S., Lester, J. D., Howe, E. J., Arandjelovic, M., Stanford, C. B., & Vigilant, L. (2015). Genetic censusing identifies an unexpectedly sizable population of an endangered large mammal in a fragmented forest landscape. *BMC Ecology, 15*(1), 21. https://doi.org/10.1186/s12898-015-0052-x

McConkey, K. R. (2018). Seed dispersal by primates in Asian habitats: From species, to communities, to conservation. *International Journal of Primatology, 39*(3), 466–492.

McKinney, T. (2015). A classification system for describing anthropogenic influence on nonhuman primate populations. *American Journal of Primatology, 77*(7), 715–726.

McLennan, M. R. (2008). Beleaguered chimpanzees in the agricultural district of Hoima, Western Uganda. *Primate Conservation, 23*(1), 45–54.

McLennan, M. R., & Hockings, K. J. (2016). The aggressive apes? Causes and contexts of great ape attacks on local persons. In F. M. Angelici (Ed.), *Problematic wildlife* (pp. 373–394). Springer.

McLennan, M. R., Hyeroba, D., Asiimwe, C., et al. (2012). Chimpanzees in mantraps: Lethal crop protection and conservation in Uganda. *Oryx, 46*(4), 598–603.

McLennan, M. R., Spagnoletti, N., & Hockings, K. J. (2017). The implications of primate behavioral flexibility for sustainable human–primate coexistence in anthropogenic habitats. *International Journal of Primatology, 38*(2), 105–121.

McLennan, M. R., Lorenti, G. A., Sabiiti, T., & Bardi, M. (2020). Forest fragments become farmland: Dietary response of wild chimpanzees (Pan troglodytes) to fast-changing anthropogenic landscapes. *American Journal of Primatology, 82*(4), e23090. https://doi.org/10.1002/ajp.23090

McLennan, M. R., Hintz, B., Kiiza, V., et al. (2021). Surviving at the extreme: Chimpanzee ranging is not restricted in a deforested human-dominated landscape in Uganda. *African Journal of Ecology, 59*(1), 17–28. https://doi.org/10.1111/aje.12803

Meijaard, E., Buchori, D., Hadiprakarsa, Y., et al. (2011). Quantifying killing of orangutans and human-orangutan conflict in Kalimantan, Indonesia. *PLoS One, 6*(11), e27491. https://doi.org/10.1371/journal.pone.0027491

Mittermeier, R. A., Turner, W. R., Larsen, F. W., et al. (2011). Global biodiversity conservation: The critical role of hotspots. In F. E. Zachos & J. C. Habel (Eds.), *Biodiversity hotspots: Distribution and protection of conservation priority areas* (pp. 3–22). Springer.

Mkumbukwa, A. R. (2008). The evolution of wildlife conservation policies in Tanzania during the colonial and post-independence periods. *Development Southern Africa, 25*(5), 589–600. https://doi.org/10.1080/03768350802447875

Mormile, J. E., & Hill, C. M. (2017). Living with urban baboons: Exploring attitudes and their implications for local baboon conservation and management in Knysna, South Africa. *Human Dimensions of Wildlife, 22*(2), 99–109.

Muldoon, K. M., & Goodman, S. M. (2015). Primates as predictors of mammal community diversity in the forest ecosystems of Madagascar. *PLoS One, 10*(9), e0136787. https://doi.org/10.1371/journal.pone.0136787

Nater, A., Mattle-Greminger, M. P., Nurcahyo, A., et al. (2017). Morphometric, behavioral, and genomic evidence for a new orangutan species. *Current Biology, 27*(22), 3487–3498.e10. https://doi.org/10.1016/j.cub.2017.09.047

@NatGeoUK. (2019). *'I am scared all the time': Chimps and people are clashing in rural Uganda*. National Geographic. https://www.nationalgeographic.co.uk/animals/2019/11/i-am-scared-all-time-chimps-and-people-are-clashing-rural-uganda. Accessed 2 Feb 2021.

Nekaris, K. A. I., Poindexter, S., Reinhardt, K. D., et al. (2017). Coexistence between Javan slow lorises (*Nycticebus javanicus*) and humans in a dynamic agroforestry landscape in West Java, Indonesia. *International Journal of Primatology, 38*(2), 303–320.

Nekaris, K. A. I., Campera, M., Nijman, V., et al. (2020). Slow lorises use venom as a weapon in intraspecific competition. *Current Biology, 30*(20), R1252–R1253. https://doi.org/10.1016/j. cub.2020.08.084

Norris, K., Terry, A., Hansford, J. P., & Turvey, S. T. (2020). Biodiversity conservation and the Earth system: Mind the gap. *Trends in Ecology & Evolution, 35*(10), 919–926. https://doi. org/10.1016/j.tree.2020.06.010

Osborn, F. V., & Hill, C. M. (2005). Techniques to reduce crop loss: Human and technical dimensions in Africa. In R. Woodroffe, S. Thirgood, & A. Rabinowitz (Eds.), *People and wildlife, conflict or co-existence?* (pp. 72–85). Cambridge University Press.

Pádua, S., Dietz, L. A., Souza, M. G., & Santos, G. R. (2002). In situ conservation education and the lion tamarins. In D. G. Kleiman & A. B. Rylands (Eds.), *Lion tamarins: Biology and conservation* (pp. 315–335). Smithsonian Institution Press.

Palmer, A. (2018). Kill, incarcerate, or liberate? Ethics and alternatives to orangutan rehabilitation. *Biological Conservation, 227*, 181–188.

Parathian, H. E., McLennan, M. R., Hill, C. M., Frazão-Moreira, A., & Hockings, K. J. (2018). Breaking through disciplinary barriers: Human–wildlife interactions and multispecies ethnography. *International Journal of Primatology, 39*, 749–775.

Parker, G. E., & Osborn, F. V. (2006). Investigating the potential for chilli Capsicum spp. to reduce human-wildlife conflict in Zimbabwe. *Oryx, 40*(3), 343–346.

Pedersen, A. B., & Davies, T. J. (2009). Cross-species pathogen transmission and disease emergence in primates. *EcoHealth, 6*(4), 496–508.

Pooley, S., Bhatia, S., & Vasava, A. (2020). Rethinking the study of human–wildlife coexistence. *Conservation Biology, 35*, 784–793.

Priston, N. E. C., & McLennan, M. R. (2013). Managing humans, managing macaques: Human–macaque conflict in Asia and Africa. In S. Radhakrishna, M. A. Huffman, & A. Sinha (Eds.), *The macaque connection: Cooperation and conflict between humans and macaques* (pp. 225–250). Springer.

Rambaldi, D. M., Kleiman, D. G., Mallinson, J. J. C., et al. (2002). The role of nongovernmental organizations and the International Committee for the Conservation and Management of *Leontopithecus* in lion tamarin conservation. In D. G. Kleiman & A. B. Rylands (Eds.), *Lion tamarins: Biology and conservation* (pp. 71–94). Smithsonian Institution Press.

Reed, M. S. (2008). Stakeholder participation for environmental management: A literature review. *Biological Conservation, 141*(10), 2417–2431.

Riley, E. P., & Priston, N. E. C. (2010). Macaques in farms and folklore: Exploring the human–nonhuman primate interface in Sulawesi, Indonesia. *American Journal of Primatology, 72*(10), 848–854.

Rubis, J. M., & Theriault, N. (2020). Concealing protocols: Conservation, Indigenous survivance, and the dilemmas of visibility. *Social and Cultural Geography, 21*(7), 962–984.

Ruiz-Miranda, C. R., Beck, B. B., Kleiman, D. G., et al. (2010). Re-introduction and translocation of golden lion tamarins, Atlantic Coastal Forest, Brazil: The creation of a metapopulation. In *Global re-introduction perspectives: Additional case-studies from around the globe* (pp. 225–230). IUCN/SSC Re-introduction Specialist Group.

Ruiz-Miranda, C. R., de Morai, M. M., Jr., Dietz, L. A., et al. (2019). Estimating population sizes to evaluate progress in conservation of endangered golden lion tamarins (*Leontopithecus rosalia*). *PLoS One, 14*(6), e0216664. https://doi.org/10.1371/journal.pone.0216664

Ruiz-Miranda, C. R., Vilchis, L. I., & Swaisgood, R. R. (2020). Exit strategies for wildlife conservation: Why they are rare and why every institution needs one. *Frontiers in Ecology and the Environment, 18*(4), 203–210. https://doi.org/10.1002/fee.2163

Sabuhoro, E., Wright, B., Munanura, I. E., et al. (2017). The potential of ecotourism opportunities to generate support for mountain gorilla conservation among local communities neighbouring Volcanoes National Park in Rwanda. *Journal of Ecotourism, 20*(1), 1–17. https://doi.org/1 0.1080/14724049.2017.1280043

Simberloff, D. (1998). Flagships, umbrellas, and keystones: Is single-species management passé in the landscape era? *Biological Conservation, 83*(3), 247–257.

Sodik, M., Pudyatmoko, S., Yuwono, P. S. H., & Imron, M. A. (2019). Resource selection by Javan slow loris Nycticebus javanicus E. Geoffroy, 1812 (Mammalia: Primates: Lorisidae) in a lowland fragmented forest in Central Java, Indonesia. *Journal of Threatened Taxa, 11*(6), 13667–13679.

Sodik, M., Pudyatmoko, S., Yuwono, P. S. H., et al. (2020). Better providers of habitat for Javan slow loris (Nycticebus javanicus E. Geoffroy 1812): A species distribution modeling approach in Central Java, Indonesia. *Biodiversitas Journal of Biological Diversity, 21*(5), 1890–1900. https://doi.org/10.13057/biodiv/d210515

Spehar, S. N., & Rayadin, Y. (2017). Habitat use of Bornean orangutans (*Pongo pygmaeus morio*) in an industrial forestry plantation in East Kalimantan, Indonesia. *International Journal of Primatology, 38*(2), 358–384.

Spehar, S. N., Sheil, D., Harrison, T., et al. (2018). Orangutans venture out of the rainforest and into the Anthropocene. *Science Advances, 4*(6), e1701422.

Suzuki, K., & Muroyama, Y. (2010). Topic 5: Resolution of human–macaque conflicts: Changing from top-down to community-based damage management. In N. Nakagawa, M. Nakamichi, & H. Sugiura (Eds.), *The Japanese macaques* (pp. 359–373). Springer Japan.

Temudo, M. P. (2012). 'The white men bought the forests': Conservation and contestation in Guinea-Bissau, Western Africa. *Conservation & Society, 10*(4), 354.

Thornton, S. A., Setiana, E., Yoyo, K., et al. (2020). Towards biocultural approaches to peatland conservation: The case for fish and livelihoods in Indonesia. *Environmental Science & Policy, 114*, 341–351.

Torres-Romero, E. J., & Olalla-Tárraga, M. Á. (2015). Untangling human and environmental effects on geographical gradients of mammal species richness: A global and regional evaluation. *The Journal of Animal Ecology, 84*(3), 851–860.

United Nations. (2021). *THE 17 GOALS | Sustainable Development*. https://sdgs.un.org/goals. Accessed 25 Feb 2021.

Wallis, J., Benrabah, M., Pilot, M., et al. (2020). Macaca sylvanus. In *The IUCN red list of threatened species*. IUCN. https://www.iucnredlist.org/en. Accessed 3 Dec 2020.

Waters, S., Bell, S., & Setchell, J. M. (2018). Understanding human-animal relations in the context of primate conservation: A multispecies ethnographic approach in North Morocco. *Folia Primatologica, 89*(1), 13–29.

Webber, A. D., Hill, C. M., & Reynolds, V. (2007). Assessing the failure of a community-based human-wildlife conflict mitigation project in Budongo Forest Reserve, Uganda. *Oryx, 41*(2), 177–184.

Wich, S. A., Gaveau, D., Abram, N., et al. (2012). Understanding the impacts of land-use policies on a threatened species: Is there a future for the Bornean orangutan? *PLoS One, 7*(11), e49142. https://doi.org/10.1371/journal.pone.0049142

World Bank. (2021). *Guinea-Bissau | Data*. https://data.worldbank.org/country/GW. Accessed 25 Feb 2021.

Chapter 11
Primate Tourism

Malene Friis Hansen, Stefano S. K. Kaburu, Kristen S. Morrow, and Laëtitia Maréchal

Contents

Abstract Primate tourism, where people travel and see non-human primates, is a rapidly growing activity. This chapter introduces the history and the multidimensions of primate tourism across the world. We then focus on tourism associated with wild primate viewing and assess the costs and benefits of primate tourism related to

Malene Friis Hansen and Stefano S. K. Kaburu contributed equally as joint first authors.

M. F. Hansen (✉)
Behavioural Ecology Group, Department of Biology, University of Copenhagen, Copenhagen, Denmark

The Long-Tailed Macaque Project, Broerup, Denmark

Department of Social Sciences, Oxford Brookes University, Oxford, UK

S. S. K. Kaburu
Department of Biomedical Science and Physiology, Faculty of Science and Engineering, University of Wolverhampton, Wolverhampton, UK

K. S. Morrow
Department of Anthropology, University of Georgia, Athens, GA, USA

L. Maréchal
University of Lincoln, Lincoln, UK

© The Author(s), under exclusive license to Springer Nature Switzerland AG 2023
T. McKinney et al. (eds.), *Primates in Anthropogenic Landscapes*, Developments in Primatology: Progress and Prospects, https://doi.org/10.1007/978-3-031-11736-7_11

habitat protection, revenue generation, co-existence with local communities, knowledge sharing, provisioning, health and habituation. Following this assessment, we explore the different drivers for human-primate interactions associated with primate tourism. This chapter concludes by summarising responsible primate tourism guidelines.

Keywords Health · Human-primate interactions · Management · Provisioning · Viewing primates

11.1 Primate Tourism: Definitions and History

Primate tourism is broadly defined as people travelling and viewing nonhuman primates (hereafter primates). This definition includes the multifaceted dimensions of primate tourism from captive (e.g. zoos) to semi-free ranging and wild populations, from consumptive to non-consumptive activities, from highly **anthropogenic** to less anthropogenic habitats, from incidental to targeted tourism (business, research, conservation) and from strictly managed to unmanaged settings. Primate tourism can be of great benefit to humans and primates, yet it can also incur severe costs.

Primate tourism includes hunting of particular primates and the viewing of primates. Primate hunting tourism preceded observation-based primate tourism (Russon & Susilo, 2014) and still exists today. For example, between 2015 and 2020, different African countries exported approximately 4800 hunting trophies from chacma baboons (*Papio ursinus*) (CITES Trade Database, 2020). However, primate viewing now far exceeds primate hunting as a tourist activity (Russon & Wallis, 2014). In this chapter, we will focus on primate viewing outside captive facilities. We begin with a brief overview of the history of primate viewing and then touch upon the benefits and costs of primate tourism as well as the human and primate drivers of **human-primate interactions**, before ending with suggestions for responsible primate tourism guidelines.

11.1.1 The History of Viewing Primates

Research, cultural reasons and conservation purposes are the basis of many primate tourist sites, but a growing tourism industry has made many of them hotspots for primate tourism. For example, in the 1950s, researchers used provisioning to habituate Japanese macaque (*Macaca fuscata*) groups for research purposes with these groups becoming the subject of tourist attention with the creation of the so-called monkey parks (Asquith, 1989; Knight, 2010; Kurita, 2014). Other primate tourism projects were initiated for cultural and/or religious reasons (Fuentes & Gamerl, 2005; Fuentes et al., 2007). In Bali (Indonesia), for example, Hindu temple staff began provisioning resident long-tailed macaques (*M. fascicularis*) in the 1980s, followed by a sharp increase of **provisioning** and tourism in the 1990s (Fuentes et al., 2007; Wheatley, 1999). In contrast, in Gibraltar in the 1940s, it was the British military who initiated provisioning of introduced Barbary macaques (*M. sylvanus*), which in the 1980s–1990s led to an increase in tourism and provisioning (Fuentes et al., 2007). Although provisioning was, initially, the preferred method of rapidly habituating primates to human presence, nowadays provisioning is discouraged and prohibited in many areas because of its negative effects on primates and their habitats.

Other primate tourism projects began for conservation purposes and through the establishment of protected areas (Russon & Wallis, 2014). For instance, orangutan (*Pongo* sp.) tourism, began in the 1970s in both Sumatra and Borneo with rehabilitant orangutans (Russon & Susilo, 2014), while in the late 1970s, it began in the Virunga National Park in order to protect mountain gorillas (*Gorilla beringei beringei*) from poaching and habitat destruction (Kalpers et al., 2003). National Parks where tourists could watch primates were also established in Costa Rica in the 1970s (Boza, 1993; Kauffman, 2014) and in Madagascar in the 1990s (Wright et al., 2014).

11.1.2 Different Types of Primate Tourism Management

Many primate tourism activities strive to become **ecotourism**. The International Union for Conservation of Nature (IUCN) defines ecotourism as "the environmentally responsible travel to natural areas, in order to enjoy and appreciate nature (and accompanying cultural features, both past and present) that promote conservation, have a low visitor impact and provide for beneficially active socio-economic involvement of local peoples" (Ceballos-Lascurain, 1996, 20). In other words, to be considered as ecotourism, primate tourism must be ecologically sustainable, low impact, culturally sensitive, learning-oriented and community supporting. However, the great complexity surrounding primate tourism, associated with differences in geographical areas, local cultures, the species in question and the management authority means that we need different approaches to manage the activities, the tourists and the primates, and as such, ecotourism may not always be initially achievable.

Primate tourism can be classified based on the degree of management: strictly managed, loosely managed or unmanaged. The management authority can be either governmental (e.g. Baluran National Park, East Java, Indonesia: Hansen et al., 2019) or community-based and/or private (e.g. The Community Baboon Sanctuary in Belize: Alexander, 2000). Management possibilities will vary greatly with respect to the primate habitat in which the tourism activity occurs and the species the activity focuses on.

Strictly managed primate tourism settings have consistent, strict regulations and a high degree of active management. We often find this type of tourism for great apes, where the focus lies on managing tourist behaviour to mitigate potential negative effects on apes such as **pathogen transmission**. In these projects, the number of tourists and the minimum distance allowed near an ape group as well as the time spent in proximity to a group are restricted (Williamson & Macfie, 2014). Guides are present at all times to enforce restrictions, and some groups are protected by guards. Strictly managed projects can also include some forms of **incidental primate tourism**, when tourists venture into areas with primates without the specific purpose of viewing them (Grossberg et al., 2003). This occurs, for example, when tourists visit large National Parks to watch other wildlife. Some primate tourism projects focus on **strict management** of both humans and primates. For instance, the government in Singapore manages their long-tailed macaque population in recreational areas by **culling** and other forms of population control (Sha et al., 2009), while human behaviours towards primates are strictly regulated and enforced with fines (Fuentes et al., 2008).

Loosely managed settings have some regulations and management, but they often lack consistency. Most of these projects mainly focus on managing the primates. One management initiative relies on provisioning with the goal of enticing the primates closer (Knight, 2009), to keep them in certain areas or to deter them from crop foraging (Fuentes et al., 2007). Other initiatives include translocation, culling and other forms of population control, such as **sterilisation** (Priston & McLennan, 2013). These projects may often be reactive when problems arise. For example, management strategies tend to be implemented when primates are observed approaching tourists aggressively (Priston & McLennan, 2013). Loosely managed tourism activities may also have rules and regulations in place regarding tourist behaviours, yet they lack enforcement (Hansen et al., 2019).

Finally, unmanaged projects have no management of either the tourists or the primates. This may be incidental or intentional. Incidental forms can occur, for example, in urban areas or on roads (e.g. Morrow et al., 2019; Sengupta & Radhakrishna, 2020), whereas intentional unmanaged activities can happen in cultural or religious sites, where visitors purposely visit revered primates (Medhi et al., 2007).

11.2   Benefits and Costs of Primate Tourism

With more than 60% of primate species currently threatened with extinction (Estrada et al., 2017), the growing primate tourism industry can play a crucial role for primate conservation (Russon & Wallis, 2014). The conservation value of primate tourism, however, can only be assessed if there is a thorough investigation of the benefits and costs that both primates and people experience as a result of visitors' presence at primate sites (Russon & Wallis, 2014). Accordingly, the last three decades have seen an increase in studies examining the impact of tourists on primates and vice versa, especially with the emergence of the field of **ethnoprimatology** (Fuentes & Hockings, 2010; Lee, 2010). It is worth noting, however, that benefits and costs vary in relation to the level of analysis (e.g. depending on whether the analysis is conducted at individual, population, species or ecosystem levels) and/or the duration of the effect examined (short-term vs long-term effects), and all these aspects need to be taken into consideration when performing a cost-benefit analysis. In this section, we will provide an overview of some of the main benefits and costs of primate tourism for both primates and humans.

11.2.1 Habitat Protection

One of the most effective strategies to mitigate primate habitat loss and protect primate populations is the establishment of protected areas (PAs) (Bruner et al., 2001). Accordingly, since the late 1960s and early 1970s, a growing number of primate habitats have been declared PAs. The case of the mountain gorilla of Virunga National Park offers perhaps the most striking example of the importance of PAs for the conservation and protection of primates. By 1978, there were only 252–285 mountain gorillas remaining in the park (Weber & Vedder, 1983). In 1979, the Mountain Gorilla Project was established with the goal of protecting the gorilla population through the promotion of gorilla tourism. This tourism programme prevented poaching, logging and cattle grazing inside the park, and, along with a professional veterinary programme, the gorilla population gradually increased in size reaching 480 individuals by 2010 (Goldsmith, 2014). PAs, however, can be costly to the local people, as their establishment often leads to the resettling of local communities, thereby depriving people of access to resources they have depended on for generations (Allendorf, 2007; Snyman, 2014). Furthermore, communities living adjacent to or in PAs can incur **direct costs** such as damage to crops, loss of livestock or loss of life (Hill, 2000) and **indirect costs** including increased ethnic conflict, reliance on volatile tourism markets and political marginalisation (Laudati, 2010; van der Duim et al., 2014). In this context, the revenues that come from primate tourism and the involvement of local people in tourism-related jobs play a key role in reducing the negative impact PA establishment may have on local communities (Snyman, 2014). However, it is worth noting that while the IUCN (2008) has provided a definition of the different kinds of PAs and their level of protection, the national governments in primate habitats may have their own definitions, making it difficult to generalise costs and benefits of habitat protection.

11.2.2 Revenue Generation and Human-Primate Coexistence

In addition to the direct protection of primate habitats, tourism provides an important revenue to use for primate conservation. Annual gorilla **tourism revenue** in Bwindi Impenetrable National Park, (Uganda), Volcanoes National Park (Rwanda) and Virunga National Park, for example, was estimated to be around $7.75 million US in 2000–2001 (Hatfield, 2005). In 2010, gorilla tourism in Bwindi was estimated to account for more than 80% of all foreign tourist revenues (Goldsmith, 2014). Similarly, Wright et al. (2014) estimated that lemur tourism in the region of Ranomafana National Park in Madagascar generated nearly $2 million US in funds in 2011. Income from tourism can be invested in the protection of primate species and their habitat and/or in local communities. In Kibale National Park (popular for chimpanzee tourism), the Uganda Wildlife Authority shares 20% of entrance fees with the local communities living near the park with an estimated $150,000 US distributed between 1999 and 2008 (MacKenzie, 2012). Although most of this money is spent to build schools, interviews with local people suggest that revenues are most effectively spent on building fences to protect crops against elephant incursions (MacKenzie, 2012). This highlights the importance of including local communities when deciding the allocation of funds as this can maximise their conservation effectiveness (MacKenzie, 2012). Unfortunately, not all forms of primate tourism manage to turn revenue into conservation efforts. Russon and Susilo (2014), for example, found no evidence that orangutan tourism provides substantial economic benefits to orangutan conservation. Both because of the high percentage of **monetary leakage** (i.e., the money that is not spent in the region but is used to purchase imported goods, such as insurance, advertising, and foreign employees; (Hvenegaard, 2014)) and corruption in orangutan tourism, and because of the low price of the entry fees.

Local communities can have mixed perceptions and attitudes towards coexisting with primates and/or associated tourists, and the involvement of local communities in tourist activity can offer local people incentives to protect primate populations. This is particularly important in areas where local communities traditionally depend on primates for subsistence, or where negative interactions with primates occur. For example, the Uaso Ngiro Baboon Project at Il Ngwesi Community Sanctuary in Kenya created "baboon walks" in which local Maasai communities could take tourists to see baboon groups, shifting local communities' attitudes towards baboons, and reducing human-related baboon deaths (Strum & Nightingale, 2014). Crucially, involving the Maasai community in the baboon walks exposes tourists to the local culture, providing a broader tourist experience (Strum & Nightingale, 2014). This may provide a positive quality of interaction between visitors and local people and contribute to the acceptance and tolerance of tourists by residents. To our knowledge, however, there are no studies examining to the extent the relationship between tourists and residents affects the latter's attitudes towards primate tourism.

The presence of primate populations in **unmanaged tourism** sites such as temples can also impose several economic costs to the local people living near the sites. These costs can range from damage to property, the stealing of objects and the risk of injury or death (Priston & McLennan, 2013). Population control strategies such as culling, **translocation** and/or sterilisation of primates are the measures most commonly used to limit such costs (Priston & McLennan, 2013). However, these measures are not always effective, and local communities can perceive them negatively (Saraswat et al., 2015).

11.2.3   Education: Knowledge Sharing

Knowledge sharing with tourists on the importance of protecting nature, animal biodiversity and local community culture is another important contribution of tourism to primate conservation. Educational programmes can encourage tourists to engage in behaviours that are more appropriate towards wildlife and local communities and/or increase their contribution to primate conservation. In fact, several studies have highlighted how little tourists know about the primate species they are visiting. Only 20% of tourists interviewed in Morocco, for example, knew that Barbary macaques are endangered and only 40% of interviewees recognised the health and safety risks associated with interacting with the macaques (Stazaker & Mackinnon, 2018). Additionally, there is often a mismatch between tourists' awareness of how they are supposed to behave around wildlife and how they actually behave. For instance, results from a combination of tourist interviews and direct observations in Parque Nacional de Brasília (Brazil) showed that, while 79.2% of interviewees reported that monkeys initiated human-capuchin interactions, direct observations revealed that people started 47.3% of the interactions (Sabbatini et al., 2006). Although sites characterised by incidental or opportunistic primate tourism (e.g. religious temples) generally do not offer educational programmes (Matheson, 2016), many parks do offer tourists some form of information about primate behaviour and/or conservation status. A tour guide and/or brochures or information boards provide this information. Surprisingly, to our knowledge, no studies have examined how successful such information is in changing people's attitudes towards primate conservation. Furthermore, a growing number of studies have highlighted that tour guides frequently fail to enforce park regulations leading to tourists breaking rules during their visits (Leasor & Macgregor, 2014; Nakamura & Nishida, 2009; Weber et al., 2020). Weber et al. (2020), for example, showed that in 98% of tourist visits to the gorillas of the Bwindi Impenetrable National Park, the 7-m distance rule was violated. The authors suggest various reasons why park regulations are not enforced, such as the pressure guides face to maximise the gorilla experience for tourists who have often travelled long distances and spent a lot of money to see the gorillas, especially if the guides receive tips from the tourists. Not only does the lack of law enforcement jeopardise the health and safety of both humans and wildlife, but it communicates the wrong message to the tourists: that it is acceptable to compromise wildlife safety in order to have a more enjoyable experience.

   In addition to potentially providing some form of education for visitors, primate tourism sites can also offer forms of knowledge sharing with the local communities. Crucially, this knowledge sharing can go both ways: on the one hand, these tourist sites can provide information on the diversity of local wildlife and how to protect it to local people (e.g. Kurita, 2014); on the other hand, these sites can benefit from traditional knowledge about environment, medical plants and animal behaviour (Strum & Nightingale, 2014). It is also important that conservation education programmes align with local cultural beliefs and practices. For instance, a complete ban on provisioning might not be a viable solution in cultures where people want to feed primates for religious reasons (Sengupta & Radhakrishna, 2020). Alternatively, attempts to align conservation programmes with local cultural practices can unintentionally exacerbate primate conservation issues if not appropriately, carefully and respectfully implemented (e.g. lemur hunting: Sodikoff, 2011).

11.2.4 Provisioning

Intentional provisioning by tourists and/or staff occurs frequently in many primate tourism sites, especially in loosely managed and unmanaged projects. Provisioning can affect several aspects of primates' lives, from their behaviour to their ecology and health. Anthropogenic food tends to be more caloric and more digestible than non-cultivated and non-processed food (McLennan & Ganzhorn, 2017). This can provide primate groups and/or individuals with important advantages compared to those foraging mainly on natural food. This includes higher competitive abilities (i.e. stronger and bigger physical features), especially in **high-ranking individuals** (Balasubramaniam et al., 2020a; Campbell, 2013; McKinney, 2014), and more time available for resting or social activities (Jaman & Huffman, 2013; Koirala et al., 2017). However, when provisioning is unpredictable for those primates relying on tourist provisioning, individuals spend time monitoring human activity and interacting with visitors to obtain food, which inevitably reduces the time they invest in resting or social activities (Balasubramaniam et al., 2020a; Kaburu et al., 2019; Marty et al., 2019). In the short term, these behavioural changes can alter group dynamics, potentially jeopardising primate welfare. In comparison, primate groups that inhabit or visit tourist areas tend to be larger (e.g. Fuentes et al., 2011; Hasan et al., 2013; Hansen et al., 2015, 2019, 2020) and less cohesive (Morrow et al., 2019) with higher rates of intra-group aggression, which can lead to injuries or even death (Hill, 2000; Maréchal et al., 2016a).

There seems to be a negative correlation between the **home ranges** of provisioned primates and the level of provisioning (e.g. Klegarth et al., 2017). In Baluran National Park, East Java, Indonesia, the home range of a provisioned group of long-tailed macaques was 23 times smaller than that of a non-provisioned group and negatively correlated with the number of tourists (Hansen et al., 2020). The provisioned group furthermore actively selected tourist sites, and the distribution of long-tailed macaques in the park was centred around roads and trails (Hansen et al., 2019). Given that primates play a key ecological role as seed dispersers, provisioning-induced alterations of primate movement and habitat selection can negatively affect the surrounding ecosystems (Waterman et al., 2019). In India, rhesus macaques (*M. mulatta*), for instance, shortened their daily range and reduced seed dispersal during tourist season, where they were provisioned. They also dispersed seeds on tarmac roads reducing germination success (Sengupta et al., 2015). However, at incidental tourism sites with spatially dispersed provisioning, primate home ranges and daily travel lengths may increase (Riley et al., 2021), possibly leading to higher energetic expenditures.

Finally, there is mixed evidence that access to highly caloric human food increases females' birth rates, shortening their inter-birth intervals and lowering infant mortality. For instance, provisioning led to an increase of the Japanese macaque population from about 160 individuals in 1950 to more than 2000 individuals in the 1990s (Kurita, 2014). In contrast, female pygmy marmosets (*Cebuella pygmaea*) exposed to tourists were found to have lower birth rates, smaller litter sizes and larger inter-birth intervals, compared to groups that received fewer tourist visits (de la Torre et al., 2000).

11.2.5 Health

In addition to altering primates' behaviour, human provisioning can change primate diets, by compensating for seasonal shortages in natural food availability, but it can also have detrimental effects on primate health. The high carbohydrate content and low fibre of anthropogenic foods (McLennan & Ganzhorn, 2017) is of particularly poor nutritional value for primates. Such diets can increase serum insulin and cholesterol levels (Kemnitz et al., 2002), thereby increasing the risks of cardiovascular diseases and diabetes (Sapolsky, 2014). Recent work has also shown that anthropogenic food can change gut **microbiome** and parasite load (Borg et al., 2014; Lee et al., 2019). Human provisioning can also affect primate health indirectly. For example, while a primate group can rapidly grow with supplementary feeding, exceeding the natural food availability capacity can increase the risk of starvation if provisioning is abruptly reduced or stopped (e.g. Kurita, 2014). Finally, when provisioning occurs close to roads, primates face higher risks of death by vehicle collisions (Campbell et al., 2016).

A particular concern for primate tourism is the bidirectional **pathogen transmission** between humans and primates because of their close evolutionary proximity (Melin et al., 2020), ecological overlap and close physical interactions (Muehlenbein & Wallis, 2014; Carne et al., 2017). Moreover, international tourists may introduce new pathogens for which animals and the local human population do not have immunity (Russon & Wallis, 2014). Despite the higher risk of pathogen transmission associated with primate tourism, there is no confirmed case of transmission from tourists to primates due to the difficulty in identifying the exact source of infection (Muehlenbein & Wallis, 2014). However, there are several examples of individual cases of a pathogen spread from primates to an individual person (Jones-Engel et al., 2005), or humans to primates (Goldberg et al., 2007).

Exposure to, and interactions with, tourists and other human disturbances may be stressful for primates. For example, animals that are more exposed to tourists exhibit higher **glucocorticoid** levels (Maréchal et al., 2011, 2016a; Sarmah et al., 2017). In addition, some primates exposed to tourists present physical signs of stress, including poor coat conditions such as **alopecia** (Jolly, 2009; Maréchal et al., 2016a). To deal with such associated stress, primates may use behavioural coping strategies when exposed to tourists, by displaying behavioural changes, self-directed behaviours and/or higher aggression rates towards conspecifics or tourists (Maréchal et al., 2016b).

While we often relate tourism to health concerns for primates, human presence might provide protection from predators. For example, long-tailed macaques in tourist areas have higher preference for sleeping trees near human settlements (Brotcorne et al., 2014). Tourism **management** might also provide veterinary care, which might prevent and/or reduce potential disease outbreaks for the benefits to individual and population health (Wallis & Lee, 1999).

11.2.6 Habituation

Habituation occurs when primates' fear of humans is reduced through repeated neutral contacts between primates and humans (Williamson & Feistner, 2011) and is often necessary in primate tourism in order to be able to guarantee primate sightings (Blom et al., 2004; Ando et al., 2008). Habituation may be a planned management activity leading to tourism or it may occur accidentally. This process is far from neutral (Hanson & Riley, 2018), and very complex (Ampumuza & Driessen, 2020). The mutual influences between tourists and primates on their behaviour, physiology and health described in this chapter can also occur during the habituation process. The effects of habituation are multi-directional with humans affecting primates, primates affecting humans and primates affecting primates (Ampumuza & Driessen, 2020). Importantly, the achievable level of habituation differs between individuals (Allan et al., 2020), meaning that not all individuals in a habituated primate group may be habituated or tolerant of tourists.

11.3   Human Drivers for Human-Primate Interactions Within Primate Tourism

Understanding the drivers of human-primate interactions from both the human and primate perspectives is key to developing more sustainable primate tourism for both primates and the human communities with whom they share space. From the human perspective, it is important to consider the implications of (1) tourist attitudes towards primates, (2) the different types of human-primate interactions and (3) human-animal communication.

The sociocultural background of tourists can underpin their representation of wildlife, and their attraction to specific charismatic species, generally based on size, beauty, charisma, accessibility and similarity to humans (Curtin, 2005). Although people of different nationalities may value nature-based tourism for different reasons (Cochrane, 2006), primates are generally attractive to tourists (Russon & Wallis, 2014). For example, people refer to primates in terms associated with children, highlighting the emotional attraction to primates (Russell, 1995; Knight, 2011). This attitude towards primates shapes the motivations behind the different types of human-primate interactions sought by tourists, that is, viewing, photography and provisioning.

We often base primate tourism on the concept of viewing wild primates in their natural environment, but such viewing can range from simply seeing primates, such as mere registration of the presence and visibility of the animal, to observing primates, such as attentive and cognitively involved viewing of primates (Marvin, 2005). For instance, the "ready-to-view" primate tourism in monkey parks in Japan (Knight, 2010) allows tourists to see Japanese macaques, but the short time devoted to the visits often precludes tourists' ability to watch or observe these animals in their complex social environment. This difference in viewing interaction levels can shape tourist perceptions of the animal involved (Curtin, 2010).

Photography also plays a key role in the tourist experience, with photographs representing a souvenir or a trophy to bring back home or share on social media (Russell & Ankenman, 1996; Lenzi et al., 2020). The rapid increase in primate tourism has resulted in an increasing number of photographs of people in close proximity with primates on social media and elsewhere (Lenzi et al., 2020). However, the popularisation of such photographs raises serious concerns for primate welfare and conservation. For example, showing to the public a picture including a human in proximity of a primate increases their perception that primates would be suitable pets, enhancing illegal primate trade (Ross et al., 2011). These photographs can distort the public's understanding of both the conservation status of primates and what constitutes appropriate interaction (Waters et al., 2021).

Provisioning strongly affects both these interactions, such as viewing and photography, which has a considerable impact on shaping tourist attitudes towards primates. Provisioning might alter tourist views of primates as wild animals, bringing closer interactions with primates. For example, the public views monkey parks in Japan as mega zoos. Here, they lure macaques to an open feeding station where the macaques are conditioned and controlled by humans (Knight, 2006).

Human-primate communication influences the proximity and types of human-primate interactions such as facial signalling, gestures or vocalisations. For example, people are poor at judging the emotional states of Barbary macaques based on their facial expressions and, consequently, are not able to accurately predict their subsequent behaviour (Maréchal et al., 2017). This raises serious concerns for human safety, as people can get closer to primates displaying aggressive facial expressions than expected, potentially resulting in aggressive exchanges between species. Furthermore, people are willing to closely approach macaques they perceive to be cute, young, female, trustworthy, social, healthy and non-dominant (Clark et al., 2020). However, macaques that are more dominant also drove close human-primate interactions despite human proclivity to avoid dominant individuals.

11.4   Primate Individual Factors Involved
in Human-Primate Interactions

At tourist sites where visitors interact with primates, there are strong inter-individual differences in the frequency and way animals interact with people. Several attributes seem to drive primate interactions with people, such as sex (Balasubramaniam et al., 2020b; Fuentes & Gamerl, 2005), dominance rank (Balasubramaniam et al., 2020b; Carne et al., 2017), age (Fuentes & Gamerl, 2005; McKinney, 2014), social network position (Morrow et al., 2019) and spatial position (Balasubramaniam et al., 2020b). However, such attributes might be species-specific or associated with factors inherent to each tourist site. For instance, male long-tailed macaques tend to interact with people more frequently than females (Balasubramaniam et al., 2020b; Fuentes & Gamerl, 2005), but studies have found the opposite pattern in rhesus macaques (Beisner et al., 2015) or no sex effect in white-faced capuchin (McKinney, 2014). Similarly, recent studies showed a positive relationship between centrality in social network position (i.e. the number of direct and indirect social connections individuals have in their groups) and rates of interactions with people among bonnet macaques (*M. radiata*) (Balasubramaniam et al., 2020b), but a negative relationship in Barbary and moor macaques (*M. maura*) (Carne et al., 2017; Morrow et al., 2019). We need more research to establish a comprehensive model that incorporates the complexity of all these features (e.g. species identity, dominance rank, sex, site characteristics) that accounts for these intra- and inter-specific differences in the primate attributes driving the interaction with people.

11.5   Be a Responsible Primate Tourist: Guidelines

As presented throughout this chapter, primate tourism can benefit primates and the local community, but it can also present some costs. By following the below guidelines (Fig. 11.1), tourists can help mitigate these costs: (1) to limit pathogen transmission risk, a 7-m distance between humans and primates should be maintained at any time, personal protective equipment such as mask and gloves should be worn, shoe soles should be cleaned with detergent and multiple site visitations over a short period of time limited to avoid cross site contamination, and primate provisioning by tourists or littering should be regulated; (2) to improve primate welfare, it is recommended to limit tourist group size (<10 people), to maintain low noise levels and to avoid sudden gestures that could frighten primates, and reduce visitation time and frequency; (3) to reduce illegal primate trade, it is recommended to avoid posting selfies with primates on social media; (4) to support and respect the local community and culture, it is recommended to investigate if revenues are distributed to local communities and tourist behaviour is culturally appropriate.

All these recommendations can be found in specialised guidelines such as the *Great Apes Best Practice Tourism Guidelines* (Williamson & Macfie, 2014), IUCN

Fig. 11.1 Primate tourism guidelines

Responsible image guidelines (Waters et al., 2021) and specific primate tourism site guidelines. These guidelines in addition to responsible measures followed by tourism management and researchers should reduce the costs of primate tourism.

Box 11.1 COVID-19
The **COVID-19** pandemic underlined the dangers of close encounters between primates and humans as well as the dangers of habituating primates to provisioning and then abruptly ending it. Macaques are being used for research into the spread and the treatment of the virus SARS-CoV2 and other SARS-CoV strains as they are susceptible to the virus and show symptoms of the disease (Rockx et al., 2011, 2020). Only inoculated laboratory primates were known to be infected with the virus until two western lowland gorillas (*Gorilla gorilla gorilla*) started coughing. Both gorillas and possibly the rest of the troop tested positive for SARS-CoV2 and proved that humans can infect primates, which can also show symptoms of COVID-19 (San Diego Zoo, 2021). This endangers wild primate populations in tourism areas tremendously, and some fear that they can even become virus reservoirs (Liu et al., 2020). Primate tourism decreased considerably due to intermittent lockdown in countries across the globe in March 2020 (Lappan et al., 2020). The reduction of tourism had many consequences, especially financially for tourism providers. It also affected primate groups reliant on provisioning, forcing them to increase aggressive encounters with other groups searching for food and with local people (e.g. Thailand: The Guardian, 2020). Researchers and staff in protected areas feared an increase in poaching with the lack of tourist attention and fewer people working inside primate habitats (Arif Pratiwi, personal communication, 2020; Kone, 2020). Protection efforts of primate populations may have to increase as a result of the pandemic (Liu et al., 2020).

Acknowledgements We would like to thank the editors for inviting us to contribute to the present book and two anonymous reviewers for their helpful comments that helped us improve the chapter. We are also very grateful to the governments and communities that, over the years, have given us the opportunity to conduct research on primate tourism.

References

Alexander, S. E. (2000). Resident attitudes towards conservation and black howler monkeys in Belize: The Community Baboon Sanctuary. *Environmental Conservation, 27*(4), 341–350.

Allan, A. T. L., Bailey, A. L., & Hill, R. A. (2020). Habituation is not neutral or equal: Individual differences in tolerance suggest an overlooked personality trait. *Science Advances, 6*, 28. https://doi.org/10.1126/sciadv.aaz0870

Allendorf, T. D. (2007). Residents' attitudes toward three protected areas in Southwestern Nepal. *Biodiversity and Conservation, 16*, 2087–2102.

Ampumuza, C., & Driessen, C. (2020). Gorilla habituation and the role of animal agency in conservation and tourism development at Bwindi, South Western Uganda. *Journal of Environmental Planning E Nature and Space, 4*, 4. https://doi.org/10.1177/2514848620966502

Ando, C., Takenoshita, Y., & Yamagiwa, J. (2008). Progress of habituation of western lowland gorillas and their reaction to observers in Moukalaba-Doudou National Park, Gabon. *African Study Monographs, 39*, 55–69.

Asquith, P. J. (1989). Provisioning and the study of free-ranging primates: History, effects, and prospects. *American Journal of Physical Anthropology, 32*, 129–158.

Balasubramaniam, K. N., Marty, P. R., Arlet, M. E., et al. (2020a). Impact of anthropogenic factors on affiliative behaviors among bonnet macaques. *American Journal of Physical Anthropology, 171*, 704–717.

Balasubramaniam, K. N., Marty, P. R., Samartino, S., et al. (2020b). Impact of individual demographic and social factors on human–wildlife interactions: A comparative study of three macaque species. *Scientific Reports, 10*, 1–16.

Beisner, B. A., Heagerty, A., Seil, S. K., et al. (2015). Human–wildlife conflict: Proximate predictors of aggression between humans and rhesus macaques in India. *American Journal of Physical Anthropology, 156*(2), 286–294.

Blom, A., Cipolletta, C., & Prins, H. H. T. (2004). Behavioural responses of gorillas to habituation in the Dzanga-Ndoki Natonal Park, Central African Republic. *International Journal of Primatology, 25*, 179–196.

Borg, C., Majolo, B., Qarro, M., & Semple, S. (2014). A comparison of body size, coat condition and endoparasite diversity of wild Barbary macaques exposed to different levels of tourism. *Anthrozoös, 27*, 49–63.

Boza, M. A. (1993). Conservation in action: Past, present, and future of the National Park System of Costa Rica. *Conservation Biology, 7*, 239–247.

Brotcorne, F., Maslarov, C., Wandia, I. N., et al. (2014). The role of anthropic, ecological, and social factors in sleeping site choice by long-tailed macaques (*Macaca fascicularis*). *American Journal of Primatology, 76*, 1140–1150.

Bruner, A. G., Gullison, R. E., Rice, R. E., & Da Fonseca, G. A. (2001). Effectiveness of parks in protecting tropical biodiversity. *Science, 291*(5501), 125–128.

Campbell, J. (2013). *White-faced capuchins (Cebus capucinus) of Cahuita National Park, Costa Rica: Human foods and human interactions*. Graduate Theses and Dissertations. https://lib.dr.iastate.edu/etd/13620

Campbell, L. A., Tkaczynski, P. J., Mouna, M., et al. (2016). Behavioral responses to injury and death in wild Barbary macaques (*Macaca sylvanus*). *Primates, 57*, 309–315.

Carne, C., Semple, S., MacLarnon, A., et al. (2017). Implications of tourist–macaque interactions for disease transmission. *EcoHealth, 14*, 704–717.

Ceballos-Lascurain, H. (1996). *Tourism, ecotourism and protected areas*. IUCN.

CITES Trade Database. (2020). https://trade.cites.org/. Accessed 14 Dec 2020.

Clark, L., Butler, K., Ritchie, K. L., & Maréchal, L. (2020). The importance of first impression judgements in interspecies interactions. *Scientific Reports, 10*, 1–10.

Cochrane, J. (2006). Indonesian national parks: Understanding leisure users. *Annals of Tourism Research, 33*, 979–997.

Curtin, S. C. (2005). Nature, wild animals and tourism: An experiential view. *Journal of Ecotourism, 4*, 1–15.

Curtin, S. C. (2010). The self-presentation and self-development of serious wildlife tourists. *International Journal of Tourism Research, 12*, 17–33.

de la Torre, S., Snowdon, C. T., & Bejarano, M. (2000). Effects of human activities on wild pygmy marmosets in Ecuadorian Amazonia. *Biological Conservation, 94*, 153–163.

Estrada, A., Garber, P. A., Rylands, A. B., et al. (2017). Impending extinction crisis of the world's primates: Why primates matter. *Science Advances, 3*(1), e1600946.

Fuentes, A., & Gamerl, S. (2005). Disproportionate participation by age/sex classes in aggressive interactions between long-tailed macaques (*Macaca fascicularis*) and human tourists at Padangtegal monkey forest, Bali, Indonesia. *American Journal of Primatology, 66*, 197–204.

Fuentes, A., & Hockings, K. J. (2010). The ethnoprimatological approach in primatology. *American Journal of Primatology, 72*, 841–847.

Fuentes, A., Shaw, E., & Cortes, J. (2007). Qualitative assessment of macaque tourist sites in Padangtegal, Bali, Indonesia, and the upper rock nature reserve, Gibraltar. *International Journal of Primatology, 28*, 1143–1158.

Fuentes, A., Kalchik, S., Gettler, L., et al. (2008). Characterizing human-macaque interactions in Singapore. *American Journal of Primatology, 70*, 879–883.

Fuentes, A., Rompis, A. L., Putra, I., et al. (2011). Macaque behavior at the human–monkey interface: The activity and demography of semi-free-ranging *Macaca fascicularis* at Padangtegal, Bali, Indonesia. In M. D. Gumert, A. Fuentes, & L. Jones-Engel (Eds.), *Monkeys on the edge: Ecology and management of long-tailed macaques and their interface with humans* (pp. 159–179). Cambridge University Press.

Goldberg, T. L., Gillespie, T. R., Rwego, I. B., et al. (2007). Patterns of gastrointestinal bacterial exchange between chimpanzees and humans involved in research and tourism in western Uganda. *Biological Conservation, 135*, 511–517.

Goldsmith, M. L. (2014). Mountain gorilla tourism as a conservation tool: Have we tipped the balance? In A. E. Russon & J. Wallis (Eds.), *Primate tourism: A tool for conservation?* (pp. 177–198). Cambridge University Press.

Grossberg, R., Treves, A., & Naughton-Treves, L. (2003). The incidental ecotourist: Measuring visitor impacts on endangered howler monkeys at a Belizean archaeological site. *Environmental Conservation, 30*, 40–51.

Hansen, M. F., Wahyudi, H. A., Supriyanto, S., & Damanik, A. R. (2015). The interactions between long-tailed macaques (*Macaca fascicularis*) and tourists in Baluran National Park, Indonesia. *Journal of Indonesian Natural History, 3*, 36–41.

Hansen, M. F., Nawangsari, V. A., Beest, F. M., et al. (2019). Estimating densities and spatial distribution of a commensal primate species, the long-tailed macaque (*Macaca fascicularis*). *Conservation Science and Practice, 1*, 1–12. https://doi.org/10.1111/csp2.88

Hansen, M. F., Ellegaard, S., Moeller, M. M., et al. (2020). Comparative home range size and habitat selection in provisioned and non-provisioned long-tailed macaques (*Macaca fascicularis*) in Baluran National Park, East Java, Indonesia. *Contribution to Zoology, 89*, 393–411.

Hanson, K. T., & Riley, E. P. (2018). Beyond neutrality: The human–primate interface during the habituation process. *International Journal of Primatology, 39*, 852–877.

Hasan, M. K., Aziz, M. A., Alam, S. R., et al. (2013). Distribution of rhesus macaques (*Macaca mulatta*) in Bangladesh: Inter-population variation in group size and composition. *Primate Conservation, 26*, 125–132.

Hatfield, R. (2005). *Economic value of the Bwindi and Virunga gorilla mountain forests*. African Wildlife Foundation.

Hill, C. M. (2000). Conflict of interest between people and baboons: Crop raiding in Uganda. *International Journal of Primatology, 21*, 299–315.

Hvenegaard, G. T. (2014). Economic aspects of primate tourism associated with primate conservation. In A. E. Russon & J. Wallis (Eds.), *Primate tourism. A tool for conservation?* (pp. 259–278). Cambridge University Press.

International Union for Conservation of Nature. (2008). *Guidelines for applying protected area management categories*. Dudley N (ed). Downloaded from: https://portals.iucn.org/library/sites/library/files/documents/PAG-021.pdf

Jaman, M. F., & Huffman, M. A. (2013). The effect of urban and rural habitats and resource type on activity budgets of commensal rhesus macaques (*Macaca mulatta*) in Bangladesh. *Primates, 54*, 49–59.

Jolly, A. (2009). Coat condition of ringtailed lemurs, *Lemur catta* at Berenty Reserve, Madagascar: I. differences by age, sex, density and tourism, 1996–2006. *American Journal of Primatology, 71*, 191–198.

Jones-Engel, L., Engel, G. A., Schillaci, M. A., et al. (2005). Primate-to-human retroviral transmission in Asia. *Emerging Infectious Diseases, 11*, 1028.

Kaburu, S. S., Marty, P. R., Beisner, B., et al. (2019). Rates of human–macaque interactions affect grooming behavior among urban-dwelling rhesus macaques (*Macaca mulatta*). *American Journal of Physical Anthropology, 168*, 92–103.

Kalpers, J., Williamson, E. A., Robbins, M. M., et al. (2003). Gorillas in the crossfire: Population dynamics of the Virunga mountain gorillas over the past three decades. *Oryx, 37*, 326–337.

Kauffman, L. (2014). Interactions between tourists and white-faced monkeys (*Cebus capucinus*) at Manuel Antonio National Park, Quepos, Costa Rica. In A. E. Russon & J. Wallis (Eds.), *Primate tourism. A tool for conservation?* (pp. 230–244). Cambridge University Press.

Kemnitz, J. W., Sapolsky, R. M., Altmann, J., et al. (2002). Effects of food availability on serum insulin and lipid concentrations in free-ranging baboons. *American Journal of Primatology, 57*, 13–19.

Klegarth, A. R., Hollocher, H., Jones-Engel, L., et al. (2017). Urban primate ranging patterns: GPS-collar deployments for *Macaca fascicularis* and *M. sylvanus*. *American Journal of Primatology, 79*, 1–17.

Knight, J. (2006). Monkey mountain as a megazoo: Analyzing the naturalistic claims of "wild monkey parks" in Japan. *Society and Animals, 14*, 245–264.

Knight, J. (2009). Making wildlife viewable: Habituation and attraction. *Society and Animals, 17*, 167–184.

Knight, J. (2010). The ready-to-view wild monkey: The convenience principle in Japanese wildlife tourism. *Annals of Tourism Research, 37*, 744–762.

Knight, J. (2011). *Herding monkeys to paradise: How macaque troops are managed for tourism in Japan*. Brill.

Koirala, S., Chalise, M. K., Katuwal, H. B., et al. (2017). Diet and activity of *Macaca assamensis* in wild and semi-provisioned groups in Shivapuri Nagarjun National Park, Nepal. *Folia Primatologica, 88*, 57–74.

Kone, I. (2020). *SOS African Wildlife and the COVID-19 pandemic*. Available via: https://www.youtube.com/watch?fbclid=IwAR0Z0PPyRo3HhM4K1BehtnlXwyVXMNOuXzzboZJzW96s-bJsePYq-OY9cHc&v=8VJEDz6_J4Y&feature=youtu.be&ab_channel=IUCNSOS-SaveOurSpecies. Accessed 7 Feb 2021.

Kurita, H. (2014). Provisioning and tourism in free-ranging Japanese macaques. In A. Russon & J. Wallis (Eds.), *Primate tourism. A tool for conservation?* (pp. 44–55). Cambridge University Press.

Lappan, S., Malaivijitnod, S., Radhakrishna, S., et al. (2020). The human-primate interface in the New Normal: Challenges and opportunities for primatologists in the COVID-19 era and beyond. *American Journal of Primatology, 82*, 1–12.

Laudati, A. (2010). Ecotourism: The modern predator? Implications of gorilla tourism on local livelihoods in Bwindi Impenetrable National Park, Uganda. *Environment and Planning D: Society and Space, 28*, 726–743.

Leasor, H. C., & Macgregor, O. J. (2014). Proboscis monkey tourism: Can we make it "ecotourism". In A. Russon & J. Wallis (Eds.), *Primate tourism: A tool for conservation?* (pp. 56–75). Cambridge University Press.

Lee, P. C. (2010). Sharing space: Can ethnoprimatology contribute to the survival of nonhuman primates in human-dominated globalized landscapes? *American Journal of Primatology, 72*, 925–931.

Lee, W., Hayakawa, T., Kiyono, M., et al. (2019). Gut microbiota composition of Japanese macaques associates with extent of human encroachment. *American Journal of Primatology, 81*(12), e23072.

Lenzi, C., Speiran, S., & Grasso, C. (2020). "Let me take a selfie": Implications of social media for public perceptions of wild animals. *Society and Animals, 1*, 1–20.

Liu, Z.-J., Qian, X.-K., Hong, M.-H., et al. (2020). Global view on virus infection in non-human primates and implication for public health and wildlife conservation. *Zoological Research, 42*, 626–632. https://doi.org/10.1101/2020.05.12.089961

MacKenzie, C. A. (2012). Trenches like fences make good neighbours: Revenue sharing around Kibale National Park, Uganda. *Journal for Nature Conservation, 20*, 92–100.

Maréchal, L., Semple, S., Majolo, B., et al. (2011). Impacts of tourism on anxiety and physiological stress levels in wild male Barbary macaques. *Biological Conservation, 144*, 2188–2193.

Maréchal, L., Semple, S., Majolo, B., & MacLarnon, A. (2016a). Assessing the effects of tourist provisioning on the health of wild Barbary macaques in Morocco. *PLoS One, 11*(5), e0155920.

Maréchal, L., MacLarnon, A., Majolo, B., & Semple, S. (2016b). Primates' behavioural responses to tourists: Evidence for a trade-off between potential risks and benefits. *Scientific Reports, 6*, 1–1.

Maréchal, L., Levy, X., Meints, K., & Majolo, B. (2017). Experience-based human perception of facial expressions in Barbary macaques (*Macaca sylvanus*). *PeerJ, 5*, e3413.

Marty, P. R., Beisner, B., Kaburu, S. S. K., et al. (2019). Time constraints imposed by anthropogenic environments alter social behaviour in long-tailed macaques. *Animal Behaviour, 150*, 157–165.

Marvin, G. (2005). Seeing, looking, watching, observing non-human animals. *Society and Animals, 13*, 1–13.

Matheson, M. D. (2016). Primate tourism. In A. Fuentes (Ed.), *The international encyclopaedia of primatology* (3rd ed., pp. 1–8). Wiley.

McKinney, T. (2014). Species-specific responses to tourist interactions by white-faced capuchins (*Cebus imitator*) and mantled howlers (*Alouatta palliata*) in a Costa Rican Wildlife Refuge. *International Journal of Primatology, 35*, 573–589.

McLennan, M. R., & Ganzhorn, J. U. (2017). Nutritional characteristics of wild and cultivated foods for chimpanzees (Pan troglodytes) in agricultural landscapes. *International Journal of Primatology, 38*(2), 122–150.

Medhi, R., Chetry, D., Basavdatta, C., & Bhattacharjee, P. C. (2007). Status and diversity of temple primates in Northeast India. *Primate Conservation, 22*, 135–138.

Melin, A. D., Janiak, M. C., Marrone, F., et al. (2020). Comparative ACE2 variation and primate COVID-19 risk. *Communications Biology, 3*, 1–9.

Morrow, K. S., Glanz, H., Ngakan, P. O., & Riley, E. P. (2019). Interactions with humans are jointly influenced by life history stage and social network factors and reduce group cohesion in moor macaques (*Macaca maura*). *Scientific Reports, 9*, 1–12.

Muehlenbein, M. P., & Wallis, J. (2014). Considering risks of pathogen transmission associated with primate-based tourism. In A. Russon & J. Wallis (Eds.), *Primate tourism. A tool for conservation?* (pp. 278–287). Cambridge University Press.

Nakamura, M., & Nishida, T. (2009). Chimpanzee tourism in relation to the viewing regulations at the Mahale Mountains National Park, Tanzania. *Primate Conservation, 24*, 85–90.

Priston, N. E., & McLennan, M. R. (2013). Managing humans, managing macaques: Human–macaque conflict in Asia and Africa. In S. Radhakrishna, M. A. Huffman, & A. Sinha (Eds.), *The macaque connection* (pp. 225–250). Springer.

Riley, E. P., Shaffer, C. A., Trinidad, J. S., et al. (2021). Roadside monkeys: Anthropogenic effects on moor macaque (*Macaca maura*) ranging behavior in Bantimurung Bulusaraung National Park, Sulawesi, Indonesia. *Primates, 62*, 477–489.

Rockx, B., Feldmann, F., Brining, D., et al. (2011). Comparative pathogenesis of three human and zoonotic SARS-CoV strains in cynomolgus macaques. *PLoS One, 6*, 1–9. https://doi.org/10.1371/journal.pone.0018558

Rockx, B., Kuiken, T., Herfst, S., et al. (2020). Comparative pathogenesis of COVID-19, MERS, and SARS in a nonhuman primate model. *Science, 368*, 1012–1015. https://doi.org/10.1126/science.abb7314

Ross, S. R., Vreeman, V. M., & Lonsdorf, E. V. (2011). Specific image characteristics influence attitudes about chimpanzee conservation and use as pets. *PLoS One, 6*(7), e22050.

Russell, C. L. (1995). The social construction of orangutans: An ecotourist experience. *Society and Animals, 3*, 151–170.

Russell, C. L., & Ankenman, M. J. (1996). Orangutans as photographic collectibles: Ecotourism and the commodification of nature. *Tourism Recreation Research, 21*, 71–78.

Russon, A. E., & Susilo, A. (2014). Orangutan tourism and conservation: 35 years' experience. In A. E. Russon & J. Wallis (Eds.), *Primate tourism. A tool for conservation?* (pp. 76–97). Cambridge University Press.

Russon, A. E., & Wallis, J. (2014). Reconsidering primate tourism as a conservation tool: An introduction to the issues. In A. E. Russon & J. Wallis (Eds.), *Primate tourism. A tool for conservation?* (pp. 3–18). Cambridge University Press.

Sabbatini, G., Stammati, M., Tavares, M. C. H., et al. (2006). Interactions between humans and capuchin monkeys (*Cebus libidinosus*) in the Parque Nacional de Brasília, Brazil. *Applied Animal Behaviour Science, 97*, 272–283.

San Diego Zoo. (2021). Available via: https://zoo.sandiegozoo.org/pressroom/news-releases/gorilla-troop-san-diego-zoo-safari-park-test-positive-covid-19. Accessed 20 Jan 2021.

Sapolsky, R. M. (2014). Some pathogenic consequences of tourism for non-human primates. In A. E. Russon & J. Wallis (Eds.), *Primate tourism. A tool for conservation?* (pp. 147–155). Cambridge University Press.

Saraswat, R., Sinha, A., & Radhakrishna, S. (2015). A god becomes a pest? Human-rhesus macaque interactions in Himachal Pradesh, northern India. *European Journal of Wildlife Research, 61*, 435–443.

Sarmah, J., Hazarika, C. R., Berkeley, E. V., et al. (2017). Non-invasive assessment of adrenocortical function as a measure of stress in the endangered golden langur. *Zoo Biology, 36*, 278–283.

Sengupta, A., & Radhakrishna, S. (2020). Factors predicting provisioning of macaques by humans at tourist sites. *International Journal of Primatology, 41*, 471–485.

Sengupta, A., McConkey, K. R., & Radhakrishna, S. (2015). Primates, provisioning and plants: Impacts of human cultural behaviours on primate ecological functions. *PLoS One, 10*, 1–13. https://doi.org/10.1371/journal.pone.0140961

Sha, J. C. M., Gumert, M. D., Lee, B. P. Y. H., et al. (2009). Status of the long-tailed macaque *Macaca fascicularis* in Singapore and implications for management. *Biodiversity and Conservation, 18*, 2909–2926.

Snyman, S. (2014). Assessment of the main factors impacting community members' attitudes towards tourism and protected areas in six southern African countries. *Koedoe, 56*(2), 1–12.

Sodikoff, G. (2011). Totem and taboo reconsidered; endangered species and moral practice in Madagascar. In G. Sodikoff (Ed.), *The anthropology of extinction: Essays on culture and species death* (pp. 68–86). Indiana University Press.

Stazaker, K., & Mackinnon, J. (2018). Visitor perceptions of captive, endangered Barbary macaques (*Macaca sylvanus*) used as photo props in Jemaa El Fna Square, Marrakech, Morocco. *Anthrozoös, 31*, 761–776.

Strum, S. C., & Nightingale, D. L. M. (2014). Baboon ecotourism in the larger context. In A. E. Russon & J. Wallis (Eds.), *Primate tourism. A tool for conservation?* (pp. 155–176). Cambridge University Press.

The Guardian. (2020). *Mass monkey brawl highlights coronavirus effect on Thailand tourism.* Available via: https://www.theguardian.com/world/2020/mar/13/fighting-monkeys-highlight-effect-of-coronavirus-on-thailand-tourism. Accessed 14 Dec 2020.

van der Duim, R., Ampumuza, C., & Ahebwa, W. M. (2014). Gorilla tourism in Bwindi Impenetrable National Park, Uganda: An actor-network perspective. *Society & Natural Resources, 27*, 588–601.

Wallis, J., & Lee, D. R. (1999). Primate conservation: The prevention of disease transmission. *International Journal of Primatology, 20*, 803–826.

Waterman, J. O., Campbell, L. A., Maréchal, L., et al. (2019). Effect of human activity on habitat selection in the endangered Barbary macaque. *Animal Conservation, 23*, 373–385.

Waters, S., Setchell, J. M., Maréchal, L., et al. (2021). *IUCN best practice guidelines for responsible images of non-human primates.* IUCN Primate Specialist Group. https://human-primate-interactions.org/resources/

Weber, A. W., & Vedder, A. (1983). Population dynamics of the Virunga gorillas: 1959–1978. *Biological Conservation, 26*, 341–366.

Weber, A., Kalema-Zikusoka, G., & Stevens, N. J. (2020). Lack of rule-adherence during mountain gorilla tourism encounters in Bwindi Impenetrable National Park, Uganda, places gorillas at risk from human disease. *Frontiers in Public Health, 8*, 1. https://doi.org/10.3389/fpubh.2020.00001

Wheatley, B. P. (1999). *The sacred monkeys of Bali.* Waveland Press Inc.

Williamson, E. A., & Feistner, A. T. C. (2011). Habituating primates: Processes, techniques, variables and ethics. In J. M. Setchell & D. J. Curtis (Eds.), *Field and laboratory methods in primatology* (2nd ed., pp. 33–50). Cambridge University Press.

Williamson, E. A., & Macfie, E. J. (2014). Guidelines for best practice in great ape tourism. In A. E. Russon & J. Wallis (Eds.), *Primate tourism. A tool for conservation?* (pp. 292–310). Cambridge University Press.

Wright, P. C., Andriamihaja, B. J., King, S., et al. (2014). Lemurs and tourism in Ranomafana National Park, Madagascar: Economic boom and other consequences. In A. E. Russon & J. Wallis (Eds.), *Primate tourism. A tool for conservation?* (pp. 123–146). Cambridge University Press.

Chapter 12
Shared Ecologies, Shared Futures: Using the Ethnoprimatological Approach to Study Human-Primate Interfaces and Advance the Sustainable Coexistence of People and Primates

Erin P. Riley, Luz I. Loría, Sindhu Radhakrishna, and Asmita Sengupta

Contents

Abstract Ethnoprimatology is a research approach used to study the diverse ways that human and other primates' lives and livelihoods intersect. Our objective in this chapter is to illustrate the use and value of ethnoprimatology in studying human-primate interfaces across both time and space. We begin by clarifying how human-primate interfaces occur across a gradient from rural to urban landscapes and illustrate how the ethnoprimatological approach is implemented. We showcase how ethnoprimatology's theoretical and methodological landscape has expanded since

E. P. Riley (✉)
Department of Anthropology, San Diego State University, San Diego, CA, USA
e-mail: epriley@sdsu.edu

L. I. Loría
Facultad de Ciencias Agropecuarias, Universidad de Panamá, Panama City, Panama

Fundación Pro-Conservación de los Primates Panameños (FCPP), Panama City, Panama

S. Radhakrishna
School of Natural Sciences and Engineering, National Institute of Advanced Studies, Bengaluru, Karnataka, India

A. Sengupta
Ashoka Trust for Research in Ecology and the Environment, Bengaluru, Karnataka, India

its emergence in the late 1990s to include the integration of frameworks and tools from the natural and social sciences and the humanities. To illustrate the practice of ethnoprimatology, we highlight research conducted on the human-primate interface across three geographies of human-primate encounters: tourism sites, urban and peri-urban settings, and agroecosystems. We conclude by showing how the human-primate interconnections uncovered by ethnoprimatology have important implications for conservation, management of human-primate interfaces, and the sustainable coexistence of multispecies communities in the contemporary era.

Keywords Ethnoprimatology · Tourism · Urban spaces · Agroecosystems · Mixed methods · Conservation · Anthropology · Human-wildlife conflict · Sympatry · Community ecology

12.1 Introduction

12.1.1 What Are 'Human-Primate Interfaces'?

In many areas around the world, humans and nonhuman primates (primates henceforth) live in close proximity and regularly overlap in their use of space and the resources available in the environment. In some areas, these overlapping existences are relatively recent phenomena. For example, the free-ranging population of rhesus macaques (*Macaca mulatta*) that people encounter along the banks of the Silver River, Florida, USA, dates back to the 1930s, when its founding members were first introduced to the area (Riley & Wade, 2016). In other regions, however, human-primate coexistence, or what is referred to as 'sympatry' in ecological terms, has considerable time depth. Archaeological evidence from Gabon, for example, indicates that three hominoid genera (*Pan, Gorilla,* and *Homo*) coexisted for a period of at least 60,000 years, likely competing for similarly favoured plant foods (Tutin & Oslisly, 1995). Similarly, literary analysis of Tamil poetry reveals how the cohabitation of people and primates in southern India dates back thousands of years (Radhakrishna, 2018). These shared existences constitute human-primate interfaces: the intersections of the lives and livelihoods of humans and primates.

12.1.2 What Is Ethnoprimatology?

Given the growing reality and ubiquity of human-primate interfaces globally, and the fact that the world's remaining primate taxa are primarily threatened with extinction as a result of human activities (Estrada et al., 2017), in the past few decades, primatologists have increasingly prioritised the human-primate interface as a research focus (Dore et al., 2017; Fuentes & Wolfe, 2002; Paterson & Wallis, 2005).

Ethnoprimatology is an approach used to study human-primate interfaces across both space and time. It focuses on examining the multifaceted ways that people and primates interconnect ecologically, such as how they overlap in their resource use patterns (Hockings et al., 2020; Riley, 2007), and their cultural interconnections. The latter includes uncovering how primates feature in peoples' folklore, customs, religion, and worldviews, and, in turn, how these cultural factors shape people's perceptions and behaviour towards primates (Anand et al., 2018; Waters et al., 2019). Being the brainchild of anthropologists Leslie Sponsel (1997) and Bruce Wheatley (1999), ethnoprimatology is an inherently anthropological approach, drawing from sociocultural, biological, and environmental anthropology. At the same time, it also draws theoretical, conceptual, and methodological inspiration from the natural sciences (behavioural ecology, community ecology, ethology, evolutionary biology, ecology) as well as other social sciences (political ecology, geography, science and technology studies). Also inherent to the ethnoprimatological approach is recognising humans as part of nature (rather than separate from it) and viewing primate responses to human influences as integral and interesting components of their ecological flexibility (Fuentes, 2012; Riley, 2013).

12.1.3 Objectives of This Chapter

Our objective in this chapter is to illustrate the use and value of ethnoprimatology in studying human-primate interfaces. We begin by clarifying where human-primate interfaces occur and address how the ethnoprimatological approach is implemented. We showcase how it was used initially, and how, since its emergence, its theoretical and methodological landscape has expanded. To illustrate the practice of ethnoprimatology, we highlight research conducted on the human-primate interface across a diverse array of settings.

As a comparatively nascent field, ethnoprimatology's conceptual contributions are multifold. It enables a more nuanced understanding of primate behavioural flexibility, provides insights into the causes and consequences of human-primate **sympatry**, and further uncovers what it means to be human *and* what it means to be a primate. We further highlight the applied significance of ethnoprimatology by showing how the human-primate interconnections this approach uncovers have important implications for conservation, management of human-primate interfaces, and the sustainable coexistence of multispecies communities in the contemporary era, including public health concerns during the COVID-19 pandemic and beyond.

12.2   Studying Human-Primate Interfaces: Where and How

Humans and primates interface within a wide array of contexts, most of which exist across a gradient from rural to urban landscapes (Fig. 12.1). This gradient ranges from areas where forested land or other forms of native primate habitat (e.g. savanna, fynbos, mangroves) border agricultural and/or livestock areas to peri-urban contexts where primates spatially overlap with human-dominated landscapes, such as commercial and residential buildings (e.g. the Cape Peninsula, South Africa; Fehlmann et al., 2017; Chap. 8, this volume). The gradient also extends to urban centres, where primates live in relic forested areas (e.g. Tijuca forest of Rio de Janeiro, Brazil; Cunha et al., 2006) and cities, such as Delhi, India, where original forms of primate habitat are scarce, leading to primates living within the city centre, utilising anthropogenic substrates as dwellings and for transport (e.g. buildings and electrical wires, respectively) and relying predominantly on **anthropogenic** foods (e.g. from **provisioning** and/or raiding of homes or food stalls).

Recognising the mutual interests of human ecologists and primate ecologists, Sponsel (1997) defined 'ethnoprimatology' as an integrative anthropological approach aimed at studying the human-primate interface. Much of the earliest ethnoprimatological work was conducted by scholars trained in sociocultural and ecological anthropology (e.g. Shepard, 2002; Cormier, 2003). Accordingly, the methods used were primarily ethnographic, whereby a researcher lives in and among communities, engages in participant observation, and documents people's perceptions of and attitudes towards human-primate interfaces via in-depth interviews with key informants and rich descriptions. Today, ethnographic methods remain a critical component of the ethnoprimatological toolkit (Dore et al., 2017, 2018).

Theory and methods in the study of animal behaviour (ethology) and in ecology were also important to early ethnoprimatology, and primatologists who began adopting the ethnoprimatological approach in the early 2000s showcased how ethology, ecology, and ethnography could be integrated and used to address key questions about the causes and consequences of human-primate sympatry (Hardin &

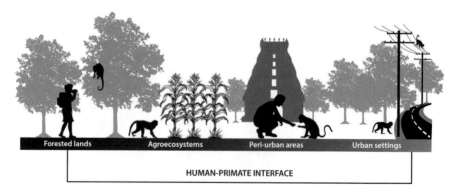

Fig. 12.1 The human-primate interface across rural to urban landscapes

Remis, 2006; Loudon et al., 2006; Riley, 2007). Today's ethnoprimatology is characterised by even greater theoretical and methodological diversity (Riley, 2018). Borrowing from evolutionary biology, ethnoprimatology has used the **niche construction** framework to explore how humans shape the ecology, sociality, and evolutionary trajectories of primates (Ellwanger & Lambert, 2018; Fuentes, 2010). Practitioners of ethnoprimatology have also recognised that evolutionary theory is not the only theoretical framework that is useful for studying human-primate interfaces (Malone et al., 2014; Parathian et al., 2018; Radhakrishna & Sengupta, 2020; Riley, 2018). For example, recent work has integrated ethnoprimatology with perspectives and concepts in human-animal studies, multispecies anthropology, and science studies to explore the intersubjectivity between researchers and their primate study subjects during the habituation process (Hanson & Riley, 2018), to illustrate how narratives about primates are emblematic of local critiques of externally driven nature conservation initiatives (Sousa et al., 2017), and to demonstrate how understanding the multispecies communities, including what people know about wildlife and their perceptions of them, can inform conservation practices (Jost Robinson & Remis, 2018; Waters et al., 2018).

The methodological toolkit of ethnoprimatology has also expanded and now incorporates a diverse array of techniques, including stable isotope analysis (Loudon et al., 2014) and nutritional ecology (Riley et al., 2013) to examine how humans and their activities affect primate feeding ecology and dietary patterns, spatial ecology/GIS (Klegarth et al., 2017) and camera trap technology (Zak & Riley, 2017) to examine overlap in resource and habitat use between people and primates, biodemographic modelling to monitor hunting sustainability (Shaffer et al., 2018), disease ecology to examine human-primate pathogen exchange (Muehlenbein, 2016), **social network analysis** (Beisner et al., 2016; Morrow et al., 2019) to understand how humans affect primate social networks, discourse analysis (Radford et al., 2018), and literary analysis (Radhakrishna, 2018) to uncover people's perceptions of and attitudes towards primates and their relationships with them. Importantly, the ethnoprimatological approach is not just about studying anthropogenic impacts on primates; rather, it prioritises *both* human and primate perspectives to understand how they perceive one another and behave *together*.

12.3 Human-Primate Interfaces

In the following sections, we review extant ethnoprimatological studies to showcase the range of theoretical and methodological approaches that characterise research on the human-primate interface. We consider three geographies of human-primate encounters, namely, tourism sites, urban and peri-urban settings, and **agroecosystems**, as settings provide illuminating examples of the diverse ways in which humans and primates mutually affect each other's lives and behaviours. While some ethnoprimatological studies report the influence of humans and primates on each other, others focus on understanding the impacts faced by one group and provide insights into their potential effect on the other group.

12.3.1   The Human-Primate Interface in Tourist Settings

Humans and primates influence each other in a multitude of ways across different forms of primate tourism (Chap. 11, this volume), making these interfaces particularly relevant for examination through the lens of ethnoprimatology. Practices such as **translocation**, provisioning, and range restriction are common in primate tourism settings (Berman et al., 2007; Sengupta & Radhakrishna, 2020). These procedures, alongside interactions with visitors in managed or incidental primate tourism sites (Grossberg et al., 2003), can influence primate morphology, physiology, behaviour, and ecology.

Provisioned primates often have poor coat conditions (Jolly, 2009), scars and wounds (Westin, 2007), and larger body sizes (Maréchal et al., 2016). Additionally, such primates are characterised by higher physiological stress (Behie et al., 2010) and higher parasitic loads (Borg et al., 2014; Junge et al., 2011) and are more prone to various pathogens that are of human origin (Devaux et al., 2019). Tourist-exposed primates are further known to increase self-scratching (an indicator of anxiety; Maréchal et al., 2011), alter social interactions (Berman et al., 2007), change ranging patterns (de la Torre et al., 2000), and reduce feeding rates when observer party size increases (Klailova et al., 2010). Studies show that both humans and primates can initiate interactions. Bonnet macaques (*M. radiata*), for instance, mostly initiated interactions in the context of seeking food and repeated certain behaviours which were associated with being given food by tourists thereby creating feedback loops for human-primate interactions (Sengupta & Radhakrishna, 2018a). One such feedback loop is the 'robbing and bartering' behaviour which long-tailed macaques (*M. fascicularis*) and rhesus macaques display in Indonesia (Brotcorne et al., 2017) and India (Kaburu et al., 2019), respectively. In these scenarios, the macaques rob visitors of objects such as spectacles and return them in exchange for food (Brotcorne et al., 2017). Additionally, particular tourist behaviours can elicit specific primate responses. For example, when tourists 'pointed' at Tibetan macaques (*M. thibetana*), they mostly responded with a 'facial threat', whereas when tourists 'slapped the railing', the macaques responded with a 'lunge/ground slap' (McCarthy et al., 2009). At Sabah, Malaysia, macaques were more likely to exhibit aggressive responses, albeit low intensity (e.g. open-mouth threats), when tourists stared at them compared to when they took pictures (Gilhooly et al., 2021). Sound decibel levels on tourist viewing platforms were positively correlated with the display of threats by Tibetan macaques (Ruesto et al., 2010).

The nature of human-primate interactions could be a function of primate species involved or the age/sex category of individuals. For instance, in Costa Rica, white-faced capuchin monkeys (*Cebus imitator*) initiated more frequent and variable interactions with humans than mantled howler monkeys (*Alouatta palliata*)

(McKinney, 2014). Similarly, long-tailed macaques in Bali engaged in more aggressive encounters than Barbary macaques (*M. sylvanus*) in Gibraltar (Fuentes et al., 2007). Across different species of macaques though, adult males initiated most interactions (Fuentes & Gamerl, 2005; Gilhooly et al., 2021; Maibeche et al., 2015).

In incidental primate tourism settings, owing to the predictable availability of calorie-rich anthropogenic food, provisioned primates alter their ranging patterns. For instance, in Indonesia, the home range of a provisioned group of long-tailed macaques was 10.62 ha, whereas that of an unprovisioned group was 249.9 ha. The daily path length of the former was also shorter and was negatively correlated with the number of tourists in the study area (Hansen et al., 2020). In other contexts, where **provisioning** is irregular and offered from moving vehicles along roads, for example, home ranges and daily travel distances have been shown to increase (e.g. Riley et al., 2021; Sha & Hanya, 2013). Provisioning also alters diets of primates involved in incidental tourism. In India, fruits comprised 28% of the diet of a group of rhesus macaques when provisioned by tourists in comparison to 70% of the diet when the area was closed to visitors (Sengupta & Radhakrishna, 2018b). Changes in diet and ranging also affected their ecological functions – provisioned rhesus macaques dispersed fewer seeds and deposited them in areas unconducive for germination compared to when they were unprovisioned (Sengupta et al., 2015).

Several studies have assessed the impact of primates on humans in tourism settings. Aggression is common towards humans and has been documented in various macaque species (Gilhooly et al., 2021; Hsu et al., 2009; Maréchal et al., 2016). Such aggressive encounters often lead to injury and, in some cases, have resulted in human deaths (Zhao & Deng, 1992). Macaques also take people's belongings by force (Sha et al., 2009). Additionally, humans are threatened by disease transmission from primates. Several cases of tourists contracting pathogens from primates have been documented in Asia and Africa (e.g. Dunay et al., 2018; Jones-Engel et al., 2006).

The attitudes and behaviour of people towards primates and **primate tourism** vary and are a function of various cultural or socio-economic/demographic factors (Loudon et al., 2006). Humans are known to initiate interactions by actively provisioning primates (Sengupta & Radhakrishna, 2020) and for taking photographs of/ with primates (Sengupta & Radhakrishna, 2018a; Fig. 12.2). Men mostly initiate these interactions and also tend to be the primary recipients of aggressive behaviour of macaques (Hsu et al., 2009). Ethnic and cultural background is another important factor shaping human perceptions of primates and their associated behaviour. Tourists from non-primate habitat countries, for instance, provisioned bonnet macaques to watch these wild, exotic primates up close. In contrast, Indian tourists at the same site were driven by their religious belief to feed the macaques (Sengupta & Radhakrishna, 2020) (Fig. 12.3). Residents of tourist sites may also have different perceptions of primates than visitors. In Singapore, residents reported more problems related to long-tailed macaques and had more negative perceptions of them than tourists who mostly had neutral views (Sha et al., 2009).

Studies guided by ethnoprimatology have most importantly revealed that particular primate behaviours influence humans to engage in behaviours such as

Fig. 12.2 The human-primate interface at the Ubud Monkey Forest, a managed primate tourism setting. (Photo credit: Tanumay Datta)

Fig. 12.3 Provisioning bonnet macaques in a tourist site in Bengaluru City, India. (Credit: Ninaad Kulshreshtha)

provisioning thereby clearly showing that they indeed mutually impact each other's lives. For instance, in India, 43 of 80 tourist respondents said that they provisioned rhesus macaques as they feared their aggressive behaviours or felt sorry seeing them make 'begging' gestures (Sengupta & Radhakrishna, 2018b). Ampumuza and Driessen (2021) have further argued that primates are not passive players in the process of habituation to humans, as is commonly understood. Using the Actor Network Theory, they suggested that just as gorillas habituated to other gorillas, they habituated to humans as well. This argument and the observation that high-ranking macaque individuals and/or individuals better connected in their own grooming network initiated more interactions with humans (Balasubramaniam et al., 2020) show that the role of animal agency and individual motivations of primates needs further examination in primate tourism settings.

12.3.2   Urban and Peri-urban Human-Primate Encounters

Human-primate interactions in urban and peri-urban spaces typically occur within city parks, temple spaces, suburban forest edges, and human residential areas. Urban landscapes offer rich incentives for primates in the form of abundant food sources (Maibeche et al., 2015), refuge from predators (Malik et al., 1984), and a resource buffer in times of environmental threats (Waite et al., 2007). They also present challenges for the survival of primates in the form of increased risk perception (Mikula et al., 2018), anthropogenic disturbance (Scheun et al., 2015), **pathogen transmission** and higher parasitic loads (Huemer et al., 2002; Thatcher et al., 2018), death from vehicle collisions, electrocution (Lokschin et al., 2007; Pragatheesh, 2011), and lethal retaliation (Jones-Engel et al., 2011; Radhakrishna & Shankar, 2016). While some generalist primate species have a long history of voluntarily associating with human habitats (Richard et al., 1989), other specialist species have more recently begun to exhibit synanthropic trends due to anthropogenic pressures on their habitat (Maibeche et al., 2015). Humans coexisting with primates in cityscapes face the possibilities of damage to property and material possessions, kitchen garden depredations, disturbance in public spaces, potential disease transmission and physical injury and sometimes even death due to the activities of primate individuals (Hoffman & O'Riain, 2011; Jones-Engel et al., 2008; Radhakrishna & Sinha, 2010; Yeo & Neo, 2010). Studies on human-primate interactions in urban spaces have addressed people's attitudes towards primates (Leite et al., 2011), the positive or negative aspects of human-primate interactions (Suzin et al., 2017), the impact of human interactions on primate ecology and behaviour (Pinheiro et al., 2018), the consequences of human provisioning for primate population growth and range expansion (Maibeche et al., 2015; Malik et al., 1984), and the costs of such interactions for human and primate health (Kowalewski et al., 2011).

Traditional primatology views primates' use of urban spaces as non-natural and an aberrational behaviour that is forced by human-caused degradation or loss of primate habitats. In contrast, ethnoprimatology, with its emphasis on the mutual influences of humans and primates on each other, is a particularly useful approach to investigate human-primate interactions in urban spaces, as it brings to the fore the multiple ways in which human and primate lives are modified by each other. For example, primate species adapted to cityscapes usually have larger group sizes relative to rural or forest-dwelling groups, and this is understood to be related to their access to provisioned food (Hasan et al., 2013; Ilham et al., 2017). Feeding on anthropogenic food also leads to behavioural changes such as increased aggression and altered activity budgets (Thatcher et al., 2018) and diminished health status (Shively et al., 2009). In their study on human-primate conflict in urban contexts, Beisner et al. (2016) adopted an ethnoprimatological approach to examine how human behaviour provokes aggression in rhesus macaques and vice versa. They concluded that aggressive behaviours by humans and macaques occur in response to the other party's actions and that such conflict interactions can only be mitigated through a holistic understanding of both participants' behaviours.

As in tourist settings, provisioning is a critical element of human-primate interactions in urban contexts: while 'nuisance marmoset' cases in Brazil were often the result of people voluntarily feeding marmosets (Goulart et al., 2010), rhesus macaques on the streets of Delhi in India display bipedal begging behaviour that incites people to offer them food (Barua & Sinha, 2019). The critical role of provisioning in driving human-primate interactions in urban contexts was clearly visible during the **COVID-19** pandemic, when abrupt cessation of provisioning in many regions due to travel restrictions led to primate groups moving away from their urban feeding sites (Lappan et al., 2020).

Apart from provisioning, human impacts on primate ecologies are most keenly felt when primates are introduced in or translocated to non-native areas through human intervention. Such introductions, either due to pet-keeping practices or as a conflict mitigation strategy, have led to human-primate conflicts or resulted in spillover effects on other primate ecologies in urban contexts (Radhakrishna & Sinha, 2011). For example, the translocation of rhesus macaques, to reduce conflict intensities, into the range of bonnet macaques in southern India led to the displacement of endemic bonnet macaques from anthropogenic spaces in those areas (Kumar et al., 2011). Human impacts on urban primates may also have hidden dimensions like noise or light pollution (Duarte et al., 2011; Scheun et al., 2015; Chap. 8, this volume).

Possibly because traditional primatology tends to treat primates' presence in human areas as non-natural, human-primate interactions within urban spaces are usually treated as conflict scenarios (Lee & Chan, 2011). Consequently, less attention has generally been paid to the benefits humans may derive from living with urban primates. Yet it is well documented that people receive religious satisfaction from feeding monkeys (Radhakrishna, 2017). Additionally, they also derive pleasure from the interactions and personal well-being from feeling that they are helping animals by providing them with food. Residents who interacted with capuchin monkeys in an urban forest fragment in Brazil expressed positive feelings towards them and liked feeding them (Suzin et al., 2017). Similarly, Leite et al. (2011) observe that people who interacted with urban marmosets in a city park in Brazil considered the species' presence important as they provided entertainment for children. Despite their many frustrations with the Barbary macaque, Gibraltar residents describe the species as 'ours', indicating a sense of connection with Barbary macaques and pride in their coexistence (Radford et al., 2018). Urban primates also serve as inspiration for human art and culture; historical and contemporary art and literature attest that primates continue to alter human lives and the metropolises they inhabit (Radhakrishna, 2018), much as they are transformed by the human spaces that they have made their own.

12.3.3 The Human-Primate Interface in Agroecosystems

The clearing of land for agricultural development and livestock farming across primate range countries has created another context in which humans and primates interface. In response to anthropogenic habitat change, many primates adjust their feeding, foraging, and ranging strategies to utilise resources provided in these altered landscapes (Galán-Acedo et al., 2019). One of the most common behavioural responses of primates is crop feeding, the consumption of cultivated foods in **agroecosystems** (Fig. 12.4), which in turn, can lead to resource competition and conflict with humans (Hill, 2015; Humle & Hill, 2016). Ethnoprimatologists study this interface by exploring *both* the primate and human perspective, that is, examining the ecological and nutritional factors driving crop feeding by primates, and people's perceptions, attitudes, and behaviour towards primate crop feeders (Riley, 2007).

Viewed through an ethnoprimatological lens, crop feeding is an excellent example of highly adaptive behaviour and a foraging strategy in its own right. Entering agricultural gardens and feeding on crops is a risky enterprise from the perspective of the primates as they may face retaliation from farmers (Hill, 2017). Critical to examining this flexible foraging strategy is understanding the trade-offs at play. For

Fig. 12.4 Camera trap photograph of a white-faced capuchin carrying a maize cob in a monocrop garden, Chiriquí province, Panama. (Photo credit: Luz I. Loría)

primates living in and around agroecosystems, crop feeding may provide an alternative source of food during periods of wild food scarcity (e.g. de Freitas et al., 2008; Hockings et al., 2009). Primates may also be attracted to crops regardless of seasonal patterns of wild food availability due to the nutritional and energetic benefits of cultivated foods, namely, they tend to be higher in digestible carbohydrates and lower in insoluble fibre (Bryson-Morrison et al., 2020; Riley et al., 2013).

From the human perspective, ethnoprimatological research has revealed considerable variation in the nature of interactions that ensue from overlapping resource use in agroecosystems, ranging from benign to neutral to negative. For example, while farmers in Brazil perceive that bearded capuchins (*C. libidinosus*) consume very little of their crops and, hence, do not harass them (Spagnoletti et al., 2017), in Indonesia, farmers consider orangutans (*Pongo* sp.) to be dangerous and are only tolerant towards crop feeding if they do not perceive the orangutans to be a physical threat (Campbell-Smith et al., 2010). Severe negative reactions to primate crop feeding often involve various forms of harassment, such as shouting, stone throwing, chasing with dogs, attaching a marker to crop feeding primates and then releasing them, and even killing them (Hill, 2018; Waters et al., 2019). On the other hand, in some cases, the co-use of crops leads to some advantages for people. For example, in Cantanhez National Park, Guinea-Bissau, chimpanzees (*Pan troglodytes verus*) only consume the fruit of cashew, an important cash crop, and discard the seed facilitating the harvesting of the valuable nut by the farmers (Hockings & Sousa, 2012).

In terms of what shapes peoples' perceptions, attitudes, and behaviour, socioeconomic factors have been found to be important. People are generally less tolerant of crop feeding behaviour when the cultivated food is a cash crop (Hill & Webber, 2010). When primates feed on non-commercial crops, however, they may be able to persist alongside people, such as the case of brown howler monkeys (*A. guariba*) in Porto Alegre, Brazil (Chaves & Bicca-Marques, 2017). In Ethiopia, local people who experienced relatively more crop damage displayed more negative attitudes towards Bale monkeys (*Cercopithecus djamdjamensis*) compared to people inhabiting the community where crop feeding was infrequent (Mekonnen et al., 2020). However, the intensity or severity of crop feeding need not always explain people's willingness to tolerate primate crop feeders. For example, Hardwick et al. (2017) found that farmers' attitudes towards the Buton macaque (*M. ochreata brunnescens*) varied across three communities of Sulawesi, Indonesia, whereby the least wealthy community used lethal control methods more frequently than the wealthier communities even when crop feeding by macaques was less severe. These authors also found a link between wealth and conservation education, such that farmers in the wealthiest community where lethal control appeared to be uncommon were aware of the macaque's ecological role. Similarly, landowners of the Curú Wildlife Refuge in Costa Rica recognise white-faced capuchins as key members of the forest community, thereby leading them to accept some crop damage by capuchins (McKinney, 2011).

Only quantifying crop losses, however, cannot explain what the losses mean for farmers and why some of them retaliate against primates that feed on crops while others do not (Hill, 2017). Thus, the ethnoprimatological approach encourages the exploration of other factors that influence people's attitudes and behaviour towards primate crop feeders, such as folklore, religious beliefs, and political ecological contexts (Dore, 2018). For example, in Lore Lindu National Park, Indonesia, villagers do not kill Tonkean macaques (*M. tonkeana*) in retaliation for crop feeding because of their cultural conceptions of how they should respond to monkeys encountered in their gardens (Riley, 2010). Balinese people consider long-tailed macaques as sacred due to the prominent role Hanuman, the monkey god, plays in the Hindu epic 'Ramayana' (Wheatley, 1999). As a result, the macaques are protected; however, when the macaques leave sacred spaces, such as temples, and feed on cultivated food in neighbouring farms, their perceived sacredness and associated protection wanes (Peterson et al., 2015), and sometimes they are shot by farmers (Schillaci et al., 2010). The underlying political ecological context of crop feeding scenarios can also explain why farmer-primate conflict exists and persists. On the Caribbean island of St. Kitts, Dore et al. (2018) found that rates of contact between farmers and green monkeys (*Chlorocebus sabeus*) increased because of the shift from sugar cane estates to small-scale farms. Kittitian farmers' negative perceptions of the monkeys derive from the latter's crop feeding behaviour, and yet farmers are not willing to invest in mitigation strategies because they do not own the land (Dore, 2018).

Primates have also been shown to use elements of agroecosystems that result in benign interactions between people and primates. For instance, polycultures and agroforestry systems may become temporary or permanent habitats for primates that feed on the native vegetation growing around or among crops rather than the crops itself. Central American farmers, howler monkeys and white-faced capuchins peaceably coexist in shade coffee plantations, where the primates feed on the leaves and fruits of shade trees without damaging the crop (Loría & Méndez-Carvajal, 2017; Williams-Guillén et al., 2006). Similarly, Javan slow lorises (*Nycticebus javanicus*) use chayote bamboo frames for foraging on insects and moving across open agricultural fields in Cipaganti, Indonesia (Nekaris et al., 2017). Since the lorises' consumption of insects on bamboo frames may help minimise the presence of insect pests in chayote crops, the Cipaganti farmers perceived them as useful and tolerated these primates.

12.4   Summary and Conclusions

Ethnoprimatology recognises humans as integral components of primate community ecology and has made the 'human-nonhuman primate community' its central focus. As we showcased in this chapter, the ethnoprimatological approach strives to investigate human-primate interfaces from *both* human and primate perspectives to understand how humans and other primates are co-shaping each other's ecology, sociality, and evolutionary trajectories.

From the outset, natural history has been an important component of ethnopri-matology's toolkit (as it should for all ethological investigations; see Altmann & Altmann, 2003; Tinbergen, 1963). Ethnoprimatological research on the human-primate interface has also been theoretically rigorous; it has drawn from behav-ioural ecology, community ecology, and life history theory to examine, for example, the causes and consequences of human-primate sympatry (Hockings et al., 2012; Morrow et al., 2019). Moreover, by viewing humans as integral components of pri-mates' community ecology, ethnoprimatology is aligned with other evolutionary sciences studying how human niche construction activities act as a major evolution-ary force on wildlife species across both time and space (Boivin et al., 2016; Sih, 2013). Given its focus on the human dimensions of primate ecology and behaviour, ethnoprimatology has also benefited from perspectives from the social sciences and humanities, thereby making it a methodologically and theoretically diverse field of study (Riley, 2018).

The conceptual contributions that stem from ethnoprimatological research are manifold. In addition to recognising the role of humans in effecting evolutionary change, ethnoprimatological research has highlighted the need to rethink the terms we use to characterise ecological relationships that arise from human-primate sym-patry (Riley, 2018). Primates that live in proximity to humans have long been described as 'commensal' (Maréchal & McKinney, 2020). A commensal relation-ship is defined as when one species benefits while the other is unaffected. However, it is becoming increasingly clear that neither humans nor primates are unaffected by their ecological overlap. Rather, as we have shown in this chapter, both humans and primates can be negatively or positively impacted. Because words matter in the practice of science, sympatry or **synanthropy** – that is, cohabiting with humans – constitute more accurate terms to characterise human-primate interfaces (Maréchal & McKinney, 2020; Riley, 2018).

12.4.1 Why Does Studying the Human-Primate Interface Matter?

In the current era in which human-primate sympatry is becoming increasingly wide-spread, solely relying on traditional forest and wildlife protection approaches that emphasise human exclusion, ignore anthropogenic environments, or discount com-munities' perspectives can no longer be an effective strategy for conserving pri-mates (Chapman & Peres, 2001; Riley, 2019). With its focus on both people and primates, ethnoprimatological research has assisted in the development of socially just conservation measures, the mitigation of human-primate interfaces, and inform-ing efforts to ensure that people and primates can coexist in the future (McLennan et al., 2017; Nekaris et al., 2017; Riley, 2019). For example, by integrating elements from ethnoprimatology and **multispecies ethnography**, Parathian et al. (2018) were able to appreciate the mutuality and complexity of the relationships Nalu people and chimpanzees have in Guinea-Bissau, providing insight to inform the development of socially inclusive measures (see also Remis & Hardin, 2009). Similarly, in their work on the human-macaque interface in north Morocco, Waters et al. (2018, 2019) focused on understanding local shepherds' knowledge and perceptions towards sympatric Barbary macaques. The researchers' immersive methodological approach enabled trust building with the shepherds, who in turn became more accepting of information shared by the researchers, ultimately altering how they interacted with the macaques. For example, upon learning about Barbary macaques' ecological role in the forest and their endangered status, shepherds stopped persecuting them during encounters.

Finally, beyond the major anthropogenic activities, such as expanding agricultural landscapes and increasing urbanisation, that affect the ability of primates to successfully coexist with humans, in the twenty-first century and beyond, zoonotic disease transmission will continue to be a major concern. The COVID-19 pandemic has resulted in shifting human-primate interfaces worldwide. Stay-at-home orders, travel restrictions, and the closures of protected areas and other tourist sites have meant fewer tourists and, hence, fewer human-macaque encounters at those sites. While some of the more positive outcomes of this pandemic-induced shift (e.g. reduced rates of provisioning; reduced risk of zoonotic exchange) may only be temporary, it is possible that the COVID-19 pandemic has made communities more receptive to messaging about the risk provisioning, and other encounters with primates pose for human-primate aggression and human-primate disease transmission (Balasubramaniam et al., 2020; Lappan et al., 2020). What is certain though is that ethnoprimatology, with its attention to both primates and people, will continue to offer place-based, culturally sensitive, and impactful recommendations to advance human-primate coexistence moving forward.

References

Altmann, S. A., & Altmann, J. (2003). The transformation of behaviour field studies. *Animal Behaviour, 65*, 413–423.

Ampumuza, C., & Driessen, C. (2021). Gorilla habituation and the role of animal agency in conservation and tourism development at Bwindi, south western Uganda. *Environment and Planning E: Nature and Space.* https://doi.org/10.1177/2514848620966502

Anand, S., Binoy, V. V., & Radhakrishna, S. (2018). The monkey is not always a god: Attitudinal differences toward crop-raiding macaques and why it matters for conflict mitigation. *Ambio, 47*, 711–720.

Balasubramaniam, K. N., et al. (2020). Impact of individual demographic and social factors on human–wildlife interactions: A comparative study of three macaque species. *Scientific Reports, 10*, 21991. https://doi.org/10.1038/s41598-020-78881-3

Barua, M., & Sinha, A. (2019). Animating the urban: An ethological and geographical conversation. *Social & Cultural Geography, 20*, 1160–1180.

Behie, A. M., Pavelka, M. S. M., & Chapman, C. A. (2010). Sources of variation in fecal cortisol levels in howler monkeys in Belize. *American Journal of Primatology, 72*, 600–606.

Beisner, B. A., Finn, K. R., Boussina, T., et al. (2016). Network position and human presence in Barbary macaques of Gibraltar. *American Journal of Physical Anthropology, 159*, 91–91.

Berman, C. M., Li, J., Ogawa, H., Ionica, C., & Yin, H. (2007). Primate tourism, range restriction, and infant risk among Macaca Thibetana at Mt. Huangshan, China. *International Journal of Primatology, 28*, 1123–1141.

Boivin, N. L., et al. (2016). Ecological consequences of human niche construction: Examining long-term anthropogenic shaping of global species distributions. *Proceedings of the National Academy of Sciences, 113*, 6388–6396.

Borg, C., Majolo, B., Qarro, M., & Semple, S. (2014). A comparison of body size, coat condition and endoparasite diversity of wild Barbary macaques exposed to different levels of tourism. *Anthrozoös, 27*, 49–63.

Brotcorne, F., et al. (2017). Intergroup variation in robbing and bartering by long-tailed macaques at Uluwatu temple (Bali, Indonesia). *Primates, 58*, 505–516.

Bryson-Morrison, N., Beer, A., Gaspard Soumah, A., et al. (2020). The macronutrient composition of wild and cultivated plant foods of West African chimpanzees (*Pan troglodytes verus*) inhabiting an anthropogenic landscape. *American Journal of Primatology, 82*. https://doi.org/10.1002/ajp.23102

Campbell-Smith, G., Simanjorang, H. V. P., Leader-Williams, N., & Linkie, M. (2010). Local attitudes and perceptions toward crop-raiding by orangutans (*Pongo abelii*) and other nonhuman primates in northern Sumatra, Indonesia. *American Journal of Primatology, 72*, 866–876.

Chapman, C. A., & Peres, C. A. (2001). Primate conservation in the new millennium: The role of scientists. *Evolutionary Anthropology, 10*, 16–33.

Chaves, Ó. M., & Bicca-Marques, J. C. (2017). Crop feeding by brown howlers (*Alouatta guariba clamitans*) in forest fragments: The conservation value of cultivated species. *International Journal of Primatology, 38*, 263–281.

Cormier, L. A. (2003). *Kinship with monkeys: The Guaja foragers of eastern Amazonia*. Columbia University Press.

Cunha, A. A., Vieira, M. V., & Grelle, C. E. V. (2006). Preliminary observations on habitat, support use and diet in two non-native primates in an urban Atlantic Forest fragment: The capuchin monkey (*Cebus* sp.) and the common marmoset (*Callithrix jacchus*) in the Tijuca forest, Rio de Janeiro. *Urban Ecosystem, 9*, 351–359.

de la Torre, S., Snowdon, C. T., & Bejarano, M. (2000). Effects of human activities on wild pygmy marmosets in Ecuadorian Amazonia. *Biological Conservation, 94*, 153–163.

Devaux, C. A., Mediannikov, O., Medkour, H., & Raoult, D. (2019). Infectious disease risk across the growing human-nonhuman primate interface: A review of the evidence. *Frontiers in Public Health, 7*. https://doi.org/10.3389/fpubh.2019.00305

Dore, K. M. (2018). Ethnoprimatology without conservation: The political ecology of farmer–green monkey (*Chlorocebus sabaeus*) relations in St. Kitts, West Indies. *International Journal of Primatology, 39*, 918–944.

Dore, K. M., Riley, E. P., & Fuentes, A. (2017). *Ethnoprimatology: A practical guide to research at the human–nonhuman primate interface*. Cambridge University Press.

Dore, K. M., Radford, L., Alexander, S., & Waters, S. (2018). Ethnographic approaches in primatology. *Folia Primatologica, 89*, 5–12.

Duarte, M. H., Vecci, M. A., Hirsch, A., & Young, R. J. (2011). Noisy human neighbours affect where urban monkeys live. *Biology Letters, 7*(6), 840–842.

Dunay, E., Apakupakul, K., Leard, S., Palmer, J. L., & Deem, S. L. (2018). Pathogen transmission from humans to great apes is a growing threat to primate conservation. *EcoHealth, 15*, 148–162.

Ellwanger, A. L., & Lambert, J. E. (2018). Investigating niche construction in dynamic human-animal landscapes: Bridging ecological and evolutionary timescales. *International Journal of Primatology, 39*, 797–816.

Estrada, A., et al. (2017). Impending extinction crisis of the world's primates: Why primates matter. *Science Advances, 3*, e1600946.

Fehlmann, G., O'Riain, M. J., Kerr-Smith, C., & King, A. J. (2017). Adaptive space use by baboons (*Papio ursinus*) in response to management interventions in a human-changed landscape. *Animal Conservation, 20*, 101–109.

Freitas, C. H. D., Setz, E. Z. F., Araújo, A. R. B., & Gobbi, N. (2008). Agricultural crops in the diet of bearded capuchin monkeys, *Cebus libidinosus spix* (primates: Cebidae), in forest fragments in southeast Brazil. *Revista Brasileira de Zoologia, 25*, 32–39.

Fuentes, A. (2010). Naturalcultural encounters in Bali: Monkeys, temples, tourists, and ethnoprimatology. *Cultural Anthropology, 25*, 600–624.

Fuentes, A. (2012). Ethnoprimatology and the anthropology of the human-primate interface. *Annual Review of Anthropology, 41*, 101–117.

Fuentes, A., & Gamerl, S. (2005). Disproportionate participation by age/sex classes in aggressive interactions between long-tailed macaques (*Macaca fascicularis*) and human tourists at Padangtegal monkey forest, Bali, Indonesia. *American Journal of Primatology, 66*, 197–204.

Fuentes, A., & Wolfe, L. (2002). *Primates face to face: The conservation implications of human-nonhuman primate interconnections.* Cambridge University Press.

Fuentes, A., Shaw, E., & Cortes, J. (2007). Qualitative assessment of macaque tourist sites in Padangtegal, Bali, Indonesia, and the Upper Rock Nature Reserve, Gibraltar. *International Journal of Primatology, 28*, 1143–1158.

Galán-Acedo, C., Arroyo-Rodríguez, V., Andresen, E., et al. (2019). The conservation value of human-modified landscapes for the world's primates. *Nature Communications, 10*, 152.

Gilhooly, L. J., Burger, R., Sipangkui, S., & Colquhoun, I. C. (2021). Tourist behavior predicts reactions of macaques (*Macaca fascicularis* and *M. nemestrina*) at Sepilok orang-utan rehabilitation centre, Sabah, Malaysia. *International Journal of Primatology, 42*, 349–368.

Goulart, V. D., Teixeira, C. P., & Young, R. J. (2010). Analysis of callouts made in relation to wild urban marmosets (*Callithrix penicillata*) and their implications for urban species management. *European Journal of Wildlife Research, 56*, 641–649.

Grossberg, R., Treves, A., & Naughton-Treves, L. (2003). The incidental ecotourist: Measuring visitor impacts on endangered howler monkeys at a Belizean archaeological site. *Environmental Conservation, 30*, 40–51.

Hansen, M. F., et al. (2020). Comparative home range size and habitat selection in provisioned and non-provisioned long-tailed macaques (*Macaca fascicularis*) in Baluran National Park, East Java, Indonesia. *Contributions to Zoology, 89*, 393–411.

Hanson, K. T., & Riley, E. P. (2018). Beyond neutrality: The human–primate interface during the habituation process. *International Journal of Primatology, 39*, 852–877.

Hardin, R., & Remis, M. J. (2006). Biological and cultural anthropology of a changing tropical forest: A fruitfall collaboration across subfields. *American Anthropologist, 108*, 273–285.

Hardwick, J. L., Priston, N. E. C., Martin, T. E., et al. (2017). Community perceptions of the crop-feeding Buton macaque (*Macaca ochreata brunnescens*): An ethnoprimatological study on Buton Island, Sulawesi. *International Journal of Primatology, 38*, 1102–1119.

Hasan, M. K., et al. (2013). Distribution of rhesus macaques (*Macaca mulatta*) in Bangladesh: Inter-population variation in group size and composition. *Primate Conservation, 26*, 125–132.

Hill, C. M. (2015). Perspectives of "conflict" at the wildlife–agriculture boundary: 10 years on. *Human Dimensions of Wildlife, 20*(4), 296–301.

Hill, C. M. (2017). Primate crop feeding behavior, crop protection, and conservation. *International Journal of Primatology, 38*, 385–400.

Hill, C. M. (2018). Crop foraging, crop losses, and crop raiding. *Annual Review of Anthropology, 47*, 377–394.

Hill, C. M., & Webber, A. D. (2010). Perceptions of nonhuman primates in human–wildlife conflict scenarios. *American Journal of Primatology, 72*, 919–924.

Hockings, K. J., & Sousa, C. (2012). Differential utilization of cashew—a low-conflict crop—by sympatric humans and chimpanzees. *Oryx, 46*, 375–381.

Hockings, K. J., Anderson, J., & Matsuzawa, T. (2009). Use of wild and cultivated foods by chimpanzees at Bossou, Republic of Guinea: Feeding dynamics in a human-influenced environment. *American Journal of Primatology, 71*, 636–646.

Hockings, K. J., Anderson, J., & Matsuzawa, T. (2012). Socioecological adaptations by chimpanzees, pan troglodytes verus, inhabiting an anthropogenically impacted habitat. *Animal Behaviour, 83*, 801–810.

Hockings, K. J., Parathian, H., Bessa, J., & Frazão-Moreira, A. (2020). Extensive overlap in the selection of wild fruits by chimpanzees and humans: Implications for the management of complex social-ecological systems. *Frontiers in Ecology and Evolution, 8*. https://doi.org/10.3389/fevo.2020.00123

Hoffman, T. S., & O'Riain, M. J. (2011). The spatial ecology of chacma baboons (*Papio ursinus*) in a human-modified environment. *International Journal of Primatology, 32*, 308–328.

Hsu, M. J., Kao, C., & Agoramoorthy, G. (2009). Interactions between visitors and Formosan macaques (*Macaca cyclopis*) at Shou-shan Nature Park, Taiwan. *American Journal of Primatology, 71*, 214–222.

Huemer, H. P., Larcher, C., Czedik-Eysenberg, T., et al. (2002). Fatal infection of a pet monkey with human herpesvirus. *Emerging Infectious Diseases, 8*, 639–642.

Humle, T., & Hill, C. (2016). People–primate interactions: Implications for primate conservation. In S. Wich & A. J. Marshall (Eds.), *An introduction to primate conservation* (pp. 219–240). Oxford University Press.

Ilham, K., Rizaldi, Nurdin, J., & Tsuji, Y. (2017). Status of urban populations of the long-tailed macaque (*Macaca fascicularis*) in West Sumatra, Indonesia. *Primates, 58*, 295–305.

Jolly, A. (2009). Coat condition of ring-tailed lemurs, *Lemur catta* at Berenty Reserve, Madagascar: I. Differences by age, sex, density and tourism, 1996–2006. *American Journal of Primatology, 71*, 191–198.

Jones-Engel, L., et al. (2006). Temple monkeys and health implications of commensalism, Kathmandu, Nepal. *Emerging Infectious Diseases, 12*, 900–906.

Jones-Engel, L., et al. (2008). Diverse contexts of zoonotic transmission of simian foamy viruses in Asia. *Emerging Infectious Diseases, 14*, 1200–1208.

Jones-Engel, L., Engel, G. A., Gumert, M. D., & Fuentes, A. (2011). Developing sustainable human–macaque communities. In M. D. Gumert, A. Fuentes, & L. Jones-Engel (Eds.), *Monkeys on the edge: Ecology and management of long-tailed macaques and their interface with humans* (pp. 295–327). Cambridge University Press.

Jost Robinson, C. A., & Remis, M. J. (2018). Engaging holism: Exploring multispecies approaches in ethnoprimatology. *International Journal of Primatology, 39*, 776–796.

Junge, R. E., Barrett, M. A., & Yoder, A. D. (2011). Effects of anthropogenic disturbance on indri (*Indri indri*) health in Madagascar. *American Journal of Primatology, 73*, 632–642.

Kaburu, S. S., et al. (2019). Rates of human–macaque interactions affect grooming behavior among urban-dwelling rhesus macaques (*Macaca mulatta*). *American Journal of Physical Anthropology, 168*, 92–103.

Klailova, M., Hodgkinson, C., & Lee, P. C. (2010). Behavioral responses of one western lowland gorilla (*Gorilla gorilla gorilla*) group at Bai Hokou, Central African Republic, to tourists, researchers and trackers. *American Journal of Primatology, 72*, 897–906.

Klegarth, A. R., et al. (2017). Urban primate ranging patterns: GPS-collar deployments for *Macaca fascicularis* and *M. sylvanus*. *American Journal of Primatology, 79*. https://doi.org/10.1002/ajp.22633

Kowalewski, M. M., Salzer, J. S., Deutsch, J. C., et al. (2011). Black and gold howler monkeys (*Alouatta caraya*) as sentinels of ecosystem health: Patterns of zoonotic protozoa infection relative to degree of human–primate contact. *American Journal of Primatology, 73*, 75–83.

Kumar, R., Radhakrishna, S., & Sinha, A. (2011). Of least concern? Range extension by rhesus macaques (*Macaca mulatta*) threatens long-term survival of bonnet macaques (*M. radiata*) in peninsular India. *International Journal of Primatology, 32*, 945–959.

Lappan, S., Malaivijitnond, S., Radhakrishna, S., Riley, E. P., & Ruppert, N. (2020). The human–primate interface in the new normal: Challenges and opportunities for primatologists in the covid-19 era and beyond. *American Journal of Primatology, 82*. https://doi.org/10.1002/ajp.23176

Lee, B. P., & Chan, S. (2011). Lessons and challenges in the management of long-tailed macaques in urban Singapore. In M. D. Gumert, A. Fuentes, & L. Jones-Engel (Eds.), *Monkeys on the edge: Ecology and management of long-tailed macaques and their interface with humans* (p. 307). Cambridge University Press.

Leite, G. C., Duarte, M. H. L., & Young, R. J. (2011). Human–marmoset interactions in a city park. *Applied Animal Behaviour Science, 132*, 187–192.

Lokschin, L. X., Rodrigo, C. P., Hallal Cabral, J. N., & Buss, G. (2007). Power lines and howler monkey conservation in Porto Alegre, Rio Grande do Sul, Brazil. *Neotropical Primates, 14*, 76–80.

Loría, L. I., & Méndez-Carvajal, P. G. (2017). Uso de habitat y patrón de actividad del mono cariblanco (*Cebus imitator*) en un agroecosistema cafetalero en la provincia de chiriquí, panamá. *Tecnociencia, 19*, 61–78.

Loudon, J. E., Sauther, M. L., Fish, D. D., Hunter-Ishikawa, M., & Ibrahim, Y. J. (2006). One reserve, three primates: Applying a holistic approach to understand the interconnections among ring-tailed lemurs (*Lemur catta*), verraux's sifaka (*Propithecus verreauxi*), and humans (*Homo sapiens*) at Beza Mahafaly Special Reserve, Madagascar. *Ecological and Environmental Anthropology, 2*, 54–74.

Loudon, J. E., Grobler, J. P., Sponheimer, M., et al. (2014). Using the stable carbon and nitrogen isotope compositions of vervet monkeys (*Chlorocebus pygerythrus*) to examine questions in ethnoprimatology. *PLoS One, 9*, e100758.

Maibeche, Y., Moali, A., Yahi, N., & Menard, N. (2015). Is diet flexibility an adaptive life trait for relictual and peri-urban populations of the endangered primate *Macaca sylvanus*? *PLoS One, 10*, e0118596.

Malik, I., Seth, P. K., & Southwick, C. H. (1984). Population growth of free-ranging rhesus monkeys at Tughlaqabad. *American Journal of Primatology, 7*, 311–321.

Malone, N., Wade, A. H., Fuentes, A., et al. (2014). Ethnoprimatology: Critical interdisciplinarity and multispecies approaches in anthropology. *Critique of Anthropology, 34*, 8–29.

Maréchal, L., Semple, S., Majolo, B., Qarro, M., Heistermann, M., & MacLarnon, A. (2011, 2011/09/01/). Impacts of tourism on anxiety and physiological stress levels in wild male Barbary macaques. *Biological Conservation, 144*(9), 2188–2193.

Maréchal, L., & McKinney, T. (2020). The (mis)use of the term "commensalism" in primatology. *International Journal of Primatology, 41*, 1–4.

Maréchal, L., Semple, S., Majolo, B., & MacLarnon, A. (2016). Assessing the effects of tourist provisioning on the health of wild Barbary macaques in Morocco. *PLoS One, 11*, e0155920.

McCarthy, M. S., Matheson, M. D., Lester, J. D., et al. (2009). Sequences of Tibetan macaque (*Macaca thibetana*) and tourist behaviors at Mt. Huangshan, China. *Primate Conservation, 24*, 145–151.

McKinney, T. (2011). The effects of provisioning and crop-raiding on the diet and foraging activities of human-commensal white-faced capuchins (*Cebus capucinus*). *American Journal of Primatology, 73*, 439–448.

McKinney, T. (2014). Species-specific responses to tourist interactions by white-faced capuchins and mantled howlers in a Costa Rican wildlife refuge. *International Journal of Primatology, 35*, 573–589.

McLennan, M. R., Spagnoletti, N., & Hockings, K. J. (2017). The implications of primate behavioral flexibility for sustainable human–primate coexistence in anthropogenic habitats. *International Journal of Primatology, 38*, 105–121.

Mekonnen, A., Fashing, P. J., Bekele, A., & Stenseth, N. C. (2020). Use of cultivated foods and matrix habitat by bale monkeys in forest fragments: Assessing local human attitudes and perceptions. *American Journal of Primatology, 82*. https://doi.org/10.1002/ajp.23074

Mikula, P., Šaffa, G., Nelson, E., & Tryjanowski, P. (2018). Risk perception of vervet monkeys *Chlorocebus pygerythrus* to humans in urban and rural environments. *Behavioural Processes, 147*, 21–27.

Morrow, K. S., Glanz, H., Ngakan, P. O., & Riley, E. P. (2019). Interactions with humans are jointly influenced by life history stage and social network factors and reduce group cohesion in moor macaques (*Macaca maura*). *Scientific Reports, 9*, 20162. https://doi.org/10.1038/s41598-019-56288-z

Muehlenbein, M. P. (2016). Disease and human/animal interactions. *Annual Review of Anthropology, 45*, 395–416.

Nekaris, K. A. I., Poindexter, S., Reinhardt, K. D., et al. (2017). Coexistence between Javan slow lorises (*Nycticebus javanicus*) and humans in a dynamic agroforestry landscape in West Java, Indonesia. *International Journal of Primatology, 38*, 303–320.

Parathian, H. E., McLennan, M. R., Hill, C. M., et al. (2018). Breaking through disciplinary barriers: Human–wildlife interactions and multispecies ethnography. *International Journal of Primatology, 39*, 749–775.

Paterson, J., & Wallis, J. (2005). *Commensalism and conflict: The human-primate interface. Special topics in primatology* (Vol. 4). ASP.

Peterson, J. V., Riley, E. P., & Putu Oka, N. (2015). Macaques and the ritual production of sacredness among Balinese transmigrants in South Sulawesi, Indonesia. *American Anthropologist, 117*, 71–85.

Pinheiro, T., Guevara, R., & Aparecida, M. (2018). Behavioral changes of free-living squirrel monkeys (*Saimiri collinsi*) in an urban park. *Primate Conservation, 32*, 89–94.

Pragatheesh, A. (2011). Effect of human feeding on the road mortality of rhesus macaques on national highway-7 routed along Pench Tiger Reserve, Madhya Pradesh, India. *Journal of Threatened Taxa, 3*, 1656–1662.

Radford, L., Alexander, S., & Waters, S. (2018). On the rocks: Using discourse analysis to examine relationships between Barbary macaques (*Macaca sylvanus*) and people on Gibraltar. *Folia Primatologica, 89*, 30–44.

Radhakrishna, S. (2017). Cultural and religious aspects of primate conservation. In A. Fuentes (Ed.), *The international encyclopedia of primatology* (pp. 1–8). Wiley.

Radhakrishna, S. (2018). Primate tales: Using literature to understand changes in human–primate relations. *International Journal of Primatology, 39*, 878–894.

Radhakrishna, S., & Sengupta, A. (2020). What does human-animal studies have to offer ethology? *Acta Ethologica, 23*, 193–199.

Radhakrishna, S., & Shankar, R. T. R. (2016). Get the monkey off the back, op-ed. *The Tribune*, p. 9. Http://www.Tribuneindia.Com/news/comment/get-the-monkey-off-the-back/284023.Html

Radhakrishna, S., & Sinha, A. (2010). Dr. Jekyll and Mr. Hyde: The strange case of human-macaque interactions in India (commentary). *Current Conservation, 4*, 39–40.

Radhakrishna, S., & Sinha, A. (2011). Less than wild? Commensal primates and wildlife conservation. *Journal of Biosciences, 36*, 749–753.

Remis, M. J., & Hardin, R. (2009). Transvalued species in an African forest. *Conservation Biology, 23*, 1588–1596.

Richard, A. F., Goldstein, S. J., & Dewar, R. E. (1989). Weed macaques: Evolutionary implications of macaque feeding ecology. *International Journal of Primatology, 10*, 569–594.

Riley, E. P. (2007). The human-macaque interface: Conservation implications of current and future overlap and conflict in Lore Lindu National Park, Sulawesi, Indonesia. *American Anthropologist, 109*, 473–484.

Riley, E. P. (2010). The importance of human-macaque folklore for conservation in Lore Lindu National Park, Sulawesi, Indonesia. *Oryx, 44*, 235–240.

Riley, E. P. (2013). Contemporary primatology in anthropology: Beyond the epistemological abyss. *American Anthropologist, 115*, 411–422.

Riley, E. P. (2018). The maturation of ethnoprimatology: Theoretical and methodological pluralism. *International Journal of Primatology, 39*, 705–729.

Riley, E. P. (2019). *The promise of contemporary primatology*. Routledge.

Riley, E. P., & Wade, T. W. (2016). Adapting to Florida's riverine woodlands: The population status and feeding ecology of the Silver River rhesus macaques and their interface with humans. *Primates, 57*, 195–121.

Riley, E. P., Tolbert, B., & Farida, W. R. (2013). Nutritional context explains the attractiveness of cacao to crop raiding Tonkean macaques. *Current Zoology, 59*, 160–169.

Riley, E. P., Shaffer, C. A., Trinidad, J. S., et al. (2021). Roadside monkeys: Anthropogenic effects on moor macaque (*Macaca maura*) ranging behavior in Bantimurung Bulusaraung National Park, Sulawesi, Indonesia. *Primates, 62*, 477–489.

Ruesto, L. A., Sheeran, L. K., Matheson, M. D., et al. (2010). Tourist behavior and decibel levels correlate with threat frequency in Tibetan macaques (*Macaca thibetana*) at Mt. Huangshan, China. *Primate Conservation, 2010*, 99–104.

Scheun, J., Bennett, N. C., Ganswindt, A., & Nowack, J. (2015). The hustle and bustle of city life: Monitoring the effects of urbanisation in the African lesser bushbaby. *The Science of Nature, 102*, 57. https://doi.org/10.1007/s00114-015-1305-4

Schillaci, M. A., Engel, G. A., Fuentes, A., et al. (2010). The not-so-sacred monkeys of Bali: A radiographic study of human-primate commensalism. In S. Gursky & J. Supriatna (Eds.), (pp. 249–256). Springer.

Sengupta, A., & Radhakrishna, S. (2018a). *For the love of selfies, monkeys and food: Initiation and recurrence of human-primate encounters in Bengaluru, southern India*. Paper presented at the 27th International Primatological Society Congress, 19th–25th August 2018, Nairobi, Kenya.

Sengupta, A., & Radhakrishna, S. (2018b). The hand that feeds the monkey: Mutual influence of humans and rhesus macaques (*Macaca mulatta*) in the context of provisioning. *International Journal of Primatology, 39*, 817–830. https://doi.org/10.1007/s10764-018-0014-1

Sengupta, A., & Radhakrishna, S. (2020). Actors predicting provisioning of macaques by humans at tourist sites. *International Journal of Primatology, 41*, 471–485.

Sengupta, A., McConkey, K. R., & Radhakrishna, S. (2015). Primates, provisioning and plants: Impacts of human cultural behaviours on primate ecological functions. *PLoS One, 10*, e0140961.

Sha, J. C. M., & Hanya, G. (2013). Diet, activity, habitat use, and ranging of two neighboring groups of food-enhanced long-tailed macaques (*Macaca fascicularis*). *American Journal of Primatology, 75*, 581–592.

Sha, J. C. M., Gumert, M. D., Lee, B. P. Y.-H., et al. (2009). Macaque–human interactions and the societal perceptions of macaques in Singapore. *American Journal of Primatology, 71*, 825–839.

Shaffer, C., Milstein, M., Suse, P., et al. (2018). Integrating ethnography and hunting sustainability modelling for primate conservation in an indigenous reserve in Guyana. *International Journal of Primatology, 39*, 945–968.

Shively, C. A., Register, T. C., & Clarkson, T. B. (2009). Social stress, visceral obesity, and coronary artery atherosclerosis: Product of a primate adaptation. *American Journal of Primatology, 71*, 742–751.

Shepard, G. H. (2002). Primates in Matsingenka subsistence and world view. In A. Fuentes & L. Wolfe (Eds.), Primates Face to Face: Conservation Implications of Human-Nonhuman Interconnections (pp. 101–136). Cambridge University Press.

Sih, A. (2013). Understanding variation in behavioural responses to human-induced rapid environmental change: A conceptual overview. *Animal Behaviour, 85*, 1077–1088.

Sousa, J., Hill, C. M., & Ainslie, A. (2017). Chimpanzees, sorcery and contestation in a protected area in Guinea-Bissau. *Social Anthropology, 25*, 364–379.

Spagnoletti, N., Cardoso, T. C. M., Fragaszy, D., & Izar, P. (2017). Coexistence between humans and capuchins (*Sapajus libidinosus*): Comparing observational data with farmers' perceptions of crop losses. *International Journal of Primatology, 38*, 243–262.

Sponsel, L. E. (1997). The human niche in Amazonia: Explorations in ethnoprimatology. In W. G. Kinzey (Ed.), *New world primates: Ecology, evolution, and behavior* (pp. 143–165). Aldine Gruyter.

Suzin, A., Back, J. P., Garey, M. V., & Aguiar, L. M. (2017). The relationship between humans and capuchins (*Sapajus* sp.) in an urban green area in Brazil. *International Journal of Primatology, 38*, 1058–1071.

Thatcher, H. R., Downs, C. T., & Koyama, N. F. (2018). Using parasitic load to measure the effect of anthropogenic disturbance on vervet monkeys. *EcoHealth, 15*, 676–681.

Tinbergen, N. (1963). On aims and methods of ethology. *Zeitschrift für Tierpsychologie, 20*, 410–433.

Tutin, C. E. G., & Oslisly, R. (1995). Homo, pan and gorilla: Co-existence over 60,000 years at lope in central Gabon. *Journal of Human Evolution, 28*, 597–602.

Waite, T. A., Chhangani, A. K., Campbell, L. G., et al. (2007). Sanctuary in the city: Urban monkeys buffered against catastrophic die-off during Enso-related drought. *EcoHealth, 4*, 278–286.

Waters, S., Bell, S., & Setchell, J. M. (2018). Understanding human-animal relations in the context of primate conservation: A multispecies ethnographic approach in North Morocco. *Folia Primatologica, 89*, 13–29.

Waters, S., El Harrad, A., Bell, S., & Setchell, J. M. (2019). Interpreting people's behavior toward primates using qualitative data: A case study from North Morocco. *International Journal of Primatology, 40*, 316–330.

Westin, J. L. (2007). *Effects of tourism on the behavior and health of red howler monkeys (Alouatta seniculus) in Suriname* (Ph.D. Dissertation). University of Michigan, Ann Arbor.

Wheatley, B. (1999). *The sacred monkeys of Bali*. Waveland Press.

Williams-Guillén, K., McCann, C., Martínez Sánchez, J. C., & Koontz, F. (2006). Resource availability and habitat use by mantled howling monkeys in a Nicaraguan coffee plantation: Can agroforests serve as core habitat for a forest mammal? *Animal Conservation, 9*, 331–338.

Yeo, J.-H., & Neo, H. (2010). Monkey business: Human–animal conflicts in urban Singapore. *Social & Cultural Geography, 11*, 681–699.

Zak, A. A., & Riley, E. P. (2017). Comparing the use of camera traps and farmer reports to study crop feeding behavior of moor macaques (*Macaca maura*). *International Journal of Primatology, 38*, 224–242.

Zhao, Q. K., & Deng, Z. Y. (1992). Dramatic consequences of food handouts to (*Macaca thibetana*) at Mount Emei, China. *Folia Primatologica, 58*, 24–31.

Part III
Primates in Captivity

Chapter 13
Perspectives on the Continuum of Wild to Captive Behaviour

Michelle A. Rodrigues, Partha Sarathi Mishra, and Michelle Bezanson

Contents

Abstract Historically, we have characterized the environments primates live in and that we observe them in as 'wild' and 'captive'. We assumed that primates exhibited 'natural' behaviour in the 'wild', whereas 'captive' behaviour was an 'unnatural' artefact of human management. This view later expanded to consider the influence of anthropogenic activities on wild primates, and the distinctions of free-ranging, provisioned animals and enriched naturalistic environments. However, these dichotomies remain. Here, we suggest primate environments should be considered a continuum, ranging from intact primary forest to heavily managed captive environments. Across this continuum, the behaviour we observe is a flexible response to ecological conditions shaped by human presence. Rather than viewing wild behaviour as natural, and captive behaviour as an artificial reaction to human management, recogni-

M. A. Rodrigues (✉)
Department of Social and Cultural Sciences, Marquette University, Milwaukee, WI, USA
e-mail: Michelle.Rodrigues@marquette.edu

P. S. Mishra
Srishti Manipal Institute of Arts Design and Technology, Govindapura, Karnataka, India

M. Bezanson
Department of Anthropology, Santa Clara University, Santa Clara, CA, USA

tion of how humans shape primates' environments across contexts improves our understanding of primate behaviour and evolution. Even the "wildest" primates' lives are altered and influenced by humans, and the behaviour of intensely managed captive primates reflects species-level behavioural flexibility and individual responses. Beyond the indirect effects of anthropogenic pressures and management, the behaviour we observe is responsive to human observation and interaction. Whether the presence of researchers following primates, the presence of local people encountering primates in daily activity, or humans caring for primates in managed environments, these observer effects influence behaviour. Recognizing how observations are shaped by anthropogenic and observer effects across the continuum of primate habitats allows us to identify how co-evolutionary pressures shape the evolution of primate socioecology and socioecological flexibility of primates.

Keywords Anthropogenic effects · Observer effects · Human-primate interface · Ethnoprimatology · Captivity

13.1   Introduction

Throughout their range, primates occur in habitats affected by a continuum of anthropogenic pressures. These pressures can be both direct and indirect, and we refer to them as indirect **anthropogenic effects** throughout this chapter. These anthropogenic effects are any influence from humans or human activity on habitat, forest structure, food sources, food availability, encroachment, hunting activity, or any results of human behaviour that reduce habitat size or function (Fuentes, 2006, 2012; Fuentes & Hockings, 2010; Malone et al., 2014; McKinney, 2015; Riley, 2019). They also include direct human pressures, or **observer effects**, including direct human interactions, observation, and habituation. Here, we consider anthropogenic pressures as anthropogenic effects, whereas the impact of human/researcher observation is considered observer effects. These interrelated pressures shape primate behavioural ecology, and those occurring over longer periods of time act as selective pressures.

Historically, we often assumed that wild primates existed apart from humans, in forests deep and wild enough to be minimally affected by human presence (Garland, 2008; Haraway, 1989; Jost Robinson & Remis, 2018). Such assumptions echo the development of North American primatology within anthropology, where similar assumptions about 'uncontacted' human groups represented a perspective that separated 'nature' from 'culture' (Haraway, 1989). While wild behaviour was considered 'natural' apart from human influences, the recognition that captivity shaped primate behavioural responses led to conclusions that such behaviour was 'unnatural' (Fedigan & Strum, 1999; Haraway, 1989; Strum & Fedigan, 2000). However, these assumptions gave way to greater recognition of how primates evolved in locations where they coexist with human presence. For example, there are reports of chimpanzees (*Pan troglodytes*) living in forests bordering human villages in West Africa dating back to the 1600s (Sept & Brooks, 1995).

With the rise of **ethnoprimatology** in the past 20 years, there is an increased awareness of how anthropogenic pressures shape wild primates (Estrada, 2006; Fuentes, 2012; Fuentes & Hockings, 2010; Malone et al., 2014; Riley, 2019; Riley & Fuentes, 2011; Sponsel, 1997). While wild and captive primate populations are traditionally considered two separate domains of research, here we suggest they are better understood as a continuum with continuous, overlapping anthropogenic and observer effects (Fig. 13.1). We now recognize that free-living primates in their natural ranges have significant interaction through human provisioning and observation; such circumstances overlap with the dynamics of free-ranging primates under managed human care outside their ranges. For example, from the monkeys' perspective, what are the distinctions between free-ranging macaques (*Macaca* spp.) provisioned at an Indian temple for the past century and those free-ranging on Cayo Santiago, Puerto Rico? While the earliest captive settings were limited to free-ranging introduced populations and laboratory, pet, or zoo environments where primates' lives were constrained to deeply 'unnatural' settings (Coe, 1989; Sampaio et al., 2020), the modern range of captive settings range from restrictively managed environments to those where primates live in complex ecosystems managed to mirror 'natural' social dynamics (Coe, 1989; Hosey, 2005). While such settings may

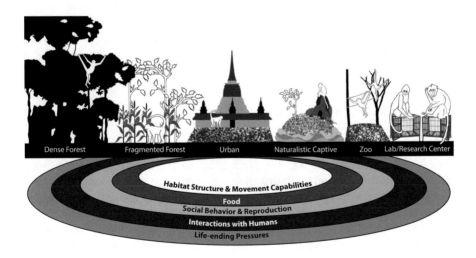

Fig. 13.1 The continuum of primate habitats, from dense forests to lab/research centres, and how anthropogenic and observer effects influence primate lives. Within habitats, anthropogenic and observer effects influence primate lifeways in varying ways (radiating circles). For example, the degree to which non-human primate individual responds to **human interactions** *is dependent on previous exposure. Human influence on* **habitat structure** *can impact* **primate movement capabilities** *from activities including forest use/selective logging to the extreme of human-built structures that house primates in research labs*

differ from wild ecosystems in many ways, rather than considering behaviours observed in these settings as artefacts, they should be viewed as flexible responses reflecting species' range of adaptability.

Keeping primates in human-managed settings has a long history, from provisioning of temple monkeys, to keeping infants as pets after hunting adults, and to assembling menageries (Coe, 1989; Fuentes & Wolf, 2002; Sampaio et al., 2020; Sponsel, 1997). Capturing primate infants begins with violence and trauma (Bradshaw et al., 2008; Crailsheim et al., 2020; Edes et al., 2016). The ability to survive this trauma and reproduce under novel circumstances favoured individuals who could adjust to new environments. The earliest captive environments were limited by a lack of understanding of wild primate behaviour to inform husbandry practices (Haraway, 1989; Rowell, 2000; Strum & Fedigan, 2000). Human-driven selection chose which species were most desirable or suitable to manage in captivity. For example, rhesus macaques (*Macaca mulatta*) may have become the 'model monkey' for biomedical research because their co-evolutionary history with humans and behavioural and ecological flexibility made them well-suited to thrive in often adverse conditions (Ahuja, 2013). Capture and transport of primates were shaped by priorities for biomedical models, ease of capture, entertainment draw, and size/ease of convenience (Ahuja, 2013; Coe, 1989; Haraway, 1989; Sampaio et al., 2020). This human-managed selection parallels existing co-evolutionary relationships between humans and non-human primates in their historical ranges.

Here we consider primates across the broad environmental categories: (1) primates in the 'wild', (2) primates who live among humans, and (3) primates in human-managed settings, with consideration for how these broad categories are continuous. We review how these overlapping environmental categories shape primate lives.

13.2   Primates in 'the Wild'

One might start a section on primates in the wild by questioning 'what is wild and does it truly exist?' We, however, are not going to address the meaning of wild because at this time, we can argue that all free-ranging primates face varying degrees of anthropogenic influence. This may be in the form of daily encounters with many humans and their tools/technology or seldom and brief encounters. Tydecks et al. (2016) identified 1268 biological field stations in 120 countries with the majority occurring where human impact is low to intermediate. Bezanson and McNamara (2019) identified 349 different field sites that were described in a review of 5 years of primatology publications. These ranged from forests without research facilities to well-established field sites with more than 40 years of documented research. In this section, we address how field sites fit on a continuum of anthropogenic influence in free-ranging primates. We explore the effects of research stations, infrastructure, and the influence of human observers.

13.2.1 The Field Site

Field researchers often begin their research plan thinking about a particular biome or location where they will work for weeks to years. We imagine a site with very few human interactions as we nobly set out to the 'wilds' to follow elusive primates. Will it be hot, humid, hilly, rocky, buggy, cold, snaky, swampy, sunny, rainy, dark, light, and what stresses might these impose on human comfort? When we complete our work, we move to site descriptions in our publications. Many field primatologists rely on the classification of biomes to describe the research site and for ecosystem mapping. Biomes are units of ecological classification based on vegetation, rainfall, and faunal and floristic communities that characterize the area (Bailey, 2004; Holdridge, 1947). These regions are controlled by climate; given how humans alter climate and landscapes, Ellis and Ramankutty (2008) suggested we might better understand ecological zones by considering anthropogenic pressures. Their resulting term 'anthromes' better captures the areas where we work and study primates. Anthromes are anthropogenic biomes and mosaic environments that integrate human activities into ecological classifications (Ellis & Ramankutty, 2008). Some examples of anthromes are 'populated forests', 'remote forests', 'populated rangelands', 'rainfed villages', and 'sparse trees'. Moving forward, primatologists might better describe field sites with this system to communicate how primate groups are impacted by human activity. The combination of habitat structure and anthropogenic activities influences resources, survival, and reproduction of individuals, generations, populations, and ultimately a world with 'wild' spaces.

A field site is an area of land kept in a natural state to promote research, education, and conservation and, in some cases, may involve tourism (Brussard, 1982). Some field sites are characterized by a healthy infrastructure with multiple housing units, a laboratory, teaching spaces, electricity, warm water, and opportunities to communicate via internet or phone. These field sites experience more researcher and education traffic and may experience greater conservation efforts (Bezanson & McNamara, 2019: Junker et al., 2020). Other field sites may be more rustic without services, and these sites experience less researcher and education traffic (Bezanson & McNamara, 2019). All field sites exist in proximity to human communities which vary with regard to human influence (Martin et al., 2014).

13.2.2 The Role of Research Teams, Learning Teams, Field Station Personnel, and Neighbouring Human Communities

When we first arrive to a research site, our expectations are met with the reality that a research site involves many humans with whom we must communicate, share resources, plan, manage, work, and all while (hopefully) interrogating the effects of our presence (e.g. Kutsukake, 2013; MacClancy & Fuentes, 2013; Malone et al.,

2010; Nelson et al., 2017; Riley & Bezanson, 2018; Strier, 2010). The first step of the research process when studying primates that have not been observed by previous researchers or subject to ongoing investigations is habituation. **Habituation** is a term for the degree of individual primate adjustment to human observation with the goal of all study subjects eventually behaving naturally or neutrally in the presence of researchers (Crofoot et al., 2010; Gazagne et al., 2020; Rasmussen, 1991; Tutin & Fernandez, 1991; van Krunkelsven et al., 1999; Williamson & Feistner, 2003). Scholars of the habituation process suggest that 'natural' behaviour or neutrality may never be met in the presence of researchers (Alcayna-Stevens, 2016; Hanson & Riley, 2018; Strier, 2013). For example, Hanson and Riley (2018) found that a group of moor macaques studied for more than 30 years did not act neutrally in the presence of researchers. This means that the macaques exhibited human-directed behaviours such as threatening, avoiding, approaching, monitoring, or quickly leaving the area where observers were present. Some primates also adjust activity budget, home ranges; exhibit varying degrees of stress during the habituation process; and influence the behaviour of other animals that are not the study target (Green & Gabriel, 2020; Hanson & Riley, 2018; Jack et al., 2008; McDougall, 2012; Rasmussen, 1991). Primates habituated to researcher presence are also habituated to all human presence. A lack of fear can potentially lead to poaching, crop foraging, poor nutrition, and pathogen exchange (e.g. Hockings et al., 2010; Malone et al., 2010; McLennan & Hill, 2010; Strier, 2013; Westin, 2017; Williamson & Fawcett, 2008). While many primatologists examine the influence of the habituation process, there are fewer studies on effects of the field site itself, including researcher traffic, student traffic, and infrastructure (Bezanson & McNamara, 2019; Bezanson et al., 2013; Fedigan, 2010; Strier, 2010).

When researchers travel to a site, they may bring research supplies, personal items, and other humans (e.g. assistants, students, family members). The research station might hire a few to many employees depending on the size of the station and the number of people temporarily calling the area home. Field stations vary in infrastructure, number of employees, funding, accessibility to the public, and local human community involvement (Bezanson & McNamara, 2019). Larger and longer-term field sites can promote long-term conservation programmes, train national and international students, safeguard the area from illegal activities, and provide predictable support for adjacent communities (Laurance, 2013; Lwanga & Isabirye-Basuta, 2008; Trevelyan & Nuttman, 2008). Key examples of successful long-term research sites are the Biological Station of Caratinga/Reserva Particular do Patrimônio Natural – Feliciano Miguel Abdalla (EBC/RPPN-FMA) in Brazil (Strier & Boubli, 2006), Ranomafana National Park in Madagascar (Wright & Andriamihaja, 2002), and Kibale National Park in Uganda (Wrangham, 2008). In addition to predictable support for adjacent communities, researchers working at these sites can often integrate local conservation programmes for community members. For example, Franquesa-Soler et al. (2020) describe a children's education art programme for 12 different communities in Southern Mexico. Larger, longer-term sites often attract eco-tourists who bring income and predictability. However, disproportionate support for some community members can introduce conflict among community members (e.g. Shoo and Songorwa 2013; Siswanto 2015). Field station traffic influences primates (and other species) by introducing trails, housing structures, trash, human foods (composting), field trash (e.g. trail markers, flagging tape, monitors), and pathogens (Bezanson et al., 2013; Fedigan, 2010; Riley & Bezanson, 2018; Strier, 2010). In fact, Laurance (2013) suggests that in some cases, field station or science safeguarding has the potential to cause more harm than good. Because of these risks, recognition of the impact of research team presence and careful consideration of the potential positive and negative effects of conducting field research should be part of the research planning process.

13.3 Primates Among Humans

For primates living alongside humans, anthropogenic pressures shape daily life. Anthropogenic activities that induce novel conditions change socioecological conditions and thereby the behaviour of individuals (Buskirk & Steiner, 2009). The behavioural responses are rapid, while the costs are low and have a major impact on the fitness of individuals and thereby the viability of the population (Tuomainen & Candolin, 2011). If the responses are adaptive and improve the fitness of individuals, they can protect a species from decline and extinction as well as provide more time for genetic adaptation (Pigliucci and Pigliucci, 2001; Tuomainen & Candolin, 2011; West-Eberhard, 1983).

13.3.1 Primates Living in Fragmented Habitats

Fragmentation of forests is a direct consequence of human presence which impacts primate populations by altering the habitat (Almeida-Rocha et al., 2017). Human intervention creates fragments in primate habitats which influence their population, activity budget, and social organization (Almeida-Rocha et al., 2017; Benchimol & Peres, 2014; Ram et al., 2003; also see McKinney, 2015 for review). Changes in the landscape and habitat can either be sudden (Cowlishaw & Dunbar, 2000; Chapman & Peres, 2001; Velankar et al., 2016) or gradual over time (Baranga et al., 2012; Marsh, 2003; Singh, 2019). Prominently, the presence of humans influences the feeding behaviour among primates which depends on the distribution and availability of food (Marsh, 2003). Provisioning, crop foraging, snatching human food, and garbage foraging are the major sources of food for primates in the anthropogenic landscapes (*Alouatta*: Asensio et al., 2009; Pozo-Montuy et al., 2013; *Cebus*: McKinney, 2011; *Macaca*: Kumara et al., 2009; Proffitt et al., 2018; Riley et al. 2021; *Trachypithecus*: Nijman & Nekaris, 2010; *Papio*: Nowak & Lee, 2013).

In many circumstances, behavioural flexibility determines the survivability of primates in a fragmented landscape, for example, urban areas (Marty et al., 2020; Ralainasolo et al., 2008). However, there is some discordance in the results found in studies under naturalistic conditions. In some studies, specialized diets helps primates survive in the fragmented landscape (Boyle & Smith, 2010), whereas in other studies, novel behavioural traits and flexibility in otherwise generalized diets were beneficial (Reader & Laland, 2003).

Most studies focus on behavioural plasticity in feeding behaviour of primates (reviewed in Nowak & Lee, 2013) and innovations to extract food from otherwise difficult sources (*Sapajus*: Moura & Lee, 2004; great apes: Watts, 2008, *Macaca*: Malaivijitnond et al., 2007; Pal et al., 2018; Tan et al., 2016). These behaviours are attributed to larger brains which help in adapting to altered environments ('cognitive buffer hypothesis') (Sol, 2009). However, the relationship between brain size and innovation is not straightforward. In some lemurs, despite the small brain to body ratios, there is behavioural flexibility (Cameron & Gould, 2013; Soma, 2006). Nowak and Lee (2013) emphasize that behavioural flexibility is not novel and is the 'norm rather than the exception'. There is a lack of information across taxa for comparative analysis of behavioural plasticity, and with increases in anthropogenic impact threatening primates with extinction, it is critically important that we better understand this plasticity. Ethnoprimatological studies provide an excellent opportunity to study conditions under which the behavioural ecology and behavioural flexibility in primates change under the influence of human-induced changes (Dore et al., 2018; Fuentes, 2012; McKinney & Dore, 2018). It presents a holistic approach for human-primate interaction affected by anthropogenic food.

13.3.2 Impacts of Shared Landscapes on Primate Behaviour

Adaptability of primates puts them at crossroads with humans, placing constraints on behaviour. Despite studies attempting to understand this relationship, there is still little information on the impact humans have on complex social behaviour in primates. A few recent studies investigated the effect of human presence on grooming strategies of macaques, but its impact on social relationships and fitness consequences is unknown (Balasubramaniam et al. 2020a, b; Beisner et al., 2015; Marty et al., 2019).

Proximity to humans causes time constraints on primates' social behaviours and reciprocity (Kaburu et al., 2019). Grooming is one of the most common of the social interactions and is beneficial for the primates by removing ectoparasites, lowering stress levels, and maintaining social cohesion (Seyfarth, 1977; Silk et al., 2009, 2010). The presence of humans can change activity and grooming patterns based on the balance of cost and benefits (Balasubramaniam et al., 2020a, b; Dhawale et al., 2020; Kaburu et al., 2019; Maréchal et al., 2011; Marty et al., 2019). Human presence increases the chances for sociality as more 'free time' is available with high-calorie anthropogenic food (McLennan & Ganzhorn, 2017; Riley et al., 2013). Also, there is higher grooming to cope with social stresses from living near humans (social stress hypothesis: Aureli & Yates, 2010; Maréchal et al., 2011; Schino et al., 1988).

Strong bonds among the primates can calm social and environmental stress (Silk et al., 2009, 2010; Young et al., 2014). Human influences reduce the strength of bonding among the individuals and increase aggressive behaviour by changing competitive regimes in the group, changing the social network, and altering mother-infant interactions (Holzner et al., 2019; Morrow et al., 2019). Therefore, living alongside humans constrains social behaviour in primates by changing the socio-ecological parameters and group structure.

13.4   Primates in Human-Managed Settings

Human-managed settings range along a continuum from free-ranging provisioned groups to naturalistic social groups in enriched-but-enclosed settings at zoos, sanctuaries, and research centres, to those in limited social groups and/or constrained captive environments, including those held as pets, at roadside zoos, or in laboratory environments where animals may be single or pair-housed (Baker, 2007; Coe, 1989; Fuentes & Wolf, 2002; Hosey, 2005; Sponsel, 1997). While the span of non-human primates living sympatrically with humans overlaps with those living in human-managed settings (Fig. 13.1), we focus this section on non-human primates that are (1) held in free-ranging settings where they are provisioned and receive veterinary care and (2) under human management providing animals with adequate social companionship, veterinary care, appropriate environmental conditions, and enrichment. The continuum of primates living in human-managed settings extends to those living in environments which do not meet the basic physical and social needs of primates (Hosey, 2005). While primates under enriched human-managed settings may exhibit the range of primates' behavioural, ecological, and social flexibility (Hosey, 2005; Rodrigues & Boeving, 2019), such impoverished environments likely exceed their range of adaptive coping abilities. On the 'wilder' side of the continuum are rehabilitated primates in sanctuaries or reintroduced populations receiving provisioning and medical/welfare interventions and free-ranging introduced primate colonies such as those on Cayo Santiago, Puerto Rico or St. Catherine's Island, Georgia (Beck, 2018; Hosey, 2005). On the managed side of this continuum are primates living in socially housed groups in zoos, sanctuaries, and laboratories (Baker, 2007; Coe, 1989; Hansen et al., 2018; Hosey, 2005).

13.4.1   Constraints of Human Management

Hosey (2005) describes zoo environments as those with primates living under human management, restricted or enclosed spaces, and large numbers of unfamiliar human visitors. 'Human management' includes feeding, veterinary care, and may include management of group size/composition and reproduction (Hosey, 2005). These constraints vary across human-managed conditions but overlap with those in free-ranging conditions. In McKinney's (2015) anthropogenic classification system, three of the four domains, landscape, diet, and human-non-human interface, overlap with these criteria. Primates living in natural landscapes experience less space restrictions, but increasingly fragmented forests may become restricted spaces (Benchimol & Peres, 2014; Estrada et al., 2017). Furthermore, provisioning occurs across wild and human-sympatric contexts, and at some sites, veterinary interventions occur (McKinney, 2015; Robbins et al., 2011). The soft-release stages of reintroductions and permanence of rehabilitated primates released onto islands (Beck, 2018; Chap. 15, this volume) occupy middle ground between 'wild' and 'captive'. Finally, exposure to unfamiliar humans is frequent for primates living at tourist sites and within cities (Hosey, 2005; Kaburu et al., 2019; Maréchal et al., 2011). Understanding the range of primate behavioural flexibility across anthropogenic pressures as part of the animals' ecological environment allows for insight into the full range of primate adaptive capabilities and niche construction (Day et al., 2003; McKinney, 2015).

13.4.2   Managing Primates' Needs

Increased understanding of primates' social, dietary, and space needs, in conjunction with increasing welfare standards, altered captive environments, though this process has often been trial-and-error. For example, Yerkes assumed chimpanzees were monogamous and housed them in nuclear family units (Haraway, 1989). Conversely, there were struggles breeding golden lion tamarins (*Leontopithecus rosalia*) in captivity until caretakers realized housing in pairs reduced stress (Kleiman, 1979, 1983). Similarly, infant chimpanzees were removed to nursery groups, assuming the standard of care would be better from humans. This practice was changed when humans recognized this practice caused harm and that the social transmission of mothering skills had generational effects (Davenport & Menzel, 1963; Chernus, 2008; Freeman & Ross, 2014). More recent studies shed light on managing primates in captivity. For example, female black and gold howler monkeys (*Aloutta caraya*) have better reproductive success under human management when housed in larger 'family' groups, rather than pairs, and when they can regularly hear the howls of familiar conspecifics (Farmer et al., 2011; Pastor-Nieto, 2015).

13.4.3   What Is 'Natural' Behaviour?

In human management, 'naturalistic', wild-like behaviours are viewed as ideal. However, 'unnatural' behaviour may not correspond to decreased welfare (Hosey, 2005). Some primates, such as Sulawesi macaques (*Macaca nigra*), have comparable activity budgets between wild and zoo-housed primates. Other primates under human-managed care have more time to allocate to socializing, due to decreased

Box 13.1 Flexibility of Chimpanzees and Bonobos in Managed Care

Rodrigues and Boeving (2019) examined captive chimpanzees and bonobos living in human-managed settings, considering the impact of sex, origin, group tenure, and kinship on social network structure. We found grooming networks were equally strong within-and-between sexes in both species. Among the bonobos, wild-born bonobos were most central in grooming networks, and group tenure correlated with strength, a social network statistic that sums individuals' association with other individuals. A similar role for wild-born chimpanzees occurred in some studies (Leve et al., 2016). Among factors examined, none accounted for the chimpanzee grooming structure. While studies on individual groups under human management cannot fully address large questions, in conjunction with other studies on *Pan* social dynamics in zoos and the wild, this approach allows us to explore primates' range of adaptive flexibility while teasing out how individual- and group-characteristics impact network social dynamics.

Research on behavioural differences between chimpanzees and bonobos emphasized contrasts between the two species, particularly among social relationships. However, the emphasis on female social relationships in bonobos was critiqued as an artefact of human management, as initial research emphasizing female-female relationships emerged from zoo settings (Parish, 1994, 1996). Further research on bonobos in these settings found female-female social engagement increases when new females join the group but decreases after the group stabilizes (Stevens et al., 2006). Data from wild populations increased our knowledge of *Pan* flexibility across sites, emphasizing male-female and male-male bonds in wild bonobos (Furuichi & Ihobe, 1994; Stanford, 1998), and strength of female-female bonds in wild chimpanzees (Lehmann & Boesch, 2009; Wakefield, 2008, 2013). However, studying social dynamics in human-managed populations allows us to examine malleability of social bonds in environments where the ecological pressures that shape or limit social bonding in wild animals are comparable for both species.

travel and ranging needs (Majolo et al., 2008). Therefore, it is valuable to see the range of behaviours that emerge in captivity as part of species' wider behavioural repertoires. For example, studies on captive ape social dynamics yield insight into the adaptive flexibility of these primates (Rodrigues & Boeving, 2019; see Box 13.1). Research on human-caretaker dynamics also indicates human! caregivers have an active role in captive primate social networks (Funkhouser et al., 2020).

13.5   Conclusion

For primates living in their 'natural' ranges, species' behavioural ecology is influenced by anthropogenic and observer effects and may reflect a long history of interactions with humans. For primates living within human-managed care, their ability to thrive in these settings is influenced by species' range of flexibility and these co-evolutionary histories. Primate behavioural ecology, including feeding ecology, social behaviour, and reproduction, are shaped by coexistence with humans. Even at the most remote field sites, the 'wild' has the potential to transition into a new anthropogenic landscape or anthrome. Further research is needed to understand how these human impacts shape primate behaviour, especially the social behaviour of primates living among human settlements. Furthermore, given that our knowledge of wild primate behaviour, especially long-term, intensive study, is biased towards a subset of species and field sites (Bezanson & McNamara, 2019), it is possible that some behaviours seen in captivity are part of species' larger behavioural repertoire, or emerging cultural traditions. For primates living under human management, caretakers and visitors may also be social influences that impact behaviour (Chap. 16, this volume). Fully understanding primate behavioural ecology requires that we recognize ourselves as part of primates' ecological and social networks.

Acknowledgements We thank the co-editors of this volume, particularly Tracie McKinney and Siân Waters, for inviting us to contribute to this volume and for their careful editing. We thank our peer and editor reviewers for their constructive comments. MAR is grateful for discussions about captive behavioural flexibility with Emily Boeving, who co-authored the case study research on grooming networks in captive chimpanzees and bonobos.

References

Ahuja, N. (2013). Macaques and biomedicine: Notes on decolonization, polio, and changing representation of Indian Rhesus in the United States, 1930–1960. In S. Radhakrishna, M. A. Huffman, & A. Sinha (Eds.), *The macaque connection: Cooperation and conflict between humans and macaques* (pp. 71–91). Springer Science & Business Media LLC. https://doi.org/10.1007/978-1-4614-3967-7

Alcayna-Stevens, L. (2016). Habituating field scientists. *Social Studies of Science, 46*(6), 833–853. https://doi.org/10.1177/0306312716669251

Almeida-Rocha, J. M., Peres, C. A., & Oliveira, L. C. (2017). Primate responses to anthropogenic habitat disturbance: A pantropical meta-analysis. *Biological Conservation, 215*, 30–38. https://doi.org/10.1016/j.biocon.2017.08.018

Asensio, N., Korstjens, A. H., & Aureli, F. (2009). Fissioning minimizes ranging costs in spider monkeys: A multiple-level approach. *Behavioral Ecology and Sociobiology, 63*(5), 649–659. https://doi.org/10.1007/s00265-008-0699-9

Aureli, F., & Yates, K. (2010). Distress prevention by grooming others in crested black macaques. *Biology Letters, 6*(1), 27–29. https://doi.org/10.1098/rsbl.2009.0513

Bailey, R. G. (2004). Identifying ecoregion boundaries. *Environmental Management, 34*(1), S14–S26. https://doi.org/10.1007/s00267-003-0163-6

Baker, K. (2007). Enrichment and primate centers: Closing the gap between research and practice. *Journal of Applied Animal Welfare Science, 10*(1), 49–54. https://doi.org/10.1080/10888700701277618

Balasubramaniam, K. N., Marty, P. R., Arlet, M. E., Beisner, B. A., Kaburu, S. S., Bliss-Moreau, E., Kodandaramaiah, U., & McCowan, B. (2020a). Impact of anthropogenic factors on affiliative behaviors among bonnet macaques. *American Journal of Physical Anthropology, 171*(4), 704–717. https://doi.org/10.1002/ajpa.24013

Balasubramaniam, K. N., Marty, P. R., Samartino, S., Sobrino, A., Gill, T., Ismail, M., Saha, R., Beisner, B. A., Kaburu, S. S., & Bliss-Moreau, E. (2020b). Impact of individual demographic and social factors on human-wildlife interactions: A comparative study of three macaque species. *Scientific Reports, 10*(1), 1–16. https://doi.org/10.1038/s41598-020-78881-3

Baranga, D., Basuta, G. I., Teichroeb, J. A., & Chapman, C. A. (2012). Crop raiding patterns of solitary and social groups of red-tailed monkeys on cocoa pods in Uganda. *Tropical Conservation Science, 5*(1), 104–111. https://doi.org/10.1177/194008291200500109

Beck, B. B. (2018). *Unwitting travelers: A history of primate reintroduction*. Salt Water Media.

Beisner, B. A., Heagerty, A., Seil, S. K., Balasubramaniam, K. N., Atwill, E. R., Gupta, B. K., Tyagi, P. C., Chauhan, N. P., Bonal, B. S., & Sinha, P. R. (2015). Human–wildlife conflict: Proximate predictors of aggression between humans and rhesus macaques in India. *American Journal of Physical Anthropology, 156*(2), 286–294. https://doi.org/10.1002/ajpa.22649

Benchimol, M., & Peres, C. A. (2014). Predicting primate local extinctions within "real-world" forest fragments: A pan-neotropical analysis. *American Journal of Primatology, 76*(3), 289–302. https://doi.org/10.1002/ajp.22233

Bezanson, M., & McNamara, A. (2019). The what and where of primate field research may be failing primate conservation. *Evolutionary Anthropology: Issues, News, and Reviews,* (November 2018), evan.21790. https://doi.org/10.1002/evan.21790

Bezanson, M., Stowe, R., & Watts, S. M. (2013). Reducing the ecological impact of field research. *American Journal of Primatology, 75*(1), 1–9. https://doi.org/10.1002/ajp.22086

Boyle, S. A., & Smith, A. T. (2010). Can landscape and species characteristics predict primate presence in forest fragments in the Brazilian Amazon? *Biological Conservation, 143*(5), 1134–1143. https://doi.org/10.1016/j.biocon.2010.02.008

Bradshaw, G. A., Capaldo, T., Lindner, L., & Grow, G. (2008). Building an inner sanctuary: Complex PTSD in chimpanzees. *Journal of Trauma and Dissociation, 9*(1), 9–34. https://doi.org/10.1080/15299730802073619

Brussard, P. F. (1982). The role of field stations in the preservation of biological diversity. *Bioscience, 32*, 327–330. https://doi.org/10.2307/1308849

Buskirk, J. V., & Steiner, U. K. (2009). The fitness costs of developmental canalization and plasticity. *Journal of Evolutionary Biology, 22*(4), 852–860. https://doi.org/10.1111/j.1420-9101.2009.01685.x

Cameron, A., & Gould, L. (2013). Fragment-adaptive behavioural strategies and intersite variation in the ring-tailed lemur (*Lemur catta*) in South-Central Madagascar. In L. K. Marsh & C. A. Chapman (Eds.), *Primates in fragments: Complexity and resilience* (pp. 227–243). Springer. https://doi.org/10.1007/978-1-4614-8839-2_16

Chapman, C. A., & Peres, C. A. (2001). Primate conservation in the new millennium: The role of scientists. *Evolutionary Anthropology: Issues, News, and Reviews, 10*(1), 16–33. https://doi.org/10.1002/1520-6505(2001)10:1<16::AID-EVAN1010>3.0.CO;2-O

Chernus, L. A. (2008). Separation/abandonment/isolation trauma: What we can learn from our nonhuman primate relatives. *Journal of Emotional Abuse, 8*(4), 469–492. https://doi.org/10.1080/10926790802480364

Coe, J. C. (1989). Naturalizing habitats for captive primates. *Zoo Biology, 8*(1S), 117–125. https://doi.org/10.1002/zoo.1430080512

Cowlishaw, G., & Dunbar, R. I. M. (2000). *Primate conservation biology*. University of Chicago Press.

Crailsheim, D., Stüger, H. P., Kalcher-Sommersguter, E., & Llorente, M. (2020). Early life experience and alterations of group composition shape the social grooming networks of former pet and entertainment chimpanzees (*Pan troglodytes*). *PLoS One, 15*(1), 1–26. https://doi.org/10.1371/journal.pone.0226947

Crofoot, M. C., Lambert, T. D., Kays, R., & Wikelski, M. C. (2010). Does watching a monkey change its behaviour? Quantifying observer effects in habituated wild primates using automated radiotelemetry. *Animal Behaviour, 80*(3), 475–480. https://doi.org/10.1016/j.anbehav.2010.06.006

Davenport, R. K., & Menzel, E. W. (1963). Stereotyped behavior of the infant chimpanzee. *Archives of General Psychiatry, 8*, 99–104. https://doi.org/10.1001/archpsyc.1963.01720070101013

Day, R. L., Laland, K. N., & Odling-Smee, J. (2003). Rethinking adaptation: The niche-construction perspective. *Perspectives in Biology and Medicine, 46*(1), 80–95. https://doi.org/10.1353/pbm.2003.0003

Dhawale, A. K., Kumar, M. A., & Sinha, A. (2020). Changing ecologies, shifting behaviours: Behavioural responses of a rainforest primate, the lion-tailed macaque *Macaca silenus*, to a matrix of anthropogenic habitats in southern India. *PLoS One, 15*(9), e0238695. https://doi.org/10.1371/journal.pone.0238695

Dore, K. M., Radford, L., Alexander, S., & Waters, S. (2018). Ethnographic approaches in primatology. *Folia Primatologica, 89*(1), 5–12. https://doi.org/10.1159/000485693

Edes, A. N., Wolfe, B. A., & Crews, D. E. (2016). Rearing history and allostatic load in adult western lowland gorillas (*Gorilla gorilla gorilla*) in human care. *Zoo Biology, 35*(2), 167–173. https://doi.org/10.1002/zoo.21270

Ellis, E. C., & Ramankutty, N. (2008). Putting people in the map: Anthropogenic biomes of the world. *Frontiers in Ecology and the Environment, 6*(8), 439–447. https://doi.org/10.1890/070062

Estrada, A. (2006). Human and non-human primate co-existence in the Neotropics: A preliminary view of some agricultural practices as a complement for primate conservation. *Ecological and Environmental Anthropology, 2*, 17–29. Retrieved from http://digitalcommons.unl.edu/icwdmeea/3

Estrada, A., Garber, P. A., Rylands, A. B., Roos, C., Fernandez-Duque, E., Di Fiore, A., et al. (2017). Impending extinction crisis of the world's primates: Why primates matter. *Science Advances*, (229). https://doi.org/10.1126/sciadv.1600946

Farmer, H. L., Plowman, A. B., & Leaver, L. A. (2011). Role of vocalisations and social housing in breeding in captive howler monkeys (*Alouatta caraya*). *Applied Animal Behaviour Science, 134*, 177–183. https://doi.org/10.1016/j.applanim.2011.07.005

Fedigan, L. M. (2010). Ethical issues faced by field primatologists: Asking the relevant questions. *American Journal of Primatology, 72*(9), 754–771. https://doi.org/10.1002/ajp.20814

Fedigan, L. M., & Strum, S. C. (1999). A brief history of primate studies: National traditions, disciplinary origins, and stages in north American field research. In P. Dolhinhow & A. Fuentes (Eds.), *The nonhuman primates* (pp. 258–269). Mayfield Publishing Company.

Franquesa-Soler, M., Jorge-Sales, L., Aristizabal, J. F., Moreno-Casasola, P., & Serio-Silva, J. C. (2020). Evidence-based conservation education in Mexican communities: Connecting arts and science. *PLoS One, 15*(2), e0228382. https://doi.org/10.1371/journal.pone.0228382

Freeman, H. D., & Ross, S. R. (2014). The impact of atypical early histories on pet or performer chimpanzees. *PeerJ, 2*, e579. https://doi.org/10.7717/peerj.579

Fuentes, A. (2006). Human-nonhuman primate interconnections and their relevance to Anthropology. *Ecological and Environmental Anthropology (University of Georgia), 1*, 0–11. https://doi.org/10.1002/eat.20929

Fuentes, A. (2012). Ethnoprimatology and the anthropology of the human-primate interface. *Annual Review of Anthropology*. https://doi.org/10.1146/annurev-anthro-092611-145808

Fuentes, A., & Hockings, K. J. (2010). The ethnoprimatological approach in primatology. *American Journal of Primatology, 72*(10), 841–847. https://doi.org/10.1002/ajp.20844

Fuentes, A., & Wolf, L. D. (2002). *Primates face to face: The conservation implications of Human-nonhuman primate interconnections*. Cambridge University Press.

Funkhouser, J. A., Mayhew, J. A., Mulcahy, J. B. B., & Sheeran, L. K. (2020). Human caregivers are integrated social partners for captive chimpanzees. *Primates*. https://doi.org/10.1007/s10329-020-00867-6

Furuichi, T., & Ihobe, H. (1994). Variation in male relationships in bonobos and chimpanzees. *Behaviour, 130*, 211–228. https://doi.org/10.1163/156853994X00532

Garland, E. (2008). The elephant in the room: Confronting the colonial character of wildlife conservation in Africa. *African Studies Review, 51*(3), 51–74. https://doi.org/10.1353/arw.0.0095

Gazagne, E., Hambuckers, A., Savini, T., Poncin, P., Huynen, M. C., & Brotcorne, F. (2020). Toward a better understanding of habituation process to human observer: A statistical approach in *Macaca leonina* (Primates: Cercopithecidea). *Raffles Bulletin of Zoology, 68*(2020), 735–749. https://doi.org/10.26107/RBZ-2020-0085

Green, V. M., & Gabriel, K. I. (2020). Researchers' ethical concerns regarding habituating wild-nonhuman primates and perceived ethical duties to their subjects: Results of an online survey. *American Journal of Primatology, 82*(9), e23178. https://doi.org/10.1002/ajp.23178

Hansen, B. K., Fultz, A. L., Hopper, L. M., & Ross, S. R. (2018). An evaluation of video cameras for collecting observational data on sanctuary-housed chimpanzees (*Pan troglodytes*). *Zoo Biology, 37*(3), 156–161. https://doi.org/10.1002/zoo.21410

Hanson, K. T., & Riley, E. P. (2018). Beyond neutrality: The human–primate interface during the habituation process. *International Journal of Primatology, 39*(5), 852–877. https://doi.org/10.1007/s10764-017-0009-3

Haraway, D. J. (1989). *Primate visions: Gender, race, and nature in the world of modern science*. Routledge, Chapman & Hall, Inc.

Hockings, K. J., Yamakoshi, G., Kabasawa, A., & Matsuzawa, T. (2010). Attacks on local persons by chimpanzees in Bossou, Republic of Guinea: Long-term perspectives. *American Journal of Primatology, 72*(10), 887–896. https://doi.org/10.1002/ajp.20784

Holdridge, L. R. (1947). Determination of world plant formations from simple climatic data. *Science, 105*(2727), 367–368. https://www.science.org/doi/10.1126/science.105.2727.367

Holzner, A., Ruppert, N., Swat, F., Schmidt, M., Weiß, B. M., Villa, G., Mansor, A., Sah, S. A. M., Engelhardt, A., & Kühl, H. (2019). Macaques can contribute to greener practices in oil palm plantations when used as biological pest control. *Current Biology, 29*(20), R1066–R1067. https://doi.org/10.1016/j.cub.2019.09.011

Hosey, G. R. (2005). How does the zoo environment affect the behaviour of captive primates? *Applied Animal Behaviour Science, 90*(2), 107–129. https://doi.org/10.1016/j.applanim.2004.08.015

Jack, K. M., Lenz, B. B., Healan, E., Rudman, S., Schoof, V. A., & Fedigan, L. (2008). The effects of observer presence on the behavior of *Cebus capucinus* in Costa Rica. *American Journal of Primatology, 70*(5), 490–494. https://doi.org/10.1002/ajp.20512

Jost Robinson, C. A., & Remis, M. J. (2018). Engaging holism: Exploring multispecies approaches in ethnoprimatology. *International Journal of Primatology, 39*(5), 776–796. https://doi.org/10.1007/s10764-018-0036-8

Junker, J., Petrovan, S. O., Arroyo-Rodríguez, V., Boonratana, R., Byler, D., Chapman, C. A., Chetry, D., Cheyne, S. M., Cornejo, F. M., Cortés-Ortiz, L., & Cowlishaw, G. (2020). A severe lack of evidence limits effective conservation of the world's primates. *Bioscience, 70*(9), 794–803. https://doi.org/10.1093/biosci/biaa082

Kaburu, S. S. K., Beisener, B., Balasubramaniam, K. N., Marty, P. R., Bliss-Moreau, E., Mohan, L., Rattan, S. K., Arlet, M. E., Atwill, E. R., & McCowan, B. (2019). Interactions with humans impose time constraints on urban-dwelling rhesus macaques. *Behaviour, 156*(12), 1255–1282. https://doi.org/10.1163/1568539X-00003565

Kleiman, D. G. (1979). Parent-offspring conflict and sibling competition in a monogamous primate. *The American Naturalist, 114*(5), 753–760.

Kleiman, D. G. (1983). The behavior and conservation of the golden lion tamarin, *Leontopithecus rosalia. Congresso Brasileiro de Primatologia*, 35–53.

Kumara, H. N., Kumar, S., & Singh, M. (2009). Of how much concern are the 'least concern' species? Distribution and conservation status of bonnet macaques, rhesus macaques and Hanuman langurs in Karnataka, India. *Primates, 51*(1), 37. https://doi.org/10.1007/s10329-009-0168-8

Kutsukake, N. (2013). Heterogeneous ethical structures in field primatology. In J. MacClancy & A. Fuentes (Eds.), *Ethics in the field: Contemporary challenges* (pp. 84–97). Berghan Books.

Laurance, W. F. (2013). Does research help to safeguard protected areas? *Trends in Ecology & Evolution, 28*(5), 261–266. https://doi.org/10.1016/j.tree.2013.01.017

Lehmann, J., & Boesch, C. (2009). Sociality of the dispersing sex: The nature of social bonds in West African female chimpanzees, *Pan troglodytes. Animal Behaviour, 77*(2), 377–387. https://doi.org/10.1016/j.anbehav.2008.09.038

Leve, M., Sueur, C., Petit, O., Matsuzawa, T., & Hirata, S. (2016). Social grooming network in captive chimpanzees: Does the wild or captive origin of group members affect sociality? *Primates, 57*(1), 73–82. https://doi.org/10.1007/s10329-015-0494-y

Lwanga, J. S., & Isabirye-Basuta, G. (2008). Long-term perspectives on forest conservation: Lessons from research in Kibale Park. In R. W. Wrangham & E. Ross (Eds.), *Science and conservation in African forests* (pp. 63–74). Cambridge University Press.

MacClancy, J., & Fuentes, A. (Eds.). (2013). *Ethics in the field: Contemporary challenges* (Vol. 7). Berghahn Books.

Majolo, B., De Bortoli Vizioli, A., & Schino, G. (2008). Costs and benefits of group living in primates : Group size effects on behaviour and demography. *Animal Behaviour, 76*, 1235–1247. https://doi.org/10.1016/j.anbehav.2008.06.008

Malaivijitnond, S., Lekprayoon, C., Tandavanittj, N., Panha, S., Cheewatham, C., & Hamada, Y. (2007). Stone-tool usage by Thai long-tailed macaques (*Macaca fascicularis*). *American Journal of Primatology, 69*(2), 227–233. https://doi.org/10.1002/ajp.20342

Malone, N. M., Fuentes, A., & White, F. J. (2010). Ethics commentary: Subjects of knowledge and control in field primatology. *American Journal of Primatology, 72*, 779–784. https://doi.org/10.1002/ajp.20840

Malone, N., Wade, A. H., Fuentes, A., Riley, E. P., Remis, M., & Robinson, C. J. (2014). Ethnoprimatology: Critical interdisciplinarity and multispecies approaches in anthropology. *Critique of Anthropology, 34*(1), 8–29. https://doi.org/10.1177/0308275X13510188

Maréchal, L., Semple, S., Majolo, B., Qarro, M., Heistermann, M., & MacLarnon, A. (2011). Impacts of tourism on anxiety and physiological stress levels in wild male Barbary macaques. *Biological Conservation, 144*(9), 2188–2193. https://doi.org/10.1016/j.biocon.2011.05.010

Marsh, L. K. (2003). The nature of fragmentation. In L. K. Marsh (Ed.), *Primates in Fragments*. Springer. https://doi.org/10.1007/978-1-4757-3770-7_1

Martin, L. J., Quinn, J. E., Ellis, E. C., Shaw, M. R., Dorning, M. A., Hallett, L. M., Heller, N. E., Hobbs, R. J., Kraft, C. E., Law, E., & Michel, N. L. (2014). Conservation opportunities across the world's anthromes. *Diversity and Distributions, 20*(7), 745–755. https://doi.org/10.1111/ddi.12220

Marty, P. R., Beisner, B., Kaburu, S. S., Balasubramaniam, K., Bliss-Moreau, E., Ruppert, N., Sah, S. A. M., Ismail, A., Arlet, M. E., & Atwill, E. R. (2019). Time constraints imposed by anthropogenic environments alter social behaviour in longtailed macaques. *Animal Behaviour, 150*, 157–165. https://doi.org/10.1016/j.anbehav.2019.02.010

Marty, P. R., Balasubramaniam, K. N., Kaburu, S. S., Hubbard, J., Beisner, B., Bliss-Moreau, E., Ruppert, N., Arlet, M. E., Sah, S. A. M., & Ismail, A. (2020). Individuals in urban dwelling primate species face unequal benefits associated with living in an anthropogenic environment. *Primates, 61*(2), 249–255. https://doi.org/10.1007/s10329-019-00775-4

McDougall, P. (2012). Is passive observation of habituated animals truly passive? *Journal of Ethology, 30*(2), 219–223. https://doi.org/10.1007/2Fs10164-011-0313-x

McKinney, T. (2011). The effects of provisioning and crop-raiding on the diet and foraging activities of human-commensal white-faced capuchins (*Cebus capucinus*). *American Journal of Primatology, 73*(5), 439–448. https://doi.org/10.1002/ajp.20919

McKinney, T. (2015). A classification system for describing anthropogenic influence on nonhuman primate populations. *American Journal of Primatology, 77*, 715–726. https://doi.org/10.1002/ajp.22395

McKinney, T., & Dore, K. M. (2018). The state of ethnoprimatology: Its use and potential in today's primate research. *International Journal of Primatology, 39*(5), 730–748. https://doi.org/10.1007/s10764-017-0012-8

McLennan, M. R., & Ganzhorn, J. U. (2017). Nutritional characteristics of wild and cultivated foods for chimpanzees (*Pan troglodytes*) in agricultural landscapes. *International Journal of Primatology, 38*(2), 122–150. https://doi.org/10.1007/s10764-016-9940-y

McLennan, M. R., & Hill, C. M. (2010). Chimpanzee responses to researchers in a disturbed forest–farm mosaic at Bulindi, western Uganda. *American Journal of Primatology, 72*(10), 907–918. https://doi.org/10.1002/ajp.20839

Morrow, K. S., Glanz, H., Ngakan, P. O., & Riley, E. P. (2019). Interactions with humans are jointly influenced by life history stage and social network factors and reduce group cohesion in moor macaques (*Macaca maura*). *Scientific Reports, 9*(1), 1–12. https://doi.org/10.1038/s41598-019-56288-z

Moura, A. d. A., & Lee, P. C. (2004). Capuchin stone tool use in Caatinga dry forest. *Science, 306*(5703), 1909–1909. https://doi.org/10.1126/science.1102558

Nelson, R. G., Rutherford, J. N., Hinde, K., & Clancy, K. B. H. (2017). Signaling safety: Characterizing fieldwork experiences and their implications for career trajectories. *American Anthropologist, 119*(4), 710–722. https://doi.org/10.1111/aman.12929

Nijman, V., & Nekaris, K. A.-I. (2010). Testing a model for predicting primate crop-raiding using crop- and farm-specific risk values. *Applied Animal Behaviour Science, 127*(3), 125–129. https://doi.org/10.1016/j.applanim.2010.08.009

Nowak, K., & Lee, P. C. (2013). "Specialist" primates can be flexible in response to habitat alteration. In L. K. Marsh & C. A. Chapman (Eds.), *Primates in fragments: Complexity and resilience* (pp. 199–211). Springer. https://doi.org/10.1007/978-1-4614-8839-2_14

Pal, A., Kumara, H. N., Mishra, P. S., Velankar, A. D., & Singh, M. (2018). Extractive foraging and tool-aided behaviors in the wild Nicobar long-tailed macaque (*Macaca fascicularis umbrosus*). *Primates, 59*(2), 173–183. https://doi.org/10.1007/s10329-017-0635-6

Parish, A. R. (1994). Sex and food control in the "uncommon chimpanzee": How Bonobo females overcome a phylogenetic legacy of male dominance. *Ethology and Sociobiology, 15*(3), 157–179. https://doi.org/10.1016/0162-3095(94)90038-8

Parish, A. R. (1996). Female relationships in bonobos (*Pan paniscus*): Evidence for bonding, cooperation, and female dominance in a male-philopatric species. *Human Nature, 7*(1), 61–96. https://doi.org/10.1007/BF02733490

Pastor-Nieto, R. (2015). Health and welfare of howler monkeys in captivity. In M. M. Kowalewski, L. Cortés-Ortiz, & B. Urbani (Eds.), *Howler monkeys: Behavior, ecology, and conservation* (pp. 313–355). Springer Science. https://doi.org/10.1007/978-1-4939-1960-4

Pigliucci, M. (2001). *Phenotypic plasticity: Beyond nature and nurture*. John Hopkins University Press.

Pozo-Montuy, G., Serio-Silva, J. C., Chapman, C. A., & Bonilla-Sánchez, Y. M. (2013). Resource use in a landscape matrix by an arboreal primate: Evidence of supplementation in black howlers (*Alouatta pigra*). *International Journal of Primatology, 34*(4), 714–731. https://doi.org/10.1007/s10764-013-9691-y

Proffitt, T., Luncz, V. L., Malaivijitnond, S., Gumert, M., Svensson, M. S., & Haslam, M. (2018). Analysis of wild macaque stone tools used to crack oil palm nuts. *Royal Society Open Science, 5*(3), 171904. https://doi.org/10.1098/rsos.171904

Ralainasolo, F. B., Ratsimbazafy, J. H., & Stevens, N. J. (2008). Behavior and diet of the critically endangered *Eulemur cinereiceps* in Manombo forest, Southeast Madagascar. *Madagascar Conservation and Development, 3*(1), 38–43. https://doi.org/10.4314/mcd.v3i1.44134

Ram, S., Venkatachalam, S., & Sinha, A. (2003). Changing social strategies of wild female bonnet macaques during natural foraging and provisioning. *Current Science, 84*(6), 780–790. https://www.jstor.org/stable/24107582

Rasmussen, D. R. (1991). Observer influence on range use of *Macaca arctoides* after 14 years of observation. *Laboratory Primate Newsletter, 30*(3), 6–11.

Reader, S. M., & Laland, K. N. (2003). *Animal innovation* (Vol. 10). Oxford University Press.

Riley, E. P. (2019). *The Promise of contemporary primatology*. Routledge.

Riley, E. P., & Bezanson, M. (2018). Ethics of primate fieldwork: Toward an ethically engaged primatology. *Annual Review of Anthropology, 47*, 493–512. https://doi.org/10.1146/annurev-anthro-102317-045913

Riley, E. P., & Fuentes, A. (2011). Conserving social-ecological systems in Indonesia: Human-nonhuman primate interconnections in Bali and Sulawesi. *American Journal of Primatology*. https://doi.org/10.1002/ajp.20834

Riley, E. P., Tolbert, B., & Farida, W. R. (2013). Nutritional content explains the attractiveness of cacao to crop raiding Tonkean macaques. *Current Zoology, 59*(2), 160–169. https://doi.org/10.1093/czoolo/59.2.160

Riley, E. P., Shaffer, C. A., Trinidad, J. S. Morrow, K. S., Sagnotti, C., Carosi, M., Ngakan, P. O. (2021). Roadside monkeys: anthropogenic effects on moor macaque *(Macaca maura)* ranging behavior in Bantimurung Bulusaraung National Park, Sulawesi, Indonesia. *Primates, 62*, 477–489. https://doi.org/10.1007/s10329-021-00899-6

Robbins, M. M., Gray, M., Fawcett, K. A., Nutter, F. B., Uwingeli, P., Mburanumwe, I., et al. (2011). Extreme conservation leads to recovery of the Virunga mountain gorillas. *PLoS One, 6*(6). https://doi.org/10.1371/journal.pone.001978

Rodrigues, M. A., & Boeving, E. R. (2019). Comparative social grooming networks in captive chimpanzees and bonobos. *Primates, 60*(3), 191–202. https://doi.org/10.1007/s10329-018-0670-y

Rowell, T. (2000). A few peculiar primates. In S. C. Strum & L. M. Fedigan (Eds.), *Primate encounters: Models of gender, science, and society* (pp. 57–70). University of Chicago Press.

Sampaio, M. B., Schiel, N., & Souto, A. (2020). From exploitation to conservation: A historical analysis of zoos and their functions in human societies. *Ethnobiology and Conservation, 9*(2), 1–32. https://doi.org/10.15451/EC2020-01-9.02-1-32

Schino, G., Scucchi, S., Maestripieri, D., & Turillazzi, P. G. (1988). Allogrooming as a tension-reduction mechanism: A behavioral approach. *American Journal of Primatology, 16*(1), 43–50. https://doi.org/10.1002/ajp.1350160106

Sept, J. M., & Brooks, G. E. (1995). Reports of chimpanzee natural history, including tool use, in 16th- and 17th-century Sierra Leone. *International Journal of Primatology, 16*(1), 867–878. https://doi.org/10.1007/BF02736073

Seyfarth, R. M. (1977). A model of social grooming among adult female monkeys. *Journal of Theoretical Biology, 65*(4), 671–698. https://doi.org/10.1016/0022-5193(77)90015-7

Shoo, R. A., & Songorwa, A. N. (2013). Contribution of eco-tourism to nature conservation and improvement of livelihoods around Amani nature reserve, Tanzania. *Journal of Ecotourism, 12*(2), 75–89. https://doi.org/10.1080/14724049.2013.818679

Silk, J. B., Beehner, J. C., Bergman, T. J., Crockford, C., Engh, A. L., Moscovice, L. R., Wittig, R. M., Seyfarth, R. M., & Cheney, D. L. (2009). The benefits of social capital: Close social bonds among female baboons enhance offspring survival. *Proceedings of the Royal Society B: Biological Sciences, 276*(1670), 3099–3104. https://doi.org/10.1098/rspb.2009.0681

Silk, J. B., Beehner, J. C., Bergman, T. J., Crockford, C., Engh, A. L., Moscovice, L. R., Wittig, R. M., Seyfarth, R. M., & Cheney, D. L. (2010). Female chacma baboons form strong, equitable, and enduring social bonds. *Behavioral Ecology and Sociobiology, 64*(11), 1733–1747. https://doi.org/10.1007/s00265-010-0986-0

Singh, M. (2019). Management of forest-dwelling and urban species: Case studies of the lion-tailed macaque (*Macaca silenus*) and the bonnet macaque (*M. radiata*). *International Journal of Primatology, 40*(6), 613–629. https://doi.org/10.1007/s10764-019-00122-

Siswanto, A. (2015). Eco-tourism development strategy Baluran National Park in the Regency of Situbondo, East Java, Indonesia. *International Journal of Evaluation and Research in Education, 4*(4), 185–195.

Sol, D. (2009). Revisiting the cognitive buffer hypothesis for the evolution of large brains. *Biology Letters, 5*(1), 130–133. https://doi.org/10.1098/rsbl.2008.0621

Soma, T. (2006). Tradition and novelty: *Lemur catta* feeding strategy on introduced tree species at Berenty Reserve. In *Ringtailed lemur biology* (pp. 141–159). Springer.

Sponsel, L. (1997). The human niche in Amazonia: Explorations in ethnoprimatology. In W. G. Kinzey (Ed.), *New world primates: Ecology, evolution, and behavior* (pp. 143–168). Aldine De Gruyter.

Stanford, C. B. (1998). The social behavior of chimpanzees and bonobos. *Current Anthropology, 39*, 399–420. https://doi.org/10.1086/204757

Stevens, J. M. G., Vervaecke, H., De Vries, H., & Van Elsacker, L. (2006). Social structures in *Pan paniscus*: Testing the female bonding hypothesis. *Primates, 47*(3), 210–217. https://doi.org/10.1007/s10329-005-0177-1

Strier, K. B. (2010). Long-term field studies: Positive impacts and unintended consequences. *American Journal of Primatology, 72*(9), 772–778. https://doi.org/10.1002/ajp.20830

Strier, K. B. (2013). Are observational studies of wild primates really noninvasive? In J. MacClancy & A. Fuentes (Eds.), *Ethics in the field: Contemporary challenges* (pp. 67–83). Berghahn.

Strier, K. B., & Boubli, J. P. (2006). A history of long-term research and conservation of northern muriquis (*Brachyteles hypoxanthus*) at the Estação Biológica de Caratinga/RPPN-FMA. *Primate Conservation, 2006*(20), 53–63. https://doi.org/10.1896/0898-6207.20.1.53

Strum, S. C., & Fedigan, L. M. (2000). Introduction and history. In S. C. Strum & L. M. Fedigan (Eds.), *Primate encounters: Models of gender, science, and society* (pp. 1–49). University of Chicago Press.

Tan, A. W., Luncz, L., Haslam, M., Malaivijitnond, S., & Gumert, M. D. (2016). Complex processing of prickly pear cactus (*Opuntia* sp.) by free-ranging long-tailed macaques: Preliminary analysis for hierarchical organisation. *Primates, 57*(2), 141–147. https://doi.org/10.1007/s10329-016-0525-3

Trevelyan, R., & Nuttman, C. (2008). The importance of training national and international scientists for conservation research. In *Science and conservation in African forests* (pp. 88–98). Cambridge University Press.

Tuomainen, U., & Candolin, U. (2011). Behavioural responses to human-induced environmental change. *Biological Reviews of the Cambridge Philosophical Society, 86*(3), 640–657. https://doi.org/10.1111/j.1469-185X.2010.00164.x

Tutin, C.E & Fernandez, M. (1991). Responses of wild chimpanzees and gorillas to the arrival of primatologists: Behaviour observed during habituation. In *Primate responses to environmental change* (pp. 187–197). Springer.

Tydecks, L., Bremerich, V., Jentschke, I., Likens, G. E., & Tockner, K. (2016). Biological field stations: A global infrastructure for research, education, and public engagement. *Bioscience, 66*(2), 164–171. https://doi.org/10.1093/biosci/biv174

van Krunkelsven, E., Dupain, J., Van Elsacker, L., & Verheyen, R. (1999). Habituation of bonobos (*Pan paniscus*): First reactions to the presence of observers and the evolution of response over time. *Folia Primatologica, 70*(6), 365–368.

Velankar, A. D., Kumara, H. N., Pal, A., Mishra, P. S., & Singh, M. (2016). Population recovery of Nicobar Long-Tailed Macaque *Macaca fascicularis umbrosus* following a tsunami in the Nicobar Islands, India. *PLoS One, 11*(2), e0148205. https://doi.org/10.1371/journal.pone.0148205

Wakefield, M. L. (2008). Grouping patterns and competition among female *Pan troglodytes schweinfurthii* at Ngogo, Kibale National Park, Uganda. *International Journal of Primatology, 29*(4), 907–929. https://doi.org/10.1007/s10764-008-9280-7

Wakefield, M. L. (2013). Social dynamics among females and their influence on social structure in an East African chimpanzee community. *Animal Behaviour, 85*(6), 1303–1313. https://doi. org/10.1016/j.anbehav.2013.03.019

Watts, D. P. (2008). Scavenging by chimpanzees at Ngogo and the relevance of chimpanzee scavenging to early hominin behavioral ecology. *Journal of Human Evolution, 54*(1), 125–133. https://doi.org/10.1016/j.jhevol.2007.07.008

West-Eberhard, M. J. (1983). Sexual selection, social competition, and speciation. *The Quarterly Review of Biology, 58*(2), 155–183. https://doi.org/10.1086/413215

Westin J. L. (2017). Habituation to tourists: Protective or harmful? In K. M. Dore, E. P. Riley, & A. Fuentes (Eds.), Ethnoprimatology: A practical guide to research at the human-nonhuman primate interface (pp. 15–28). Cambridge University Press. https://doi. org/10.1017/9781316272466.004

Williamson, E. A., & Fawcett, K. (2008). Long-term research and conservation of the Virunga mountain gorillas. In R. Wrangham & E. Ross (Eds.), *Science and conservation in African forests: The benefits of Long-term research* (pp. 213–229). Cambridge University Press.

Williamson, E. A., & Feistner, A. T. (2003). Habituating primates: Processes, techniques, variables and ethics. In J. M. Setchel & D. J. Curtis (Eds.), *Field and laboratory methods in primatology: A practical guide* (pp. 25–39). Cambridge University Press.

Wrangham, R. W. (2008). Why the link between long-term research and conservation is a case worth making. In R. W. Wrangham & E. Ross (Eds.), *Science and conservation in African forests* (pp. 1–8). Cambridge University Press.

Wright, P. C., & Andriamihaja, B. (2002). Making a rain forest national park work in Madagascar: Ranomafana National Park and its long-term research commitment. In J. Terborgh (Ed.), *Making parks work: Strategies for preserving tropical nature* (pp. 112–136). Island Press.

Young, C., Majolo, B., Heistermann, M., Schülke, O., & Ostner, J. (2014). Responses to social and environmental stress are attenuated by strong male bonds in wild macaques. *Proceedings of the National Academy of Sciences, 111*(51), 18195–18200. https://doi.org/10.1073/pnas.1411450111

Chapter 14
The Past, Present, and Future of the Primate Pet Trade

Sherrie D. Alexander, Siân Waters, Brooke C. Aldrich, Sam Shanee, Tara A. Clarke, Lucy Radford, Malene Friis Hansen, Smitha Daniel Gnanaolivu, and Andrea Dempsey

Contents

S. D. Alexander (✉)
Department of Anthropology, University of Alabama at Birmingham, Birmingham, AL, USA

Barbary Macaque Awareness and Conservation, Tetouan, Morocco
e-mail: sdalex@uab.edu

S. Waters
Barbary Macaque Awareness and Conservation, Tetouan, Morocco

Department of Anthropology, Durham University, Durham, UK

B. C. Aldrich
Asia for Animals Coalition, Hong Kong, China

Neotropical Primate Conservation, Seaton, Cornwall, UK

S. Shanee
Neotropical Primate Conservation, Seaton, Cornwall, UK

T. A. Clarke
Department of Sociology & Anthropology, North Carolina State University, Raleigh, NC, USA

Mad Dog Initiative, Antananarivo, Madagascar

L. Radford
Department of Anthropology, Durham University, Durham, UK

Sumatran Orangutan Society, Abingdon, UK

Abstract Pet primates are those kept typically for companionship, enjoyment, and status, although their uses as pets may extend beyond these parameters. The trade in pet primates is historically rooted, with many primates playing important roles in human cultures and religions. Thus, it is not surprising that current sociocultural trends reveal an ongoing fascination with primates and their purchase as status pets. Recent reports from various regions are presented in this chapter, demonstrating the need for drastic interventions to avoid further losses. Capture of animals for the pet trade may be intentional or opportunistic and is often exacerbated by internet trade and social media. This situation is complicated by the difficulty of obtaining accurate numbers of primates bought and sold illegally. The health and welfare of primates captured or kept as pets is another area of great concern. Long-term solutions will require attention from governmental, professional, and public actors on local and international levels.

Keywords Welfare · Exploitation · Illegal · Legal · Conservation

14.1 Introduction

A pet is an animal typically 'kept for companionship, enjoyment, and status' (Fuentes, 2007:129). Non-human primates' physical and behavioural similarities to humans make them intrinsically appealing as pets. Pet keeping is diverse across cultures, and cultural conceptualizations and historical factors influence people's motivations for keeping primates as a pets (Hurn, 2012). In some cases, these animals are sought out specifically for the pet trade (Van Uhm, 2016), and in others, they are the orphaned juveniles of primates hunted for food (Cormier, 2003). There is a ready market for primates as pets in Western society and elsewhere, where they may be used as companions or as entertainers (Lee & Priston, 2005; Norconk et al., 2019; Chap. 17, this volume).

M. F. Hansen
Behavioural Ecology Group, Department of Biology, University of Copenhagen, Copenhagen, Denmark

The Long-Tailed Macaque Project, Broerup, Denmark

Department of Social Sciences, Oxford Brookes University, Oxford, UK

S. D. Gnanaolivu
Institution of Excellence, Vijnana Bhavan, University of Mysore, Mysore, India

Wildlife Information Liaison Development (WILD), Coimbatore, India

A. Dempsey
West African Primate Conservation Action, Broxbourne, UK

West African Primate Conservation Action, Accra, Ghana

Recent research shows that portrayals of primates with humans in media and photographs affect people's perceptions of animals and their conservation status (Clarke et al., 2019; Leighty et al., 2015; Lenzi et al., 2020; Ross et al., 2008, 2011; Schroepfer et al., 2011). A trend for keeping a particular species of primate as a pet has drastic repercussions for wild populations when wild-captured animals are sold into the illegal wildlife trade to feed demand (Nekaris et al., 2013; Van Uhm, 2016). Even for captive-reared primates, there remains an unacceptably high cost to the psychological and physical welfare of individual animals (Soulsbury et al., 2009).

The welfare of primates kept as pets is inevitably compromised, whether they are wild-caught or bred in captivity. Infants are routinely removed from their mothers to be kept or sold as pets (Soulsbury et al., 2009). Maternal deprivation is profoundly damaging to primates and can result in major and often irreversible psychological damage, behavioural abnormalities, immune system dysfunction, and decreased survivorship (Harlow, 1962; Law et al., 2009; Lewis et al., 2000; Pryce et al., 2004). Isolation from conspecifics is the norm and is also problematic; close relationships with kin and group mates are important for the behavioural development of young primates (Thompson, 2019).

Pet primates are typically fed poor diets that do not meet their nutritional needs, leading to nutrition-related bone disease, diabetes, and other conditions (B Aldrich pers observation; Soulsbury et al., 2009). As young individuals of most primate species mature, they are likely to become unmanageable and dangerous, at which point owners may seek to surrender their pets to zoos or sanctuaries (North American Primate Sanctuary Alliance, 2021) or may set them free, with little chance of survival (H Browning, pers communication).

Primatological and veterinary bodies have stated publicly that primates are fundamentally unsuitable to be kept as pets (British Veterinary Association, 2014; International Primatological Society, 2008). Of additional concern is the direct or close contact between pet primates and their owners that puts both at risk for disease transmission (Cormier & Jolly, 2018; Chap. 9, this volume).

While live primates are also traded for use as subjects of biomedical research and the entertainment industry, the pet trade accounts for hundreds of thousands of primates traded annually (Estrada et al., 2017; Norconk et al., 2019). Here, we define the **trade** of primates as the buying, selling, or exchange of primate species, locally, nationally, or internationally. The **illegal trade** takes place through unlawful channels or when forbidden by local or international laws. Although our use of the term **pet** is consistent with that of Fuentes (2007), Hurn (2012) and Reuter et al. (2019) broadly define pet as an animal having a human owner, no longer living in their natural habitat, and reliant on humans for food.

It is also worth noting that the definition of pet may, at times, encompass grey areas outside of 'companionship, enjoyment, and status' (Fuentes, 2007:129). Fuentes (2013):116–117 adds that pet primates may 'contribute to the income, nutritional intake, or other functional facets of the humans who own them'. Captive primates kept for tourism (Chap. 11, this volume) and entertainment (Chap. 17, this volume) may be referred to as pets, although they typically have a strictly utilitarian function.

14.2   The Past Primate Pet Trade

The longevity of pet primate keeping is revealed in the archaeological record, with the earliest evidence of long-distance trade in primates to ancient Egypt. These specimens date to the Ptolemaic period (305 BCE to 30 BCE) and included Hamadryas baboons (*Papio hamadryas*), Olive baboons (*P. anubis*), Barbary macaques (*Macaca sylvanus*), and vervet monkeys (*Chlorocebus aethiops*), none of which are found locally (Dominy et al., 2020; Groves, 2008). The additional captive breeding of baboons in ancient Egypt and removal of their large canines suggests these animals functioned as companions in addition to divine representations (Dominy et al., 2020).

Keeping primates and other exotic pets has historically been a status marker (Dominy et al., 2020; Fuentes, 2007; Hurn, 2012; Serpell, 1996). For example, Greeks were known to keep pet primates along with their preferred lap dogs. In the third century B.C.E., monkeys were common household pets among the Roman upper class (Serpell, 1996).

One of the earliest, most consistently traded species of primates is the Barbary macaque (*Macaca sylvanus*) of North Africa. Outside of Egypt, this primate makes numerous appearances throughout ancient Europe and the Mediterranean (Van Uhm, 2016). Janson (1952) describes the immense popularity of Barbary macaques and their trainers throughout Europe as early as the eleventh century. However, by the seventeenth century, other primate species appear in European art and life as the result of increased intercontinental travel (Groves, 2008).

Artistic depictions from the Japanese Middle Ages suggest that monkeys were kept as pets as early as the twelfth century, while the Samurai of sixteenth century Japan were known to possess 'pet stable monkeys" to counter diseases or misfortunes affecting horses (Mito & Sprague, 2013). Fuentes (2013) explains that pet keeping is a long-lived tradition in Southeast Asia where macaques, primarily *M. fascicularis*, *M. mulatta*, and *M. nemestrina* (MF Hansen pers. observation), are kept as companions, for entertainment, or status. In India, captive slender lorises have been integrated into traditional practices since 300 B.C.E (Venkatesan, 2018). Fishermen kept caged slender lorises (*Loris* spp.) for use as compasses through the 1980s (SD Gnanaolivu unpublished data). Similarly, astrologers and fortune tellers continue to keep slender lorises in wooden boxes, going from house to house to foretell God's word (Gnanaolivu et al. In prep).

Finally, several **Indigenous** or foraging cultures keep pet primates for various purposes such as education, adornment, and food, since these animals have been incorporated into local ecologies and worldviews. Among others, the Guajá (Cormier, 2003) and Matsigenka (Shephard, 2002) of South America, and Mentawi (Fuentes, 2002) of Indonesia, capture infant and juvenile monkeys as the by-product of subsistence hunting.

14.3   The Present Primate Pet Trade

The ongoing extraction of wild primates is unsustainable given that over half of all primate species are facing extinction or have declining populations (Estrada et al., 2017; Norconk et al., 2019). Overall numbers of wild-caught primates traded illegally are impossible to discern. As of 2017, the largest numbers of live primates traded internationally were exported from China, many of which originated in Southeast Asia (CITES Trade Database, 2021). Vietnam and Cambodia followed China with roughly half of their sales received by Japan (Norconk et al., 2019).

The most heavily traded primate species is the long-tailed macaque (*Macaca fascicularis*) (Foley & Shepherd, 2011; Shepherd, 2010; Nijman et al., 2017). In the United States, 84% of all primate imports (dead and alive) from 2010 to 2014 were long-tailed macaques, and almost half of live individuals were wild-caught (Sanerib & Uhlemann, 2020). Although most legally traded primates are destined for biomedical research, the species traded for this industry are often the same as those kept as pets, as they are afforded the least legal protection and are most easily purchased and discarded (MF Hansen pers observation).

The legality of keeping primates varies between countries and regions, adding to the complexity of trade regulation. The United States and Europe provide two examples of locations where primates are not endemic, but heavily traded (American Society of Primatologists, 2021; Endcap, 2012). The practice of keeping primates is unregulated in some US states and completely prohibited in other states. Legality even varies between municipalities. In Texas, it is legal to keep certain primates within some city limits, but not others (Born Free USA, 2021). Many US states allow at least some primates to be kept; however, their breeding, transport, and exhibition may be regulated by the US Department of Agriculture (American Society of Primatologists, 2021). Recent reports suggest many of 15,000 pet primates in the United States have likely been bred in captivity specifically for the trade (Born Free USA, 2021).

Some European countries entirely prohibit the keeping of primates as pets. For example, in Hungary, Italy, and the Netherlands, the practice is banned (Endcap, 2012). The United Kingdom's legislation is confusing. While there are currently no prohibitions on keeping any species of primate as a pet in the United Kingdom, the Animal Welfare Act, 2006, and accompanying secondary legislation, should prohibit most primate pet keeping (Animal Welfare Act, 2006; Code of Practice for the Welfare of Privately Kept Non-Human Primates, 2010). In practice, licenses continue to be issued by local authorities for primates that fall under the Dangerous Wild Animals Act. Social media posts, newspaper stories, and advertisements all demonstrate there is clearly a thriving trade in callitrichids in the United Kingdom, which requires no license or registration of any kind (Wild Futures & RSPCA, 2009).

14.3.1 Sociocultural Influences on Primate Pet Keeping and Trade

Regardless of location, sociocultural factors play an important role in the pet trade. Primate pet keeping may be more prevalent in Asia than elsewhere, with macaques regularly sold or kept as pets in China (Ni et al., 2018), Vietnam (Aldrich & Neale, 2021), and Indonesia (Nijman et al., 2017). In Indonesia, macaques are prevalent as pets. Macaques, along with other primate species such as lorises, langurs, gibbons, and orangutans, are captured as infants and kept as household pets.

In many areas of Central and South America, wildlife trade and primate pet keeping are part of local cultures and economies, and these activities are widely tolerated by the public and authorities (Antunes et al., 2016; Bodmer & Pezo, 2001; Drews, 2002; Gómez Ruiz, 2010). As previously mentioned, primate pet keeping in this region is historically associated with the integration of primates into Indigenous worldviews (Cormier, 2003, 2005).

Blair et al. (2017) speculate that African primate pet trade is under-researched, perhaps because it is considered a by-product of the wild meat trade. In a rare study of pet chimpanzee ownership, Kabasawa (2009) suggested that bushmeat hunting was not a primary factor in the trade in Sierra Leone due to the cultural and religious taboos against the consumption of chimpanzee meat among the resident population. Kabasawa (2009) found that the general perception of Sierra Leone nationals was that white expatriates were the main purchasers of infant chimpanzees, and indeed this group made up over 50% of the owners of chimpanzees that were either confiscated or brought to a sanctuary. The chimpanzees were purchased for diverse reasons, such as pity for the infant, rescuing the infant to take to a sanctuary, or simple desire for the primate as a pet. As in many cases of primate pet purchase, the purchasers were unaware that the animal would increase in size and might become very aggressive (Kabasawa, 2009).

14.3.2 Regional Reports on the Primate Pet Trade

Keeping the global trade of live primates in mind, we shift our attention to reports on the current trade. These examples move beyond sociocultural factors alone to explore socio-economic, welfare concerns, and place-based efforts to mitigate the primate pet trade. Considering the breadth of the trade, an exhaustive review is not possible in this chapter alone. Thus, the following regional reports species from five different primate groups: macaques of Asia, orangutans of Southeast Asia, lemurs of Madagascar, slender lorises of India, and the collective category of platyrrhines.

14.3.2.1 Macaques

The desire to keep macaques as pets varies greatly between Southeast Asian countries (Aldrich & Neale, 2021). Many macaque species are highly **synanthropic**, living near or around humans, and perceived as abundant due to their visibility. Consequently, there is potentially less protection given to these species (Aldrich & Neale, 2021). The perception of a primate species' rarity, and thus, significance as a status symbol, seems to affect their desirability as pets. For example, in Laos, the long-tailed macaque is not a desirable pet, because it is believed to be common and uninteresting (P Phiapalath, pers communication).

The legality of macaque keeping varies. In Indonesia, most primate species are legally protected, although enforcement is poor, and many species are sold openly in the bird markets of Java and Bali (Nijman et al., 2017). Most pet macaques are taken directly from the wild, although this has been illegal since 2009. However, in 2021, the government lifted the harvest ban for long-tailed macaques and reinstated a harvest quota. While low, the quota potentially opens up further trade and pet keeping (see Hansen et al., 2021 for review).

Long-tailed macaques and southern pig-tailed macaques are excluded from Indonesia's list of protected species (Hansen et al., 2021). Long-tailed macaques are frequently seen in cages or chains outside people's houses. Pet owners often release macaques in nearby forests when they have reached maturity or decide to keep them constrained when they are no longer easily managed. Released individuals seem to experience difficulty foraging and integrating with wild groups (MF Hansen pers observation). When asked to give up a pet macaque, owners may claim it is impossible because the macaque is a member of their family (MF Hansen pers observation).

In Vietnam, possession of a macaque of any species is illegal unless a permit has been granted by the Forest Protection Department – but these permits cannot be legally granted for the purpose of pet ownership. Yet, the practice is relatively common, and penalties are too light to serve as deterrent. Aldrich and Neale (2021) report that Vietnamese rescue centres are full beyond capacity and struggle to find space for more macaques daily. Due to this lack of capacity, the authorities regularly release macaques immediately following confiscation. This is usually done without regard for the suitability of the location or the health status of the released individuals. Of Vietnam's five macaque species, northern pig-tailed macaques (*M. leonina*) are confiscated most frequently (Aldrich & Neale, 2021).

Macaques, like all other primates, are legally protected in China. Primate confiscations in China between 2000 and 2017 included 215 individual macaques, many of which were rhesus macaques. Most of the documented seizures occurred in residential areas, although the macaques could have been kept for other purposes, such as use in traditional medicine. Rhesus macaques are bred extensively in captivity in China, but the confiscated rhesus individuals may have had captive origins (Ni et al., 2018).

14.3.2.2 Orangutans

From 2005 to 2011, more orangutans were captured from the wild for the illegal trade than chimpanzees (*Pan troglodytes*), bonobos (*P. paniscus*) or gorillas (*Gorilla* sp.) (Stiles et al., 2013). The three species of orangutan, Bornean (*Pongo pygmaeus*), Sumatran (*P. abelii*), and Tapanuli (*P. tapanuliensis*), are fully protected by law in Malaysia and Indonesia, but law enforcement is inadequate (Jonas et al., 2017). All orangutan species are also listed on Appendix I of the Convention on International Trade in Endangered Species (CITES), prohibiting their unlicensed international trade.

Between 1993 and 2016, over 400 Bornean and Sumatran orangutans combined were confiscated by Indonesian authorities, and hundreds more were voluntarily surrendered to the authorities (Nijman, 2017). High mortality rates among infants captured from the wild and difficulties confiscating all illegally traded individuals mean that the numbers officially recorded are smaller than the numbers of wild captures (Meijaard et al., 2012). Marshall et al. (2009) estimated the loss of wild orangutans to hunting alone to be over the 1% rate deemed sustainable for orangutan populations.

Deforestation and increased human accessibility to orangutan habitat contribute to the prevalence of orangutan trade, as it enables opportunistic or planned killing of adults and the capture of infants (Sherman et al., 2020). However, this does not explain the motivations underlying peoples' choices to keep orangutans as pets. Over a 10-year period from 2007, facilities caring for Bornean orangutans took in almost 700 orangutans from the pet trade; some of these animals were confiscated, while others surrendered by their owners (Sherman et al., 2020). In nearly all these cases, the owners claimed that they had found the animals alone as infants on the ground or that someone else had given them an already-orphaned animal. A study on orangutan killings in Kalimantan found that very few people had killed orangutans to sell infants into the pet trade (Freund et al., 2016). Rather, most were opportunistic killings for food, self-defence, and crop protection (Meijaard et al., 2011).

Anecdotal reports from surrenders indicate that people see an infant orangutan as another child and as part of the family (Yayasan Orangutan Sumatera Lestari team, pers communication). Others report that they took in an infant orangutan without knowing how big and strong they would become in adulthood. There is also a clear economic value to an infant orangutan since it can be sold or bartered later (Nijman, 2005).

14.3.2.3 Lemurs

Today, 98% of Madagascar's lemurs are threatened with extinction and approximately a third (31%) are listed as Critically Endangered by the International Union for the Conservation of Nature (IUCN, 2020a). Between 2010 and mid-2013, the trade impacted more than 28,000 individual lemurs. Most lemurs were wild caught by their owners and kept as pets for up to 3 years. This practice, although illegal (Mittermeier et al., 2010; Petter, 1969; UN, 1973), is widespread and occurs in both urban and rural settings. However, the prevalence of ownership tends to be higher in urban areas. Foreigners living in Madagascar, as well as local Malagasy, keep pet lemurs. Captive lemurs are typically kept in poor conditions, living in small cages with no access to food or water; they are often individually housed, or restricted by a rope tied around their waist (Reuter et al., 2016).

Captive pet lemurs are found in a variety of settings, such as restaurants, bars, hotels, and eco-lodges (Reuter & Schaefer, 2016). Businesses advertising lemurs on their premises charged more for their services than businesses that did not. Thus, the presence of, and potential for, human-lemur interactions are a draw for tourists (Reuter & Schaefer, 2017).

To date, more than 40 species of lemur are affected by the illegal trade, including Critically Endangered species such as *Propithecus verreauxi*, *Eulemur flavifrons*, and *Varecia variegata* (IUCN, 2020a; Reuter & Schaefer, 2017). Thus, since the initial large-scale documentation of the trade (Reuter et al., 2016), exploitation of wild populations has steadily increased resulting in a more urgent call for conservation action (Reuter et al., 2017). Thus, there is no evidence to suggest lemurs are being internationally traded, which is illegal as they are protected by CITES Appendix I.

One species notably impacted by the pet trade is the ring-tailed lemur (*Lemur catta*), currently listed as Endangered (LaFleur et al., 2016; LaFleur & Gould, 2020; Gould & Sauther, 2016). This species is the most reported lemur in the pet trade, due to its desirability and ability to fare well in captivity (LaFleur et al., 2019). Furthermore, multiple hot spots for pet ring-tailed lemurs correspond with their geographic range, as well as popular coastal tourist destinations. About half (49%) were kept at local businesses, such as hotels and restaurants.

Wild extraction for the pet trade from threatened populations is unsustainable. Thus, it is vital that strategic conservation initiatives (e.g. education of locals and tourists) are implemented. Continued research will be essential in informing our understanding of the extent and impacts of wild extraction and trade, and how to best ensure the long-term viability of wild lemur populations.

14.3.2.4   Lorises

While slow lorises (genus *Nycticebus*) have garnered much attention from an alarming increase in trade due to social media (Grasso et al., 2020; Lenzi et al., 2020; Nekaris et al., 2010, 2013), the slender lorises (*Loris lydekkerianus* spp.) of India are some of the least studied primates in the world. These small nocturnal strepsirrhines are endemic to the Indian subcontinent and Sri Lanka and consist of two species and two subspecies (Gnanaolivu et al., 2020).

Slender lorises are subject to numerous anthropogenic threats, such as habitat loss, logging, road kills, illegal pet trade, traditional medicine, ritual practices, and hunting (Dittus et al., 2020; Radhakrishna & Singh, 2002), and are listed as 'Near

Threatened' by the International Union for the Conservation of Nature (IUCN) Red List (Dittus et al., 2020). In India, the hunting, sale, transfer, or possession of lorises or their parts (dead or alive) without prior permission from the government, are illegal under Schedule I of Wildlife (Protection) Act, 1972. Internationally, they are protected under CITES Appendix II, prohibiting their international trade (Nekaris et al., 2010).

Slender lorises are very small and freeze when threatened, making them easy to capture and traffic. They are traded internationally, nationally within Indian borders, and locally within loris habitats (Alves et al., 2010). To date, there have been no recorded instances of international trade on the CITES database, which may be due to confusion regarding revised taxonomy of the species (Nekaris et al., 2010). However, reports of international trade appear in local and national newspapers, as well as recorded instances via individuals investigating the illegal trade in India (Alves et al., 2013; Gnanaolivu et al., In prep).

The illegal trade of slender lorises is driven by local perceptions and beliefs ranging from fortune telling to warding off bad omens (Gnanaolivu et al. In prep). While these animals are often kept caged in homes, there is no evidence of slender lorises being kept as status or companion pets in India (Ahmed, 2012; Gnanaolivu et al. in prep; Morgan & Nijman, 2020). Due to a deficit of information on slender lorises (Dittus et al., 2020) and a likelihood of species misidentification, the impact of the pet trade on slender loris populations in India remains incomplete. Thus, additional research on the socioeconomics of the slender loris trade is desperately needed.

14.3.2.5 Platyrrhines

Across South and Central America, live primates are traded and kept as pets (Duarte-Quiroga & Estrada, 2003; Daut et al., 2015; de Souza et al., 2016; Lizarralde, 2002; Maldonado, 2012; Parathian & Maldonado, 2010; Shanee et al., 2017; Stafford et al., 2016; Tirira, 2013). Although the Americas have been the focus of comparatively little research (Reuter & O'Regan, 2017), most global evaluations list hunting in the continent as a lesser threat to primates than in other global regions (Estrada et al., 2017; IUCN, 2020a, b). All Pan-American primate range countries are CITES signatories, and all have laws prohibiting or controlling the capture, trade, or keeping of primates (CITES, 2020; Svensson et al., 2016). This means that few people have the necessary permits for private ownership of primates, and so by default, the vast majority is illegal.

In comparison with wild meat, the pet trade represents a minimal contribution to subsistence economies in the Amazon (Alves et al., 2013; Bodmer & Pezo, 2001; Maldonado & Waters, 2020). Mammals in general are traded as adornment, meat, or traditional medicines, whereas most primates traded and/or confiscated across the continent are live, suggesting that they are predominantly trafficked for the pet trade or for entertainment purposes (Levacov et al., 2011; Lizarralde, 2002; Parathian & Maldonado, 2010; Shanee, 2012; Stafford et al., 2016; Shanee et al., 2017; Tirira, 2013).

The welfare and conservation implications of primate pet keeping can be severe. For example, Peres (1991) estimated that ~10 female woolly monkeys (*Lagothrix lagothricha*) are killed for every infant found in nearby markets. As well as opportunistic capture by individuals or for local trade, primates are also captured in a more organized manner. Smaller species, such as marmosets and tamarins, are caught in groups using baited traps, while larger bodied species, such as capuchins and spider and woolly monkeys, are caught before weaning, by shooting the mother and recovering the infant from the ground (Shanee et al., 2017). The most common primates kept as pets and attractions, or found in markets, are capuchins, squirrel monkeys, marmosets, and woolly monkeys (Ministerio de Medio Ambeinte y Desarrollo Sostenible, 2012; Levacov et al., 2011; Tirira, 2013; Maldonado & Waters, 2017; Oklander et al., 2020; Shanee et al., 2017).

As the human population has steadily increased across primate habitats in the Neotropics, the illegal pet trade is also increasing. Limited understanding of the trade in many areas hinders strategies to control it (Bodmer et al., 2004; Daut et al., 2015). Measures to control the trade are inadequate, with official estimates severely underestimating true volumes (Shanee, 2012; Shanee et al., 2021). Authorities are often slow to act, with very few cases resulting in punishment of offenders, even if clear evidence is presented (Nóbrega Alves et al., 2012; Shanee & Shanee, 2021).

Global trends of wildlife trafficking moving online are not mirrored in the Americas, with little evidence of online wildlife trade in the region (Demeau et al., 2019; Lavorgna, 2015; Shanee et al., 2021). Efforts to increase enforcement in the region have resulted in increased confiscations and improved record keeping. Several countries have published national anti-wildlife trafficking strategies, including Colombia, Ecuador, and Peru (Ministerio del Ambiente y Agua, 2001; Ministerio de Medio Ambeinte y Desarrollo Sostenible, 2012; Ministerio de Agricultura y Riego, 2017).

14.4 Future of the Primate Pet Trade

Many primate species face an uncertain future, with technology directly influencing the demand for these animals. The global market for primates, increasing internet sales, and the use of social media collectively play a critical role in the trade of many primate species, protected or otherwise (Nekaris et al., 2013; Waters & Harrad, 2013). The open sale of primates in markets has decreased in some countries possibly due to increased law enforcement. However, the online trade in primates on social media like Facebook, Tik Tok, and Instagram has boomed because these are sites which facilitate contact between vendors and consumers (Bergin et al., 2018; Cheyne et al., In prep; Lenzi et al., 2020; Norconk et al., 2019).

14.4.1   The Primate Pet Trade and the Internet

Slowing online sales of pet primates is contingent on understanding the dynamics of these transactions. Urban centres are key consumers of pet primates as urban middle classes use their disposable income to advertise their economic status by buying an expensive pet and have no difficulties connecting to the internet. Some Middle Eastern countries feature strongly in the online illegal primate trade. Photographs of elites and celebrities featuring Critically Endangered species like gorillas inspire favourable reactions from their followers. Glorifying endangered primate pet ownership as a status symbol and feature of wealth may have implications for increased demand by others wishing to share the public's adulation (Spee et al., 2019).

Over an 18-month study of online sales of primates and carnivores on Facebook in Thailand, 20 primate species (380 individuals) were listed for sale (Siriwat et al., 2019). Slow loris and macaques accounted for most of the posts, but non-native primates such as callitrichids and squirrel monkeys were the most expensive. Another study of Facebook and Instagram conducted over a 6-year period in Indonesia found a mean number of 155 posts advertising 650 individual gibbons for sale (Cheyne et al., In prep). A decrease in such posts from 2019 may be explained by an initiative by the Indonesian state to close Facebook wildlife trade accounts and Facebook's own initiative to prohibit wildlife trafficking on its site (Cheyne et al., In prep). Open exploitation of Barbary macaques is more common in Morocco than in Algeria due to the stricter enforcement of wildlife protection laws in the latter. This may explain higher online sales of Barbary macaques in Algeria (Bergin et al., 2018).

Social media is also a weapon to counteract the presence of posts featuring primates kept as pets or advertised illegally. Loris conservationists fought back against damaging YouTube videos of lorises being 'tickled' by posting comments on the negative welfare and conservation consequences for these pets (Nekaris et al., 2013). A Moroccan NGO, Barbary Macaque Awareness and Conservation (BMAC), countered the lack of awareness about wildlife laws prohibiting the keeping of the Endangered Barbary macaque as a pet in Morocco, by taking advantage of high Facebook use among the Moroccan public. They shared information about the macaque and the laws protecting the species as well as providing Facebook as medium, empowered Moroccans to report illegally held macaques (Waters & Harrad, 2013). This had led to a collaboration of BMAC and the Department of Forestry leading to the confiscation of numerous illegally held macaques since 2012. A positive unforeseen consequence has been that regional jurisdictions are much quicker to confiscate because they are sensitive to the Facebook posts (S Waters, unpublished data).

To mitigate the high level of online gibbon trade in Indonesia, a campaign targeting potential buyers used a website and the social media platform Instagram to target potential buyers in major cities where trade and possession of pet gibbons were known to be high. The website was deemed more successful in reaching potential buyers (Cheyne et al., In prep.). All the above activities illustrate that online information sharing is low cost and can have a wide reach among urban populations.

Box 14.1 Misuse of Publicity Photos

The following is a personal account from Andrea Dempsey on the negative repercussions of photos depicting primates in close proximity to humans. "I had been working at ZSL London Zoo for two years when we received a new Endangered primate species to our animal collection – one male and two female white-naped mangabey (*Cercocebus lunulatus*).

In April 2008, one of the females gave birth via C-section. After several hours, the female had not picked up her offspring and the decision to hand rear was made. The hand-rearing process followed a strict protocol, with minimal human interaction to avoid imprinting, abnormal behaviours, and an inability to reproduce or rear their own offspring.

The offspring, Conchita, was a good news story both for the species and for ZSL London Zoo. Thus, it was agreed one photographer would be allowed to take photographs of Conchita. A number of images were captured; yet, the photographer was not entirely satisfied and asked if I could hold Conchita to my face so I too could be incorporated into the photo. I followed his directions and with Conchita by my face, she turned and bit my nose - snap! The photo was taken and I thought nothing more of the session. I was only delighted that awareness of the species was being raised.

So it came as a genuine shock sometime later to be alerted to an American website, advertising for sale 'cappuccino monkeys' using the photo of Conchita and I. I immediately contacted the press department at ZSL London Zoo, the website, the seller, but nothing could be done. Not only do primates make poor pets, but the very trade itself is driving primates like the mangabey to extinction and here am I advertising it! This notion, compounded by the fact I had no recourse, is a lesson well learnt. This was over twelve years ago, and yet a simple Google search will still find my image being used.

As primatologists, zoo keepers, conservationists, marketing departments, and publicists we must find other ways to tell our story, to create compassion for the natural world without using images of humans and primates together, which can be used without consent and paradoxically out of context, otherwise we will be responsible for perpetuating a trade that could wipe out the primate populations we are working to save.

14.4.2 Finding Solutions to the Primate Pet Trade

Primates face numerous compounding threats, such as habitat loss, hunting, climate change, and illegal capture for the pet trade (Estrada et al., 2017; IUCN, 2020a). Understanding the longevity and complexity of primate pet keeping is critical. Due to its clandestine nature, studies of wildlife trafficking are difficult, and information which does exist is often from imperfect data (Barber-Meyer, 2010). The primate pet trade in local communities is often opportunistic, but organized trade also exists at the national and international level (Leberatto, 2016; Shanee et al., 2017).

As with other conservation activities, properly motivated communities and local leaders can be effective in the fight against wildlife trafficking (Shanee, 2013) and can lead to the custom becoming less socially acceptable (Shanee, 2012; Waters et al., 2019). The primate pet trade is driven by access to markets and online social platforms and a growing demand for wildlife from an ever-increasing human population.

In some locations, primate-based ecotourism and investigation could provide economic gain from intact primate communities, and sustainable hunting practices would help maintain local livelihoods (Bodmer & Pezo, 2001). Estrada et al. (2020) remind us that the preservation of the world's primates ultimately requires attention to human well-being, health, and security. Even so, greater changes in political and social behaviours are still needed, combined with properly trained, equipped, and funded wildlife authorities. Legislation must be cleaned up and legal loopholes closed, laws updated, and corruption must be addressed at all levels (Shanee & Shanee, 2021).

Other solutions include increasing public awareness. This includes ensuring messages from scientists, primatologists, caretakers, and tour guides coming into close contact with wild primates are clear (Norconk et al., 2019; Waters et al., 2021). Often, those with access to primates are quick to disperse images of themselves with wild or captive primates despite the misleading messages generated by such imagery (see Waters et al., 2021). As our case study from the London Zoo demonstrates (A Dempsey, pers communication), even the most innocent of photographs can have far-reaching consequences.

The development of long-term solutions requires consideration of the historical, sociocultural, political, and economic contexts of the primate pet trade (Estrada et al., 2017; Estrada et al., 2020; Norconk et al., 2019). Ultimately, extensive government and public support from local to global levels will be required. Given that 65% of all extant primate species are either endangered or threatened, our persistence and collective efforts are imperative (Estrada et al., 2017; Estrada et al., 2020).

Acknowledgements The authors would like to thank the editors for our inclusion in this important publication and the reviewers for their comments and suggestions. We owe a special thanks to range country individuals and collaborators, paid and unpaid, who provided their time and information to these researchers. Without their help, this work would be impossible. Finally, we thank in advance those reading this chapter for their support in reducing the demand for pet primates by not purchasing primates for pets and refraining from sharing images that misrepresent primates and undermine efforts to slow the unethical trade in these animals.

References

Ahmed, A. (2012). Future of Indian lorises hangs upside down. *TRAFFIC Bulletin, 15*, 8–9.

Aldrich, B. C., & Neale, D. (2021). Pet Macaques in Vietnam: An NGO's perspective. *Animals, 11*, 60. https://doi.org/10.3390/ani11010060

Alves, R. R. N., Souto, W. M. S., & Barbosa, R. R. D. (2010). Primates in traditional folk medicine: A world overview. *Mammals Review, 40*(2), 155–180.

Alves, R. R. N., Souto, W. M. S., Barboza, R. R. D., et al. (2013). Primates in traditional folk medicine: World overview. In R. R. N. Alves & I. L. Rosa (Eds.), *Animals in traditional folk medicine: Implications for conservation* (pp. 135–170). Springer. https://doi.org/10.1007/978-3-642-29026-8_8

American Society of Primatologists. (2021). How can primatologists deter ownership of pet primates in the US? History and legislation (2021). *ASP-CAN Action Letter, 2*, 4.

Animal Welfare Act. (2006). *UK public general acts chapter 45*. Legistation.gov.uk. https://www.legislation.gov.uk/ukpga/2006/45/contents.

Antunes, A. P., Fewster, R. M., Venticinque, E. M., et al. (2016). Empty forest or empty rivers? A century of commercial hunting in Amazonia. *Science Advances, 2*(10), e1600936. https://doi.org/10.1126/sciadv.1600936

Barber-Meyer, S. M. (2010). Dealing with the Clandestine nature of wildlife-trade market surveys. *Conservation Biology, 24*(4), 918–923.

Bergin, D., Atoussi, S., & Waters, S. (2018). Online trade of Barbary macaques Macaca sylvanus in Algeria and Morocco. *Biodiversity & Conservation, 27*, 531–534.

Blair, M. E., Le, M. D., & Sterling, E. J. (2017). Multidisciplinary studies of wildlife trade in primates: Challenges and priorities. *American Journal of Primatology, 79*, e22710.

Bodmer, R. E., & Pezo, E. (2001). Rural development and sustainable wildlife use in Peru. *Conservation Biology, 15*(4), 1163–1170.

Bodmer, R. E., Pezo, E., & Fang, T. G. (2004). Economic analysis of wildlife use in the Peruvian Amazon. In K. M. Silvius, R. E. Bodmer, & J. M. V. Fragoso (Eds.), *People in nature: Wildlife conservation and management in South and Central America* (pp. 191–207). Columbia University Press.

Born Free USA. (2021). Public danger, private pain: The case against the primate pet trade in the United States.

British Veterinary Association. (2014). Commons select committee enquiry: Primates as pets: written evidence. UK Parliament. https://data.parliament.uk/writtenevidence/committeeevidence.svc/evidencedocument/environment-foodand-rural-affairs-committee/primates-as-pets/written/5183.html. Accessed 20 July 2021

Cheyne., et al. (In prep). *Biological conservation*.

CITES. (2020). *Convention on International Trade in Endangered Species of wild Fauna and Flora*. http://checklist.cites.org/. Accessed 07 June 2020.

CITES Trade Database. (2021). https://trade.cites.org/. Accessed 19 April 2021.

Clarke, T. A., Reuter, K. E., LaFleur, M., & Schaefer, M. S. (2019). A viral video and pet lemurs on Twitter. *PLoS One, 14*(1), e0208577.

Code of Practice for the Welfare of Privately Kept Non-Human Primates. (2010). http://archive.defra.gov.uk/wildlife-pets/pets/cruelty/documents/primate-cop.pdf

Cormier, L. A. (2003). *Kinship with monkeys*. Columbia University Press.

Cormier, L. A. (2005). A preliminary review of neotropical primates in the subsistence and symbolism of indigenous lowland South American peoples. *Ecological and Environmental Anthropology, 2*(1), 14–31.

Cormier, L. A., & Jolly, P. E. (2018). *The primate Zoonoses: Culture change and emerging diseases*. Rutledge, Taylor, and Francis Group.

Daut, E. F., Brightsmith, D. J., & Peterson, M. J. (2015). Role of non-governmental organizations in combating illegal wildlife–pet trade in Peru. *Journal for Nature Conservation, 24*(0), 72–82. https://doi.org/10.1016/j.jnc.2014.10.005

de Souza, F. M., Ludwig, G., & Valença-Montenegro, M. M. (2016). Legal international trade in live neotropical primates originating from South America. *Primate Conservation, 30*, 1–6.

Demeau, E., Vargas-Monroy, M. E., & Jeffrey, K. (2019). El tráfico de fauna silvestre por internet: ¿un mercado comparado con el tráfico de drogas virtual? *Revista Criminalidad, 61*, 101–112.

Dittus, W., Singh, M., Gamage, S. N., Kumara, H. N., et al. (2020). Loris lydekkerianus. The IUCN Red List of Threatened Species: e.T44722A17970358. https://doi.org/10.2305/IUCN. UK.2020-3.RLTS.T44722A17970358.en.

Dominy, N. J., Ikram, S., Moritz, G. L., et al. (2020). Mummified baboons reveal the far reach of early Egyptian mariners. *eLife, 9*, e60860. https://doi.org/10.7554/eLife.60860

Drews, C. (2002). Attitudes, knowledge and wild animals as pets in Costa Rica. *Anthrozoös, 15*(2), 119–138.

Duarte-Quiroga, A., & Estrada, A. (2003). Primates as pets in Mexico City: An assessment of the species involved, source of origin, and general aspects of treatment. *American Journal of Primatology: Official Journal of the American Society of Primatologists, 61*(2), 53–60.

Endcap. (2012). *Wild Pets in the European Union Report*. https://endcap.eu/wp-content/uploads/2013/02/Report-Wild-Pets-in-the-European-Union.pd

Estrada, A., Garber, P. A., Rylands, A. B., et al. (2017). Impending extinction crisis of the world's primates: Why primates matter. *Science Advances, 3*(1).

Estrada, A., Garber, P. A., & Chaudhary, A. (2020). Current and future trends in socio-economic, demographic and governance factors affecting global primate conservation. *PeerJ, 8*, e9816. https://doi.org/10.7717/peerj.9816

Foley, K.-E., & Shepherd, C. R. (2011). Trade in long-tailed macaques (Macaca fascicularis). In M. D. Gumert, L. Jones-Engel, & A. Fuentes (Eds.), *Monkeys on the edge. Ecology and management of Long-Tailed Macaques and their interface with humans* (1st ed., pp. 20–23). Cambridge University Press.

Freund, C., Rahman, E., & Knott, C. (2016). Ten years of orangutan-related wildlife crime investigation in West Kalimantan, Indonesia. *American Journal of Primatology, 9999*, 22620.

Fuentes, A. (2002). Monkeys, humans, and politics in the Mentawai Islands: No simple solutions in a complex world. In A. Fuentes & L. D. Wolfe (Eds.), *Primates face to face: The conservation implications of human-nonhuman primate interconnections* (pp. 187–207). Cambridge University Press.

Fuentes, A. (2007). Monkey and human connections: The wild, the captive and the inbetween. In R. Cassidy & M. Malone (Eds.), *Where the wild things are now: Domestication reconsidered* (pp. 123–146). Routlede.

Fuentes, A. (2013). Pets, property, and partner: Macaques as commodities in the human-other primate interface. In S. Radhakrishna, M. A. Huffman, & A. Sinha (Eds.), *The Macaque connection: Cooperation and conflict between humans and Macaques* (pp. 107–123). Springer.

Gnanaolivu, S. D., Kumara, H. N., Singh, M., et al. (2020). Ecological determinates of Malabar Slendar Loris (Loris lydekkerianus malabaricus, Cabrera 1908) occupancy and abundance in Aralam Wildlife Sanctuary, Western Ghats, India. *International Journal of Primatology, 41*, 511–524.

Gómez Ruiz, C. (2010). Influencia de factores culturales en la tenencia de monos aulladores como mascotas y su efecto sobre los individuos en cautiverio. Revista CES Medicina Veterinaria y Zootecnia:46+.

Gould, L., & Sauther, M. L. (2016). Going, going, gone… Is the iconic ring-tailed lemur (Lemur catta) headed for imminent extirpation. *Primate Conservation, 30*(1), 89–101.

Grasso, C., Lenzi, C., Speiran, S., et al. (2020). Anthropomorphized nonhuman animals in mass media and their influences on human attitudes towards wildlife. *Society & Animals*, 1–25.

Groves, C. (2008). *Extended family: Long lost cousins; a personal look at the history of primatology*. Conservation International.

Hansen, M. F., Gill, M., Nawangsari, V. A., et al. (2021). Conservation of Long-tailed Macaques: Implications of the updated IUCN status and the CoVID-19 pandemic. *Primate Conservation, 35*, 1–11.

Harlow, H. (1962). Social deprivation in monkeys. *Scientific American, 207*, 136.

Hurn, S. (2012). *Humans and other animals: Cross cultural perspectives on human-animal interactions*. Pluto Press.

International Primatological Society. (2008). *Private ownership of nonhuman primates*. International Primatological Society.

IUCN.   (2020a).   https://www.iucn.org/news/species/202007/ almost-a-third-lemurs-and-north-atlantic-right-whale-now-critically-endangered-iucn-red-list

IUCN. (2020b). *IUCN red list of threatened species*. www.redlist.org. Accessed 07 June 2020.

Janson, H. W. (1952). *Apes and Ape Lore: In the middle ages and renaissance*. Warburg Institute.

Jonas, H., Abram, N. K., & Ancrenaz, M. (2017). *Addressing the impact of large-scale oil palm plantations on orangutan conservation in Borneo: A spatial, legal and political economy analysis*. IIED.

Kabasawa, A. (2009). Current state of the chimpanzee pet trade in Sierra Leone. *African Study Monographs, 30*, 37–54.

LaFleur, M., Clarke, T. A., Reuter, K., & Schaeffer, T. (2016). Rapid decrease in populations of wild ring-tailed lemurs (*Lemur catta*) in Madagascar. *Folia Primatologica, 87*(5), 320–330.

LaFleur, M., Clarke, T. A., Reuter, K. E., et al. (2019). Illegal trade of wild-captured Lemur catta within Madagascar. *Folia Primatologica, 90*(4), 199–214.

LaFleur, M., & Gould, L. (2020). Lemur catta. The IUCN Red List of Threatened Species 2020:   e.T11496A115565760.   https://doi.org/10.2305/IUCN.UK.2020-2.RLTS. T11496A115565760.en

Lavorgna, A. (2015). The Social Organization of Pet trafficking in cyberspace. *European Journal on Criminal Policy and Research, 21*(3), 353–370. https://doi.org/10.1007/s10610-015-9273-y

Law, A. J., Pei, Q., Walker, M., et al. (2009). Early parental deprivation in the Marmoset monkey produces long-term changes in Hippocampal expression of genes involved in synaptic plasticity and implicated in mood disorder. *Neuropsychopharmacology, 34*(6), 1381–1394.

Leberatto, A. C. (2016). Understanding the illegal trade of live wildlife species in Peru. *Trends in Organized Crime, 19*(1), 42–66. https://doi.org/10.1007/s12117-015-9262-z

Lee, P. C., & Priston, N. E. C. (2005). *Human attitudes to primates: Perceptions of pests, conflict and consequences for primate conservation. Commensalism and conflict: The human primate interface* (pp. 1–23). J.D. Paterson and J Wallis. Winnipeg, Higwell Printing.

Leighty, K. A., Valuska, A. J., Grand, A. P., et al. (2015). Impact of visual context on public perceptions of non-human primate performers. *PLoS One, 10*(2), e0118487.

Lenzi, C., Speiran, S., & Grasso, C. (2020). "Let Me Take a Selfie": Implications of social media for public perceptions of wild animals. *Society & Animals, 1*(aop), 1–20.

Levacov, D., Jerusalinsky, L., & Fialho, MdS. (2011). Levantamento dos primatas recebidos em Centros de Triagem e sua relação com o tráfico de animais silvestres no Brasil. A Primatologia no Brasil–11 Belo Horizonte, Brazil: Sociedade Brasileira de Primatologia pp. 281–305.

Lewis, M. H., Gluck, J. P., Petitto, J. M., et al. (2000). Early social deprivation in nonhuman primates: Long-term effects on survival and cell-mediated immunity. *Biological Psychiatry, 47*(2), 119–126.

Lizarralde, M. (2002). Ethnoecology of monkeys among the Barí of Venezuela: Perception, use and conservation. In A. Fuentes & L. D. Wolfe (Eds.), *Primates face to face: The conservation implications of human-nonhuman primate interconnections. Cambridge studies in biological and evolutionary anthropology* (pp. 85–100). Cambridge University Press. https://doi.org/10.1017/CBO9780511542404.009

Maldonado, A. M. (2012). *Hunting by Tikunas in the Southern Colombian Amazon. Assessing the impact of subsistence hunting by Tikunas on game species in Amacayacu National Park*. Colombian Amazon. LAP Lambert Academic Publishing GmbH & KG.

Maldonado, A. M., & Waters, S. (2017). Primate trade (Neotropics). In A. Fuentes (Ed.), *International Encyclopedia of primatology*. Wiley-Blackwell.

Maldonado, A. M., & Waters, S. (2020). Ethnoprimatology of the Tikuna in the Southern Colombian Amazon. In B. Urbani & M. Lizarralde (Eds.), *Neotropical Ethnoprimatology:*

Indigenous peoples' perceptions of and interactions with nonhuman primates (pp. 89–107). Springer International Publishing. https://doi.org/10.1007/978-3-030-27504-4_5

Marshall, A. J., Lacy, R., Ancrenaz, M., et al. (2009). Orangutan population biology, life history, and conservation. *Orangutans: Geographic variation in behavioral ecology and conservation,* 311–326.

Meijaard, E., Buchori, D., Hadiprakarsa, Y., et al. (2011). Quantifying killing of Orangutans and human-Orangutan conflict in Kalimantan, Indonesia. *PLoS One, 6*(11), e27491.

Meijaard, E., Wich, S., Ancrenaz, M., et al. (2012). Not by science alone: Why orangutan conservationists must think outside the box. *Annals of the New York Academy of Sciences Issue: The Year in Ecology and Conservation Biology.*

Ministerio de Agricultura y Riego. (2017). Estrategia nacional para reducir el trafico ilegal de fauna silvestre en el Perú 2017–2027 y su plan de acción 2017–2022. Ministerio de Agricultura y Riego, Peru.

Ministerio del Ambiente y Agua. (2001). *Política y Estrategia Nacional de Biodiversidad del Ecuador 2001–2010, Ecuador.*

Ministerio de Medio Ambeinte y Desarrollo Sostenible. (2012). *Estrategia Nacional para la prevención y control al Tráfico Ilegal de Especies Silvestres: Diagnóstico y Plan de Acción ajustado.* Ministerio de Ambiente y Desarrollo Sostenible, Columbia.

Mito, Y., & Sprague, D. S. (2013). The Japanese and Japanese monkeys: Dissonant Neighbors seeking accommodation in a shared habitat. In S. Radhakrishna, M. A. Huffman, & A. Sinha (Eds.), *The Macaque connection: Cooperation and conflict between humans and Macaques* (pp. 33–51). Springer.

Mittermeier, R. A., Louis, E. E., Jr., Richardson, M., et al. (2010). *Lemurs of Madagascar* (3rd ed.). Conservation International.

Morgan, B., & Nijman, V. (2020). Little people and Rice Magic: The internet trade in Slender Lorises in South Asia. In K. Nekaris & A. Burrows (Eds.), *Evolution, ecology and conservation of Lorises and Pottos (Cambridge studies in biological and evolutionary anthropology)* (pp. 357–360). Cambridge University Press. https://doi.org/10.1017/9781108676526

Nekaris, K. A. I., Campbell, N., Coggins, T. G., et al. (2013). Tickled to death: Analysing public perceptions of 'cute' videos of threatened species (slow lorises – Nyticebus spp.) on Web 2.0 sites. *PLoS One, 8*(7), e69215.

Nekaris, K. A. I., Shepherd, C. P., Starr, C. R., et al. (2010). Exploring cultural drivers for wildlife trade via an ethnoprimatological approach: A case study of Slender and slow lorises (Loris and Nycticebus) in South and Southeast Asia. *American Journal of Primatology, 72,* 877–886.

Ni, Q., Wang, Y., Weldon, A., et al. (2018). Conservation implications of primate trade in China over 18 years based on web news reports of confiscations. *PeerJ, 6,* e6069.

Nijman V. (2005). *Hanging in the balance: An assessment of trade in Orang-utans and Gibbons in Kalimantan, Indonesia.* TRAFFIC Southeast Asia.

Nijman, V. (2017). Orangutan trade, confiscations, and lack of prosecutions in Indonesia. *American Journal of Primatology, 79,* 22652.

Nijman, V., Spaan, D., Rode-Margono, E. J., et al. (2017). Changes in the primate trade in Indonesian wildlife markets over a 25-year period: Fewer apes and langurs, more macaques, and slow lorises. *American Journal of Primatology, 79,* e22517.

Nóbrega Alves, R. R., De Farias Lima, J. R., & Araujo, H. F. P. (2012). The live bird trade in Brazil and its conservation implications: An overview. *Bird Conservation International, 23*(1), 53–65. https://doi.org/10.1017/S095927091200010X

Norconk, M. A., Atsalis, S., Tully, G., et al. (2019). Reducing the primate pet trade: Actions for primatologists. *American Journal of Primatology, 82.* https://onlinelibrary.wiley.com/doi/10.1002/ajp.23079

North American Primate Sanctuary Alliance (2021) Position statement—Private ownership of primates. Primate Sanctuaries.

Oklander, L. I., Caputo, M., Solari, A., et al. (2020). Genetic assignment of illegally trafficked neotropical primates and implications for reintroduction programs. *Scientific Reports, 10*(1), 3676. https://doi.org/10.1038/s41598-020-60569-3

Parathian, H. E., & Maldonado, A. M. (2010). Human–nonhuman primate interactions amongst Tikuna people: Perceptions and local initiatives for resource management in Amacayacu in the Colombian Amazon. *American Journal of Primatology, 72*(10), 855–865. https://doi.org/10.1002/ajp.20816

Peres, C. A. (1991). Humboldt's woolly monkeys decimated by hunting in Amazonia. *Oryx, 25*, 89–95.

Petter, J. J. (1969). Speciation in Madagascan lemurs. *Biological Journal of the Linnean Society, 1*, 77–84.

Wild Futures & RSPCA. (2009). Primates as pets: Is there a case for regulation?. http://wildfutures.s3.amazonaws.com/wp-content/blogs.dir/1/files/2009/09/PrimatePack_minusRegulatoryOptions.pdf

Pryce, C. R., Dettling, A. C., Spengler, M., et al. (2004). Deprivation of parenting disrupts development of homeostatic and reward systems in marmoset monkey offspring. *Biological Psychiatry, 56*(2), 72–79.

Radhakrishna, S., & Singh, M. (2002). Conserving the slender loris (*Loris lydekkerianus lydekkerianus*). National Seminar on Conservation of Eastern Ghats, March 24–26, pp. 227–231.

Reuter, K. E., LaFleur, M., & Clarke, T. A. (2017). Illegal lemur trade grows in Madagascar. *Nature, 541*(7636), 157.

Reuter, K. E., LaFleur, M., Clarke, T. A., et al. (2019). A national survey of household pet lemur ownership in Madagascar. *PLoS One, 14*(5), e0216593. https://doi.org/10.1371/journal.pone.0216593

Reuter, K. E., Gilles, H., Wills, A. R., et al. (2016). Live capture and ownership of lemurs in Madagascar: Extent and conservation implications. *Oryx, 50*(2), 344–354.

Reuter, K. E., & Schaefer, M. S. (2016). Captive conditions of pet lemurs in Madagascar. *Folia Primatologica, 87*(1), 48–63.

Reuter, K. E., & Schaefer, M. S. (2017). Motivations for the ownership of captive lemurs in Madagascar. *Anthrozoös, 30*(1), 33–46.

Reuter, P., & O'Regan, D. (2017). Smuggling wildlife in the Americas: Scale, methods, and links to other organised crimes. *Global Crime, 18*(2), 77–99. https://doi.org/10.1080/17440572.2016.1179633

Ross, S. R., Lukas, K. E., & Lonsdorf, E. V. et al. (2008). Inappropriate use and portrayal of chimpanzees.

Ross, S. R., Vreeman, V. M., Lonsdorf, E. V., et al. (2011). Specific image characteristics influence attitudes about chimpanzee conservation and use as pets. *PLoS One, 6*(7), e22050.

Sanerib, T., & Uhlemann, S. (2020). *Dealing in disease: How U.S. wildlife imports fuel global pandemic risk, 2020*. https://www.biologicaldiversity.org/programs/international/pdfs/Dealing-in-Disease_Center-wildlife-imports-report-9-28-20.pdf

Schroepfer, K., Rosati, A., Chartrand, T., et al. (2011). Use of "Entertainment" Chimpanzees in commercials distorts public perception regarding their conservation status. *PLoS One.* https://doi.org/10.1371/journal.pone.0026048

Serpell, J. (1996). *In the company of animals: A study of human-animal relationships*. Cambridge University Press.

Shanee, N. (2012). Trends in local wildlife hunting, trade and control in the Tropical Andes Hotspot, Northeastern Peru. *Endangered Species Research, 19*(2), 177–186.

Shanee, N. (2013). Campesino justification for self-initiated conservation actions: A challenge to mainstream conservation. *Journal of Political Ecology, 20*, 413–428.

Shanee, N., Mendoza, A. P., & Shanee, S. (2017). Diagnostic overview of the illegal trade in primates and law enforcement in Peru. *American Journal of Primatology, 79*(11), e22516. https://doi.org/10.1002/ajp.22516

Shanee, N., & Shanee, S. (2021). Denunciafauna – Social media campaign to evaluate wildlife crime and law enforcement in Peru. *Journal of Political Ecology, 21.*

Shanee, S., Mendoza, A. P., Maldonado, A. M., Fernández-Hidalgo, L., & Svensson, M. S. (2021). Traffic and trade in owl monkeys. In E. Fernandez-Duque (Ed.), *Owl monkeys - biology, adaptive radiation, and behavioral ecology of the only Nocturnal Primate in the Americas.* Springer International Publishing.

Shepherd, C. R. (2010). Illegal primate trade in Indonesia exemplified by surveys carried out over a decade in North Sumatra. *Endangered Species Research, 11*, 201–205. https://doi.org/10.3354/esr00276

Shephard, G. H. (2002). Primates in the Matsigenka subsistence and world view. In A. Fuentes & L. D. Wolfe (Eds.), *Primates face to face: The conservation implications of human-nonhuman primate interconnections* (pp. 101–136). Cambridge University Press.

Sherman, J., Ancrenaz, M., & Meijaard, E. (2020). Shifting apes: Conservation and welfare outcomes of Bornean orangutan rescue and release in Kalimantan, Indonesia. *Journal for Nature Conservation, 55*, 125807.

Siriwat, P., Nekaris, K. A. I., & Nijman, V. (2019). The role of the anthropogenic allee effect in the exotic pet trade on Facebook in Thailand. *Journal for Nature Conservation, 51*, 125726.

Soulsbury, C. D., Iossa, G., Kennell, S., et al. (2009). The welfare and suitability of primates as pets. *Journal of Applied Animal Welfare Science, 12*, 1–20.

Spee, L. B., Hazel, S. J., Dal Grande, E., et al. (2019). Endangered exotic pets on social media in the Middle East: Presence and impact. *Animals, 9*, 480 ani/9080480.

Stafford, C. A., Alarcon-Valenzuela, J., Patiño, J., et al. (2016). Know your monkey: Identifying primate conservation challenges in an indigenous Kichwa community using an ethnoprimatological approach. *Folia Primatologica, 87*(1), 31–47. https://doi.org/10.1159/000444414

Stiles, D., Redmond, I., Cress, D., et al. (eds). (2013). *Stolen Apes – the Illicit trade in Chimpanzees, Gorillas, Bonobos and Orangutans. A Rapid Response Assessment.* United Nations Environment Programme, GRID-Arendal. www.grida.no

Svensson, M. S., Shanee, S., Shanee, N., et al. (2016). Disappearing in the night: An overview on trade and legislation of night monkeys in South and Central America. *Folia Primatologica, 87*(5), 332–348. https://doi.org/10.1159/000454803

Thompson, N. A. (2019). Understanding the links between social ties and fitness over the life cycle in primates. *Behaviour, 156*(9), 859–908.

Tirira, D. G. (2013). Tráfico de primates nativos en el Ecuador. *Boletín Técnico, Serie Zoológica, 11*, Núm 8-9.

UN. (1973). *Convention on international trade in endangered species of wild Fauna and Flora.* United Nations.

Van Uhm, D. (2016). Monkey business: The illegal trade in Barbary macaques. *Journal of Trafficking, Organized Crime, and Security, 2*(1), 36–49.

Venkatesan, S. (2018). *Veera Yuga Naygan Vel Pari.* Vikatan Prasuram (Publishers) p1408 [in Tamil].

Waters, S., & Harrad, A. E. (2013). A note on the effective use of social media to raise awareness against the illegal trade in Barbary Macaques. *African Primates, 8*, 67–68.

Waters, S., Harrad, A. E., Bell, S., & Setchell, J. M. (2019). Interpreting people's behavior towards primates using qualitative data: A case study from North Morocco. *International Journal of Primatology, 40*, 316–330.

Waters, S., Setchell, J. M., Maréchal, L., et al. (2021). *Best practice guidelines for responsible images of non-human primates.* IUCN Primate Specialist Group Section for Human-Primate Interactions.

Chapter 15
Rescue, Rehabilitation, and Reintroduction

Siobhan I. Speiran, Tephillah Jeyaraj-Powell, Laurie Kauffman, and Michelle A. Rodrigues

Contents

Abstract The rescue, rehabilitation, and reintroduction of nonhuman primates (henceforth, primates) from captive and wild settings are three sequential strategies which improve species conservation and individual welfare. Rescue refers to the removal of individuals from situations of cruelty, danger, illness, or risk which harm their well-being. Upon rescue, individuals requiring ongoing veterinary care and/or supervision undergo rehabilitation. This process differs depending on the needs of the individual primate; while those rescued from both wild and captive circumstances may undergo rehabilitation, only the former group is typically candidate for reintroduction. Reintroduction refers to the release of rehabilitated individuals into spaces where they historically ranged, with the goal of improving species conserva-

S. I. Speiran (✉)
School of Environmental Studies, Queen's University, Kingston, ON, Canada

T. Jeyaraj-Powell
Department of Psychology, University of Central Oklahoma, Edmond, OK, USA

L. Kauffman
Department of Biology, Oklahoma City University, Oklahoma City, OK, USA

M. A. Rodrigues
Department of Social and Cultural Sciences, Marquette University, Milwaukee, WI, USA

© The Author(s), under exclusive license to Springer Nature Switzerland AG 2023   267
T. McKinney et al. (eds.), *Primates in Anthropogenic Landscapes*, Developments in Primatology: Progress and Prospects, https://doi.org/10.1007/978-3-031-11736-7_15

tion. In practice, however, reintroduction often occurs in the interest of improving primate welfare, which can complicate conservation objectives. This chapter reviews the literature from the past couple of decades on primate rescue, rehabilitation, and reintroduction – emphasizing the call for continuing to develop multidisciplinary, ethical, and evidence-based "best practices."

Keywords Rescue · Rehabilitation · Reintroduction · Primate · Monkey · Sanctuary · Costa Rica

15.1 Introduction

A female monkey is swinging briskly through the trees with an infant hanging from her neck. She arrives at the outer edges of the forest and looks out at a two-lane commercial street. She can see the connecting forest in the distance, but reaching it requires crossing human territory. She observes the traffic on the road below, then looks up to the electrical wiring which connects the forest canopy to the city like a spiderweb. She mounts the transformer and grasps the bottom wire until, in a fateful moment, her prehensile tail connects with the top wire and she experiences a sudden jolt!

Dropping to the ground with her infant, a passerby observes the event and calls the police, who forward the call to the local wildlife authority. Soon after, rescuers appear on the scene, note the location and cause of the monkeys' injuries, and carefully transfer them to a rescue centre for veterinary attention and rehabilitation. Over a period of weeks, the mother and infant recover from their injuries. Upon observation, the animal care staff determine the monkeys have successfully rehabilitated and the next morning they are reintroduced in a national park near the location where they were rescued.

The rescue, rehabilitation, and reintroduction of nonhuman primates (henceforth, primates) from captive and wild settings are three sequential strategies which may improve species conservation and individual welfare (Guy et al., 2014). **Rescue** refers to the removal of individuals from situations of injury or human-related danger (Pyke & Szabo, 2017). There are inconsistent definitions for rehabilitation and reintroduction. In a broad sense, **rehabilitation** is a managed process in which ill, injured, orphaned, or displaced wild animals regain the health and skills needed for successful reintroduction and survival (International Wildlife Rehabilitation Council 2005, cited in Molony et al., 2006). Rehabilitation may also describe caring for captive wild animals that are not to be reintroduced (Guy et al., 2014). In its narrowest definition, **reintroduction** refers to the release of captive, or confiscated individuals or groups animals into wild habitats from where the species was previously extirpated (Baker, 2002; Beck et al., 2007; Beck, 2018). This may include translocations, introductions of animals outside their historic range, or the release of animals where there is already to supplement the numbers of an existing population (Beck, 2018).

The opening passage describes the processes of primate rescue, rehabilitation, and reintroduction in a relatively simplistic way; every step in the process conceals complex layers, however, and is rarely so straightforward. Imagine how it could be complicated if the monkey succumbed to her injuries, leaving her infant orphaned? Or if they were so disabling (e.g., amputation, severe burns) that she was not a candidate for release? What about primates rescued from people's homes, the wildlife trade, or the entertainment industry – those so habituated to humans that a return to the wild is difficult or impossible?

Processes of rehabilitation differ depending on the needs of the individual primate; while those rescued from both wild (e.g., road collision) and captive circumstances (e.g., confiscated pet) undergo rehabilitation, only the former group is typically candidate for reintroduction (Speiran, 2021). The manner of rehabilitation can influence the success of release, including whether the primates require dishabituation, training, or time to heal from procedures. Dishabituation refers to a process in which wild animals who are considered overly habituated to humans (i.e., so tolerant or unafraid that either party could be at risk) are trained or conditioned to a point that they can be released with less concern for potential conflict with humans or disease vectors (e.g., tuberculosis) (Wallis & Lee, 1999).

The process of reintroduction varies across taxa, regions, facilities, and between individuals. For example, the method of reintroducing a group of young, orphaned monkeys will differ from the method of releasing a single adult (Schwartz et al., 2016). Reintroductions may occur in a different location from the sanctuary if there is not enough suitable surrounding habitat, or to reduce the chance that reintroduced individuals will return to the site of rehabilitation (i.e., the sanctuary) (Konstant & Mittermeier, 1982; Speiran, 2021).

Primates rescued from the wildlife trade or entertainment industry may be significantly habituated to humans and have no experience in the wild. Such individuals tend to live in captivity forever in zoos or wildlife sanctuaries (also called rescue centers). These are conservation-focused organizations which do not breed their animals nor allow tourist interaction. Sanctuaries are a diverse, innumerous global phenomena; some operate as NGOs, others for-profit; some may be private, while others rely on tourism. Ethical, accredited sanctuaries offer tourists a chance to see primates in captivity but prevent direct interactions or "wildlife selfies." The Global Federation of Animal Sanctuaries (GFAS) accredits organizations, comparable to how the World Association of Zoos and Aquariums (WAZA) accredits their sites, providing an audited list of ethical and conservation-benefiting sanctuaries. There are 18 GFAS-accredited primate sanctuaries globally. A sanctuary does not have to have GFAS accreditation to be ethical and legitimate; there are undoubtedly fiscal, geographical, and resource-based barriers that may prevent some sites from applying for accreditation.

Sanctuaries may or may not be in the home range of an individual species; for example, there are 68 sanctuaries worldwide for great apes, which range in Africa and Asia (Arcus Foundation, 2020). Sites which perform reintroductions are likely to be located within the range distribution of the primate species, and in a similar region to where the rescue occurred. It is preferable to perform all three phases of rescue, rehabilitation, and reintroduction at the same location to avoid the added stress and welfare risks of transportation. This is not always possible, however, and there are procedures for sanctuaries to send rescued individuals to other sanctuaries whether for rehabilitation or eventual release.

The definition of sanctuary, as it is operationalized across diverse cultural and ecological contexts, can seem idealistic. There are examples of primate sanctuaries which breed endangered individuals for reintroduction; in Decolonizing Extinction (2018), Parreñas bears witness to orangutans' forced copulation in Borneo sanctuaries. While forced copulation occurs among wild populations, Parreñas reframes how this behavior is exacerbated and facilitated by the physical boundaries of a sanctuary: "forced sexual reproduction highlights the underlying violence that occurs when the response to the threat of extinction is to increase the population of an endangered or threatened species" (Parreñas, 2018: 104).

The origins of rescued primates, and the course of their rehabilitation and reintroduction, often involve uncontrollable variables which may impact the success of primate conservation goals (Karesh, 1995; Junker et al., 2017). Like all conservation actions, which do not occur in an experimental or sociocultural vacuum, these processes are subject to trade-offs and hard choices (McShane et al., 2011; Lazos-Chavero et al., 2016). In this chapter, we cover the state of research on the rescue, rehabilitation, and reintroduction strategies deployed by organizations around the world in the interest of primate conservation and/or welfare, many of which operate as NGOs under the titles of sanctuaries and rescue centers. We conclude with a discussion of gaps in the literature and future directions for research.

15.2   Rescue

The term rescue could be broadened to encompass the processes of rehabilitation and/or releasing the animal back into its natural habitat. Primate rescue organizations and sanctuaries are usually involved in all these aspects of the animal's welfare. Animal rescues may be differentiated into two types: opportunistic and targeted (Pyke & Szabo, 2017). Opportunistic rescues involve the recovery of animals when they are discovered in harmful situations. Targeted rescues are organized when natural or man-made events place animals and their habitats at risk. Generally, primates arrive at rescue centers, sanctuaries, and zoos through a variety of circumstances: injury in the wild, confiscation from illegal trade, former pets (surrendered or seized), the entertainment industry (film or circus performers), from research facilities, and captured for translocation (Sherman et al., 2020). Quite frequently, an animal may be dropped off at a rescue center, sanctuary, or zoo without any prior consultation or health checks, and the organization must deal with the animal on the spot. A health assessment and a period of quarantine to monitor health status are usually the first steps.

Many primate rescue organizations do not actively carry out rescues or, if they do, do not address *how* they carry them out on their websites. The MONA Foundation wildlife rescue center in Spain is an exception and describes the steps involved in its rescue process: (1) determine if the animal is in an illegal or in an abusive situation, (2) coordinate permits for the retrieval of the animal, (3) organize the most suitable means of transport, (4) conduct health checks prior to retrieval/rescue, and (5) repeat health checks on the day of retrieval and again while in quarantine (https://fundacionmona.org/en/nuestros-primates/).

The sources of primates in need of rescue vary based on whether it happens in a range or non-range country for that species. In range countries in Central and South America, Africa, and Asia, illegal hunting and trade produce the highest number of orphaned primates (Ross & Leinwand, 2020). In fact, primates constitute 94.8% of the legal global trade of mammals with the *Cercopithecidae* family and the *Macaca* genus being the most traded primate groups (Can et al., 2019).

15.2.1 North America/Europe

Many primates in non-range countries such as the United States or European countries enter rescue centers and sanctuaries from research, entertainment, and pet industries. For example, despite the endangered Barbary macaque (*Macaca sylvanus*) being native to north Africa, it is the most confiscated mammal in the European Union (Uhm, 2016). Rescue animals end up in sanctuaries in Belgium, France, Italy, the Netherlands, and Spain. A case study on this trade is discussed in the previous chapter (Chap. 14, this volume). Since these are non-range countries, the possibility of release or reintroduction into the wild is nonexistent.

15.2.2 Latin America

Countries in Central and South America are home to a diverse range of primates, many of which are included in the International Union for Conservation of Nature (IUCN) Red List of endangered species. Illegal trafficking of wildlife is highly profitable in these regions and is estimated to be worth 10 billion US dollars a year (Romo, 2019). Howler monkeys (*Alouatta* spp.), for example, are rescued from poachers who hunt them for meat, illegal trade, or are captured for pets (Arroyo-Rodríguez & Dias, 2009). Capuchin monkeys (*Sapajus* spp.) are the most common primate held in captivity in Brazil with 90% of them obtained through surrender from pet owners or confiscated from traders (Nascimento et al., 2013).

15.2.3 Africa

The Pan African Sanctuary Alliance (PASA) includes 23 sanctuaries in 13 countries, and many centers care for a variety of primate and non-primate species (Stokes et al., 2018). Overall, chimpanzees (*Pan troglodytes*), gorillas (*Gorilla* spp.), and bonobos (*Pan paniscus*) have high rescue rates in central regions of Africa (Ferrie et al., 2014). Chimpanzees are hunted illegally and often opportunistically; many are brought to rescue and rehabilitation centers by law enforcement officers (Ghobrial et al., 2010).

Hunting for bushmeat (i.e., meat obtained from wild animals) is one of the biggest threats to primate populations in Africa. Data from seven countries in west and central Africa indicated the hunting of 22 primate species (Fa et al., 2005). Since hunters mainly seek out adult primates, there has been a notable increase in the number of orphans arriving at rescue centers, many of whom have low survival rates, especially if they were too young to be weaned (Faust et al., 2011).

The most common species of primate brought into rescue and rehabilitation centers in South Africa are galago (*Galago moholi*), chacma baboon (*Papio ursinus*), and vervet monkey (*Chlorocebus aethiops*) (Guy & Curnoe, 2013; Wimberger et al., 2010). Vervet monkeys are often kept as pets but are equally likely to be considered pests due to human-monkey conflict. The "Monkey Helpline" organization was responsible for the rescue of more than 300 vervets in just the first half of 2010 (Smit 2010, cited in Guy & Curnoe, 2013). The Vervet Monkey Foundation took in 191 vervet monkeys over an 8.5-year span (Healy & Nijman, 2014).

15.2.4 Asia

Orangutans (*Pongo pygmaeus*) in Borneo are primarily threatened by the clearing of their forest habitats for industry, agriculture (especially for palm oil), and urbanization. Rescue efforts include translocation of wild individuals from unsuitable habitats. However, most orangutans (90–96%) at rescue centers were either surrendered or confiscated from illegal ownership. Records show at least 994 orangutans arriving at rescue centers between 2007 and 2017, although the actual numbers likely exceeded this (Sherman et al., 2020).

Several primate species in Asia are hunted and trafficked for usage in traditional medicine. In Indonesia, slow lorises (*Nycticebus* spp.) are highly desired as pets and for medicinal use (Nekaris & Moore, 2014; Nekaris et al., 2013; Nijman & Nekaris, 2014). Macaques (*M. fascicularis*, *M. nemestrina*) are also often found in high numbers at Indonesian rescue centers (Nekaris & Moore, 2014). Gibbons (*Hylobates* spp.) are another primate species that are commonly confiscated from the illegal trade as pets (Nijman et al., 2009).

The Endangered Primate Rescue Center (EPRC) in Vietnam is home to many primates confiscated by the Forest Protection Department, many of which are critically endangered or endangered. These include several species of langur (*Trachypithecus* spp., *Pygathrix* spp.), gibbon (*Nomascus* spp.), and lorises (*Nycticebus* spp.) (Nadler, 2013). Due to its affiliation with the German Primate Centre, the EPRC does blood screening for diseases and stores DNA samples of individuals for genetic analysis for individuals arriving at the Center.

Twenty-five primate species live in China, and 80% of them are threatened, mainly due to hunting, anthropogenic, and natural climate-driven habitat loss (Li et al., 2018). In contrast to the vast diversity of primates, there are disproportionately few rescue centers which house a limited number of species. This is primarily due to the slow progress in governmental efforts (Li et al., 2018). Very few, if any, of the critically endangered gibbon species (*Nomascus nasutus*, *N. hainanus*, and *N. concolor*) or black snub-nosed monkey (*Rhinopithecus strykeri*) have been housed in captivity – indicating a need for breeding programs or better conservation efforts in the wild (Li et al., 2018).

India is home to 14 species of primates (Southwick & Lindburg, 1986). Yet, there are only two state-funded primate rehabilitation facilities in the entire country; the second one opened in south India in 2020. One of the goals of this facility is to sterilize the monkeys brought to the facility especially those that are not an endangered or threatened species and then release them back out (The Indian Express, 2020).

15.3 Rehabilitation

15.3.1 What Is Primate Rehabilitation?

Primate rehabilitation includes a variety of processes aimed at improving the health, socialization, and welfare of individuals. Guy et al. (2014) present a framework for looking at rehabilitation via two goals – conservation and welfare. The goal of conservation-based rehabilitation is reintroduction, where primates are released into the wild where they can reproduce and contribute to the survival of their species. The goal of welfare-based rehabilitation is to acclimate the animals to living in captivity. In 2015, the IUCN provided best practice guidelines for translocation and rehabilitation of gibbons (*Hylobates* spp.) (Campbell et al., 2015). Protocols for conservation-based rehabilitation can be made based on IUCN guidelines for reintroduction (IUCN SSC, 2013), but because welfare goals are not conservation-oriented, there are currently no consistent protocols. The current state of animal rehabilitation overall is disparate, as organizations around the world infrequently collect systematic data on their methods and outcomes, and there are few guidelines that are widely accepted by rehabilitators.

15.3.2   What Research Has Been Done on Primate Rehabilitation?

Research on the methods and success of primate rehabilitation is disparate; it covers a wide range of species, incorporates a variety of methods, and yields inconclusive results. Published papers include orangutans (*Pongo* spp.; Russon, 2009; Yeager, 1997), chimpanzees, and bonobos (*Pan* spp.; Tutin et al., 2001; Wobber & Hare, 2011), gibbons (*Hylobates* spp.; Cheyne, 2009, *H. lar*; de Veer & Van den Bos, 2000), vervets (*C. aethiopicus*; Wimberger et al., 2010), capuchin monkeys (*Sapajus apella*; Suarez et al., 2001), squirrel monkeys (*Saimiri sciureus*; Vogel et al., 2002), and slow loris (*Nycticebus* spp.; Nekaris & Moore, 2014). Many papers do not describe their rehabilitation methods in detail. This, combined with the lack of post-release monitoring to measure reintroduction success, makes it difficult to evaluate rehabilitation outcomes. Due to the similarities between nonhuman primate species, most research focuses on major skills captive primates need to learn before reintroduction, including dishabituation from humans, socialization with conspecifics, appropriate predator responses, locomotion in natural environments, and foraging skills (Guy et al., 2014; Guy & Curnoe, 2013). There are various rates and definitions of success in preparing primates for release into the wild (Cheyne, 2009). There are two opposing conclusions that are drawn from this research: some highlight that rehabilitation plans need to be detailed enough to account for the multiple skills animals need to survive and reproduce in the wild (Cheyne, 2009; Guy & Curnoe, 2013), while others conclude that rehabilitation does not contribute to post-release success of reintroductions (Baker, 2002; Beck et al., 2007). A study of vervet (*C. aethiops*) rehabilitation and reintroduction emphasizes the importance of knowing the history of the individual animals, resocializing them in social groups that mimic wild group sizes, and providing naturalistic enclosures with natural substrates, appropriate sleeping sites, and natural food (Guy & Curnoe, 2013). In addition, the animals needed to be dishabituated from humans and trained to recognize predators.

Similarly, Cheyne's (2009) case study of gibbon (*H. lar*) release in Indonesia found that the rehabilitated gibbons came from a wide variety of backgrounds and often had social adjustment problems. Through trial-and-error and knowledge of gibbons' natural behaviors, the sanctuary found that groupings of juveniles or an older adult with a younger juvenile had better survival rates after reintroduction than did opposite-sex, same-age pairs. A bad pairing during rehabilitation could lead to animals becoming scared of conspecifics. Furthermore, pairs from captivity did not necessarily remain paired in the wild (Cheyne, 2009; de Veer & Van den Bos, 2000).

Reports on great ape rehabilitation are similar. It takes a long time period of acclimatizing animals to their natural habitats, and thus time from rescue to release lasts 2–7 years (Russon, 2009; Tutin et al., 2001). This time accounts for animals becoming socialized with conspecifics and dishabituated to humans. Russon (2009) provides an overview of orangutan (*Pongo* spp.) rehabilitation projects in Borneo and Sumatra where rehabilitation is conducted based on age classification. The infants are cared for in nurseries, young juveniles are resocialized with other orangutans in enclosures, and older juveniles are taken to "forest school." "Forest school" allows for individuals to go out into the natural environment with partial provisioning during the day, but then they are brought back to sleep in enclosures. This involves a trade-off in human interaction, in which human caretakers are more heavily involved in the forest schools, while the juveniles in socialization enclosures experience minimal human influence. Similarly, a chimpanzee project in the Congo also used age-graded approaches to form social groups on islands before their eventual release (Tutin et al., 2001). Planning started 2 years before the chimpanzees were released, and with a release site carefully chosen and each individual closely evaluated for suitability of release.

15.3.3 What Is the Future of Primate Rehabilitation?

Guy et al. (2014) reviewed primate rehabilitation and assessed its contribution to conservation. They conclude that there should be species-specific guidelines for rehabilitation, considering each species' unique natural history and behavior. These conclusions are supported by findings that orangutans did better when a younger animal was paired with an older one, and without being housed in large social groups (Russon, 2009). Additionally, the natural diets of the primates are crucial. Some animals may not need to learn how to process natural foods, while extractive foragers may need extensive learning to be competent at foraging for wild foods. The proposed model outlined by Guy et al. (2014) may be a valuable starting point, since there are currently no guidelines for welfare-based rehabilitation comparable to those completed for reintroductions.

Guy et al. (2014) evaluated 28 rehabilitation and reintroduction projects, excluding captive-bred animals from their conclusions. They found that the most serious problems were projects that lacked quarantine of newly arrived animals, collection of histories, dishabituation from humans, training to recognize predators, and long-term post-release monitoring. They did find positive interventions across the 28 projects. These interventions included putting primates into appropriate social groups, giving them medical assessments upon arrival, limiting human contact, housing animals in natural enclosures, assessing each animals' skills for independent living, providing natural foods before release, and using soft release strategies.

These studies indicate that successful primate rehabilitation is possible but takes a large effort, including sufficient time and funding (Cheyne, 2009; Guy et al., 2014; Russon, 2009; Tutin et al., 2001). There must be extensive knowledge of the species' natural history, appropriate resources for the individual animals to learn survival skills, and close evaluation of each individual's abilities and likelihood of survival. However, a broader ethical question remains whether welfare-based rehabilitations, including those that end in reintroductions, are in the best interest of both the individual rehabilitant animals, as well as whether they achieve or detract from broader conservation goals.

15.4 Reintroduction

Reintroductions have a long history across Africa, Asia, and Latin America (Kleiman, 1996; Russon, 2009; Beck, 2018). Despite their prevalence, rescue and welfare-based reintroductions are inconsistent with best practice recommendations for primate reintroductions (Baker, 2002; Beck et al., 2007). While rescue and rehabilitation occur with primates across varying contexts, including captive-born animals from the pet trade and laboratories, most primate reintroductions (95.6%) are with wild-born individuals (Beck, 2018). Many reintroductions were conducted primarily for primate welfare purposes; however, both the broader IUCN guidelines on reintroductions for plants and animals and the primate-specific IUCN guidelines emphasize that the goal of reintroduction should be to conserve wild populations by increasing their population size (Baker, 2002; IUCN SSC 2013).

Historically, such endeavors were done on a trial-and-error basis or lacked sustained monitoring efforts (Beck et al., 2007; Beck, 2018). Reintroductions have an emotional appeal to individuals and conservation organizations, reflecting the desire to ameliorate harms inflicted on individual animals and their ecosystems by human capture (Estrada et al., 2017; Beck, 2018). Furthermore, North American and European zoos often justify captive breeding with the need to have a captive reservoir for future reintroductions (Croke, 1997; Snowdon, 1989). However, the success of reintroductions is variable across sites, due to variation in approach, lack of pre-release training and/or post-release monitoring, and inconsistent evaluation of long-term success and publication. While successful reintroductions may be beneficial in terms of increasing wild population sizes (Kleiman, 1996; Beck, 2018) or in contributing to ecosystem restoration (Genes et al., 2019), it is unclear whether they benefit individual primate welfare (Beck, 2018; Guy et al., 2014; Palmer, 2018; Sherman et al., 2020).

Of primate reintroduction projects, only 43.2% of reintroduction programs have successfully met benchmarks of success, including post-release survival for at least a year, transitioning to independence from human provisioning, and integration with wild populations (Beck, 2018). Furthermore, only 14% were able to reach the more stringent conservation aim of becoming a fully self-sustaining wild population (Beck, 2018). Not all projects collect or release data on post-release outcomes. To improve outcomes, the IUCN published guidelines for best practices in planning primate reintroductions (Baker, 2002) and expanded on these guidelines specifically to address reintroduction of great apes (Beck et al., 2007) and small apes (Campbell et al., 2015; Cheyne, 2012). These guidelines provided a framework for assessing suitability of animals for release, identifying suitable sites, and planning prerelease training and post-release monitoring and evaluation. The guidelines for primate reintroductions include a "decision tree" with steps for planning a reintroduction and emphasize that "the main goal of any reintroduction effort should be to reestablish self-sustaining populations of primates in the wild and to maintain the viability of those populations" (Baker, 2002:32). They emphasize the "precautionary principle" that reintroductions can pose risks to wild populations, including the risk of introducing diseases, and thus recommend that reintroductions should not be pursued if they will not benefit this conservation aim (Baker, 2002).

15.4.1 Latin America

Reintroduced Central and South American monkey species include atelines, cebids, and callitrichids (Konstant & Mittermeier, 1982; Beck, 2018). One of the earliest was the reintroduction of Geoffroy's spider monkeys (*Ateles geoffroyi*) to Barro Colorado Island, where they were previously extirpated (Milton & Hopkins, 2006; Beck, 2018). Eighteen to nineteen confiscated individuals between one and four years old were released on the island from 1959 to 1966. Only five individuals, one male and four females survived, and only three females reproduced (Milton & Hopkins, 2006). They established a breeding population that grew to 28 individuals by the early 2000s but had limited genetic diversity due to the small founder population (Milton & Hopkins, 2006). An additional reintroduction was attempted in 1991, but none of those in the second group of monkeys survived, likely due to poor planning (Milton & Hopkins, 2006).

The most well-documented reintroduction in the Americas was the reintroduction of 146 captive-born golden lion tamarins (*Leontopithecus rosalia*) from 1983 to 2000 in Brazil's Atlantic Forest (Kleiman, 1996; Kierulff et al., 2012; Beck, 2018). Golden lion tamarins had severely declined in the wild and were extirpated from the release sites. Over ten cohorts, release strategies changed, from "hard releases" without post-release support, to "soft releases" with provisioning and supportive care as needed. However, only about 30% survived to the benchmark of 2 years, which was long enough to reproduce (Beck, 2018). Supported with 42 translocated tamarins, this population increased, and the golden lion tamarins improved their conservation status from "critically endangered" to "endangered" (Kierulf et al., 2012; Ruiz-Miranda et al., 2019). However, despite climbing to above 3000 wild individuals in 2014, the population has been threatened by yellow fever outbreaks in 2017, and the populations are now decreasing (Beck, 2018; Ruiz-Miranda et al., 2019).

15.4.2 Africa

A wide range of African primates have been reintroduced, including apes, monkeys, and strepsirrhines (Beck, 2018). However, African apes have been a focal point (Tutin et al., 2001; Beck et al., 2007; Beck, 2018). Many of the chimpanzee reintroductions overlap with rehabilitation, as rehabilitation programs release animals on islands where the animals can live in wild habitats, while still allowing for monitoring, provisioning, and separation from wild populations (Beck, 2018). Multiple projects resulted in abandonment of the animals, such as when New York Blood Center withdrew financial support of released chimpanzees (Beck, 2018).

Reintroductions of chimpanzees and bonobos (*Pan* spp.) to wild, non-island habitats have had significant challenges. For example, release of 14–16 chimpanzees into Niokolo Koba National Park from 1972 to 1977 resulted in recapture and release of surviving chimpanzees to an island due to aggression from wild chimpanzees (Beck, 2018). Fifty-three chimpanzees were released into Conkouati-Douli National Park between 1996 and 2012 (Tutin et al., 2001; Goossens et al., 2005; Beck, 2018). The initial cohorts were first released to islands for 4–10 years, allowing them to gain ecological competence. Numerous individuals were attacked by wild chimpanzees. However, at least 17 infants were born to reintroduced females (Tutin et al., 2001; Goossens et al., 2005; Beck, 2018). Chimpanzee Conservation Center in Guinea also conducted a reintroduction of 16 chimpanzees in Haut Niger National Park where there were wild chimpanzees, from 2008 to 2011 (Humle et al., 2011; Beck, 2018). Half of the first cohort survived and gained independence within a year, but there were multiple deaths, and at least one female returned to the sanctuary (Beck, 2018). During a similar time period from 2009 to 2011, Lola ya Bonobo reintroduced 16 bonobos at Ekola Ya Bonobo Democratic Republic of Congo (Beck, 2018). Most of the reintroduced bonobos acclimated to their wild environment, and four infants were born from 2011 to 2016. However, three bonobos attacked trackers monitoring them post-release and had to be returned to the sanctuary (Beck, 2018).

Like chimpanzees, gorilla (*Gorilla* spp.) reintroductions occurred as early attempts at rehabilitating individual rescued animals or releasing groups onto islands (Beck, 2018). The John Aspinall Foundation led several reintroductions, including 25 predominantly wild-born infants at Lesio-Louna and Lefini Reserves, Republic of Congo between 1996 and 2006. Most of the gorillas survived in the wild, and there were new infants born, but at least four males had to be recaptured due to human-wildlife conflict, while six to eight died or disappeared (Beck, 2018). Projet Protection des Gorilles and Aspinall Foundation also reintroduced over 65 gorillas (Beck, 2018). Twenty-five individuals of a wild-born sanctuary population were introduced in 3 cohorts in the Lefini Reserve, Republic of Congo, and 23 of another confiscated Gabonese population were reintroduced into the Bateké Plateau National Park, Gabon. These reintroductions were largely successful. However, in 2014, the Aspinall Foundation reintroduced an additional nine captive-born animals from zoos in the United Kingdom in Gabon, against the recommendations of the IUCN guidelines (Beck et al., 2007; Beck, 2018). This resulted in the death of at least half of the ten gorillas within weeks of the project, and four surviving individuals were returned to islands or sanctuaries (Beck, 2018).

15.4.3 Asia

In Asia, most of the reintroductions focused on rehabilitant orangutans (Russon, 2009; Beck, 2018). However, a wide range of other Asian primates, including gibbons, macaques, langurs, and lorises, have been reintroduced (Beck, 2018). Primate reintroductions have a deeper history in parts of Asia, from the Buddhist and Taoist traditions of *feng sheng* (Beck, 2018; Magellan, 2019). These "mercy" or "prayer releases" are compassionate acts of freeing an animal that are believed to bring positive karma to the releaser. As a result, some Asian primate reintroductions, particularly gibbons, occur on the grounds of Buddhist temples (Beck, 2018). However, such mercy releases can have detrimental effects on ecosystems (Magellan, 2019).

Early efforts to rehabilitate orangutans (*Pongo* spp.) began in the 1960s (Harrison, 1963; Russon, 2009; Beck, 2018). Due to the increasing numbers of animals and difficulty accommodating them, rehabilitation projects grew to reintroductions, and other sanctuaries built on that model (Harrison, 1963; Beck, 2018; Palmer, 2018; Russon, 2009; Sherman et al., 2020). Over 5 times as many orangutans have been reintroduced as all the African apes, across 13 programs (Palmer, 2018; Beck, 2018). These differences may be due to the large number of orangutans displaced by habitat loss, along with Indonesian laws that require reintroduction (Beck, 2018). Wild-born offspring of released orangutans comprise a large proportion of the total

---

Box 15.1 Costa Rican Monkeys
Costa Rica is a tropical country in Central America that has undergone periods of deforestation and impressive reforestation of mangroves, wet and dry rainforests (Stan & Sanchez-Azofeifa, 2019). There are four species of monkey indigenous to Costa Rica: the Panamanian white-face capuchin (*Cebus capucinus*) and mantled black howler monkey (*Alouatta palliata*), which are listed by the IUCN as of least concern, the vulnerable black-capped squirrel monkey (*Saimiri oerstedii*), and the critically endangered Geoffroy's spider monkey (*A. geoffroyi*) (Fig. 15.1).

Rescue: Records from four rescue centers located in different conservation regions around the country report electrocution as approximately 25% of the reasons that monkeys are rescued (*n* = 365 monkeys) (Speiran, 2021). Monkeys – and other arboreal animals – may be electrocuted when they climb uninsulated power lines and transformers to cross fragmented habitat. It is not uncommon for adult monkeys to cross a road or power line with an infant on their back, and become critically injured or perish, leaving the infant orphaned. Over a period of 5 years, for example, one region received 624 reports of electrocuted adult monkeys and 165 orphans (IAR Costa Rica, 2021). It is estimated that at least half of rescued monkeys succumb to their injuries

(continued)

Box 15.1 (continued)

Fig. 15.1 Native monkey species in Costa Rican sanctuaries (from left to right): capuchin, howler, squirrel, and spider monkeys. (Photos: Siobhan Speiran)

Fig. 15.2 (**a**) Howler monkey undergoing veterinary procedures at a rescue center after being injured with a machete. Photo: Proyecto Asis. (**b**) Capuchin enclosure in a Costa Rican sanctuary. (Photo: Siobhan Speiran)

naturally or are euthanized after receiving a veterinary assessment (Speiran, 2021).

Rehabilitation: Rescued monkeys who cannot be released will rehabilitate, barring any physical injuries or maladies, by adjusting to sanctuary life – and a new, captive group. Rescued monkeys from the wild with potential for reintroduction will receive veterinary care to heal from health problems and

(continued)

Box 15.1 (continued)

Fig. 15.3 Howler monkey being released after successful rehabilitation. (Photo: Proyecto Asis)

may undergo training for reintroduction (Fig. 15.2). When a monkey is ready to be reintroduced, some centers will move the monkey or troupe to an isolated area away from human disturbance.

Reintroduction: Primates are typically reintroduced where they were rescued. If the rescue site is precarious (i.e., a road, urban area, etc.), then a center will usually attempt release in the closest suitable natural area. Sometimes, centers may perform a "soft" release in which monkeys are brought to the future site of release, provisioned with food, and allowed to venture out of the enclosure into the wild when they choose to fully reintegrate (Guy et al., 2014; Beck, 2018) (Fig. 15.3).

McKinney and Schutt (2005) examined programs to rehabilitate and release previously extirpated spider monkeys (*A. geoffroyi*) in the southern Nicoya peninsula Costa Rica, which included at least 6 months of post-release monitoring. Since 1989, around 30 spider monkeys were released into a designated wildlife refuge, and some had infants and moved to different ranges from where they were reintroduced – indicating some success. Recent updates, however, indicate that the monkeys were not monitored more than a year post-release, so there is no long-term assessment of their survival at present (ibid). The study concludes that local involvement in long-term species survival programs is necessary for success and encourages similar programs for different species.

orangutan population, which reflects a long history of anthropogenic pressures shaping orangutan populations (Russon, 2009; Spehar et al., 2018). Because orangutan reintroductions occurred at such a large scale and been primarily "welfare" releases, there is considerable ethical debate over the extent to which these releases may have helped or harmed individual welfare and posed risks to conservation of wild populations (Russon, 2009; Beck, 2018; Palmer, 2018; Parreñas, 2018).

15.5 Discussion

Our survey of the literature on the rescue, rehabilitation, and reintroduction of primates globally revealed several insights that can motivate future research. The Rescue section considers how the responsive, crisis-mandated nature of wildlife rescue can lead to instantaneous decision-making. This may lead to poor record-keeping, which may explain the lack of data on primate rescues. Differing definitions of what constitutes "rescued" primates further muddies attempts to generate baseline data. Another challenge faced by rescue organizations is that they often reach capacity within 1–2 years (Nekaris & Moore, 2014). This could result in the rejection of more animals in need of rescue or a decrease in the quality of care provided.

The subsequent section considers how, in practice, rehabilitation occurs across disparate organizations with little data or overarching guidelines for support. Moreover, it is difficult to measure the success of rehabilitation because studies do not always describe the rehabilitation process in detail and post-release monitoring is not always implemented. There is a lack of consistency in reported information regarding the type of species, number of individuals, age, and sex. Finally, we posit that welfare-oriented reintroductions, which serve to only benefit the individual's welfare and not a conservation goal, are not consistent with best practice recommendations for primate reintroductions. This view is aligned with the IUCN who state the goal of reintroduction is to improve species conservation. Despite this, many reintroductions occur for welfare purposes, as in the case of orangutans (Parreñas, 2018).

Taken together, it would seem the triumvirate of Rescue, Rehabilitation, and Reintroduction, though pillars of primate conservation discourse, are under-researched, under-regulated, and fraught with ethical debates and dilemmas. The complicated nature of this field of conservation work *and* scholarship is exacerbated when there is a disconnect between academics and wildlife rehabilitators. This can occur when, for example, evidence-based, best practices are deduced through research, but do not suit local realities or are impractical to implement. The disconnect is worsened if sanctuaries operate in an ad hoc, informal manner or lack transparency about their methods, successes, and failures. It is not uncommon for rehabilitators to implement trial-and-error, or homegrown methods of rehabilitation and reintroduction, especially given the veritable lack of research on the subject. The styles and strategies employed in conservation work may be shaped by cultural contexts, social norms, financial constraints, and personal experience.

Perhaps those interested in primate welfare could collaborate to create such a centralized database for this information. Another option would be for various accrediting organizations (like GFAS, PASA, NAPSA, and others) to encourage or even require rescue centers and sanctuaries to maintain accessible records or better track individuals and their histories. The benefits of maintaining online rescue data-bases are elucidated by Pyke and Szabo (2017). Such information would be highly beneficial to conservation agencies, welfare officials, and researchers. Sharing this information with other places in the region could promote better ways of reducing harm to these primate populations (Healy & Nijman, 2014).

In primate reintroductions, there is often a mismatch in practice or goals between the IUCN-recommended best practices and individual organizations conducting reintroductions. Mirroring the ideals of *fang sheng*, or "mercy releases," reintroductions are planned for ideological or welfare-based goals, as a way of restoring wild-born animals to the wild (Magellan, 2019). Such releases are guided by compassionate empathy for the animals but ultimately serve to assuage the emotions of those responsible for welfare releases, including organizers, caretakers, and funders at the expense of wider ecosystem conservation. Without carefully planned rehabilitation, poorly planned releases may be the equivalent of "dumping," an indirect form of euthanasia where the responsibility is shifted from the rehabilitators to the natural ecosystem (Palmer, 2018). Given the risk of transmitting diseases to wild populations, risks and benefits to conserving wild populations must be the biggest consideration in considering or planning reintroductions (Baker, 2002; Beck et al., 2007; Beck, 2018).

Some outstanding questions include whether reintroduction actually benefits animal welfare, and if rehabilitation and reintroduction are the best strategies for every species or individual. There are undoubtedly conservation contexts and crises in which the laborious process of rehabilitation and reintroduction may not ultimately benefit conservation or even welfare.

15.6 Conclusion

A number of overarching themes pervade this chapter, chief among them the reality that not all rescued primates can be rehabilitated and not all rehabilitants can be reintroduced. Rehabilitation requires detailed knowledge of a species' natural history, as well as the time and resources (including space, money, and staff) for individuals to learn survival skills for eventual release. There are still questions regarding if rehabilitation increases survival, whether reintroductions serve to advance primate welfare, and whether euthanasia is an ethical alternative to lifetime in impoverished captivity (Nekaris & Moore, 2014).

There is no way to know exactly how many primates have been rescued or reside in rescue organizations. An expansive review of the literature and sanctuary websites exposed two main issues: (1) a lack of data from many rescue organizations who do not make accessible online the number of individuals and types of species they cared for, and (2) organizations vary in their definition of what is considered a "rescue" animal. A factor which confounds the estimation of the number of rescue primates is the fact that many rescue centers and sanctuaries consider retired research animals as "rescues."

Numerous ethical conundrums befall primate conservation work, especially in contexts where species and ecological loss occur so rapidly that the processes of rehabilitation and reintroduction cannot keep pace. Is rescue, rehabilitation, and release the best strategy for primate conservation? At present, the literature suggests these strategies require enormous effort for little documented benefit. The onus is on rescue organizations and sanctuaries to self-report their progress, likely leading to a bias against reporting failed rehabilitations or reintroductions. The details of the successes, failures, and trial-and-error findings are necessary, however, if researchers are to address unresolved queries.

Acknowledgments The authors would like to thank the editors of this volume, Tracie McKinney, Siân Waters, and Michelle A. Rodrigues, for the invitation to contribute, and the reviewers whose comments have strengthened the chapter. MAR is grateful for the mentorship of Benjamin Beck, whose perspectives on reintroductions have greatly shaped her thinking on the topic.

References

Arcus Foundation. (2020). *Global Ape Sanctuaries Worldwide.* https://www.arcusfoundation.org/apesvisual/ Accessed 6 Apr 2021.

Arroyo-Rodríguez, V., & Dias, P. (2009). Effects of habitat fragmentation and disturbance on Howler monkeys: A review. *American Journal of Primatology, 72,* 1–16. https://doi.org/10.1002/ajp.20753

Baker, L. R. (2002). IUCN/SSC re-introduction specialist group: Guidelines for nonhuman primate re-introductions. *Re-introduction News, 21,* 29–57.

Beck, B. B. (2018). *Unwitting travelers: A history of primate reintroduction.* Salt Water Media.

Beck, B., Walkup, K., Rodrigues, M., Unwin, S., Travis, D., & Stoinski, T. (2007). *Best practice guidelines for the reintroduction of great apes.* IUCN, IUCN SSC Primate Specialist Group (PSG).

Campbell, C. O., Cheyne, S. M., & Rawson, R. M. (2015). *Best practice guidelines for the rehabilitation and translocation of gibbons.* IUCN SSC Primate Specialist Group. https://doi.org/10.2305/iucn.ch.2015.ssc-op.51.en

Can, Ö. E., D'Cruze, N., & Macdonald, D. W. (2019). Dealing in deadly pathogens: Taking stock of the legal trade in live wildlife and potential risks to human health. *Global Ecology and Conservation, 17,* e00515. https://doi.org/10.1016/j.gecco.2018.e00515

Cheyne, S. (2009). Challenges and opportunities of primate rehabilitation – gibbons as a case study. *Endangered Species Research, 9, 159–165.* https://doi.org/10.3354/esr00216

Cheyne, S. M., Campbell, C. O., & Payne, K. L. (2012). Proposed guidelines for in situ gibbon rescue, rehabilitation and reintroduction. *International Zoo Yearbook, 46*(1), 265–281. https://doi.org/10.1111/j.1748-1090.2011.00149.x

Croke, V. (1997). *The modern ark: The story of zoos, past, present, and future.* Scribner.

de Veer, M. W., & Van den Bos, R. (2000). Assessing the quality of relationships in rehabilitating lar gibbons (*Hylobates lar*). *Animal Welfare, 9,* 223–225.

Estrada, A., Garber, P. A., Rylands, A. B., Roos, C., Fernandez-Duque, E., Di Fiore, A., Nekaris, K. A., Nijman, V., Heymann, E. W., Lambert, J. E., Rovero, F., Barelli, C., Setchell, J. M., Gillespie, T. R., Mittermeier, R. A., Arregoitia, L. V., de Guinea, M., Gouveia, S., Dobrovolski, R., Shanee, S., Shanee, N., Boyle, S. A., Fuentes, A., MacKinnon, K. C., Amato, K. R., Meyer, A. L., Wich, S., Sussman, R. W., Pan, R., Kone, I., & Li, B. (2017). Impending extinction crisis of the world's primates: Why primates matter. *Science Advances, 3*(1), e1600946. https://doi.org/10.1126/sciadv.1600946

Fa, J. E., Ryan, S. F., & Bell, D. J. (2005). Hunting vulnerability, ecological characteristics and harvest rates of bushmeat species in afrotropical forests. *Biological Conservation, 121*(2), 167–176.

Faust, L. J., Cress, D., Farmer, K. H., Ross, S. R., & Beck, B. B. (2011). Predicting capacity demand on sanctuaries for African chimpanzees (*Pan troglodytes*). *International Journal of Primatology, 32*, 849–864. https://doi.org/10.1007/s10764-011-9505-z

Ferrie, G. M., Farmer, K. H., Kuhar, C. W., Grand, A., Sherman, J., & Bettinger, T. L. (2014). The social, economic, and environmental contributions of Pan African Sanctuary Alliance primate sanctuaries in Africa. *Biodiversity and Conservation, 23*(1), 187–201. https://doi.org/10.1007/s10531-013-0592-3

Genes, L., Fernandez, F. A. S., Vas-de-Mello, F. Z., da Rosa, P., Fernandez, E., & Pires, A. S. (2019). Effects of howler monkey reintroduction on ecological interactions and processes. *Conservation Biology, 33*(1), 88–98. https://doi.org/10.1111/cobi.13188

Ghobrial, L., Lankester, F., Kiyang, J. A., Akih, A. E., de Vries, S., Fotso, R., Gadsby, E. L., Denkins, P. D., Jr., & Gonder, M. K. (2010). Tracing the origins of rescued chimpanzees reveals widespread chimpanzee hunting in Cameroon. *BMC Ecology, 10*(1), 2. https://doi.org/10.1186/1472-6785-10-2

Goossens, B., Setchell, J. M., Tchidongo, E., Dilambaka, E., Vidal, C., Ancrenaz, M., & Jamart, A. (2005). Survival, interactions with conspecifics and reproduction in 37 chimpanzees released into the wild. *Biological Conservation, 123*(4), 461–475. https://doi.org/10.1016/j.biocon.2005.01.008

Guy, A. J., & Curnoe, D. (2013). Guidelines for the rehabilitation and release of vervet monkeys. *Primate Conservation, 27*, 55–63.

Guy, A. J., Curnoe, D., & Banks, P. B. (2014). Welfare based primate rehabilitation as a potential conservation strategy: Does it measure up? *Primates, 55*(1), 139–147. https://doi.org/10.1007/s10329-013-0386-y

Harrison, B. (1963). Education to wild living of young orangutan at Bako National Park. *Sarawak Museum Journal, 11*, 222–258.

Healy, A., & Nijman, V. (2014). Pets and pests: vervet monkey intake at a specialist South African rehabilitation centre. *Animal Welfare, 23*(3), 353–360. https://doi.org/10.7120/09627286.23.3.353

Humle, T., Colin, C., Laurans, M., & Raballand, E. (2011). Group release of sanctuary chimpanzees (Pan troglodytes) in the Haut Niger National Park, Guinea, West Africa: Ranging patterns and lessons so far. *International Journal of Primatology, 32*(2), 456–473. https://doi.org/10.1007/s10764-010-9482-7

IAR Costa Rica (2021, Jan 9). *Stop the Shocks*. Retrieved April 6, 2021, from https://www.iarcostarica.org/stop-the-shocks/

International Wildlife Rehabilitation Council (2005). Available from https://www.iwrc-online.org

IUCN (International Union for Conservation of Nature) SSC (Species Survival Commission). (2013). *Guidelines for reintroductions and other conservation translocations. IUCN, Gland, Switzerland.*

Junker, J., Kühl, H. S., Orth, L., Smith, R. K., Petrovan, S. O., & Sutherland, W. J. (2017). *Primate conservation: Global evidence for the effects of interventions*. University of Cambridge.

Karesh, W. B. (1995). Wildlife rehabilitation: Additional considerations for developing countries. *Journal of Zoo and Wildlife Medicine, 26*(1), 2–9.

Kierulff, M. C. M., Ruiz-Miranda, C. R., Procópio de Oliveira, P., Beck, B. B., Martins, A., Dietz, J. M., Rambaldi, D. M., & Baker, A. J. (2012). The golden lion tamarin *Leontopithecus rosalia*: A conservation success story. *International Zoo Yearbook, 46*(1), 36–45. https://doi.org/10.1111/j.1748-1090.2012.00170.x

Kleiman, D. G. (1996). Reintroduction programs. In D. G. Kleiman, K. V. Thompson, & S. Lumpkin (Eds.), *Wild mammals in captivity: Principles and techniques* (pp. 297–305). University of Chicago Press.

Konstant, W. R., & Mittermeier, R. A. (1982). Introduction, reintroduction, and translocation of neotropical primates: Past experiences and future possibilities. *International Zoo Yearbook, 22*(1), 69–77.

Lazos-Chavero, E., Zinda, J., Bennett-Curry, A., Balvanera, P., Bloomfield, G., Lindell, C., & Negra, C. (2016). Stakeholders and tropical reforestation: Challenges, trade-offs, and strategies in dynamic environments. *Biotropica, 48*(6), 900–914. https://doi.org/10.1111/btp.12391

Li, B., Li, M., Li, J., Fan, P., Ni, Q., Lu, J., Zhou, X., Long, Y., Jiang, Z., Zhang, P., Huang, Z., Huang, C., Jiang, X., Pan, R., Gouveia, S., Dobrovolski, R., Grueter, C. C., Oxnard, C., Groves, C., Estrada, A., & Garber, P. A. (2018). The primate extinction crisis in China: Immediate challenges and a way forward. *Biodiversity and Conservation, 27*, 3301–3327. https://doi.org/10.1007/s10531-018-1614-y

Magellan, K. (2019). Prayer animal release: An understudied pathway for introduction of invasive aquatic species. *Aquatic Ecosystem Health and Management, 22*(4), 452–461. https://doi.org/10.1080/14634988.2019.1691433

McKinney, T., & Schutt, A. (2005). Spider monkey (*Ateles geoffroyi*) rehabilitation, reintroduction, and conservation at Curú Wildlife Refuge, Costa Rica. *American Journal of Physical Anthropology Sup., 40*, 149–150.

McShane, T. O., Hirsch, P. D., Trung, T. C., Songorwa, A. N., Kinzig, A., Monteferri, B., Mutekanga, D., Thang, H. V., Dammert, J. L., Pulgar-Vidal, M., Welch-Devine, M., Peter Brosius, J., Coppolillo, P., & O'Connor, S. (2011). Hard choices: Making trade-offs between biodiversity conservation and human well-being. *Biological Conservation, 144*(3), 966–972. https://doi.org/10.1016/j.biocon.2010.04.038

Milton, K., & Hopkins, M. E. (2006). Growth of a reintroduced spider monkey (Ateles geoffroyi) population on Barro Colorado Island, Panama. In A. Estrada, P. A. Garber, M. S. M. Pavelka, & L. Leucke (Eds.), *New perspectives in the study of mesoamerican primates: Distribution, ecology, behavior, and conservation* (pp. 417–435). Springer.

Molony, S. E., Dowding, C. V., Baker, P. J., Cuthill, I. C., & Harris, S. (2006). The effect of translocation and temporary captivity on wildlife rehabilitation success: An experimental study using hedgehogs (*Erinaceus europaeus*). *Biological Conservation, 130*, 530–537.

Nekaris, K. A. I., & Moore, R. S. (2014). Compassionate conservation, rehabilitation and translocation of Indonesian slow lorises. *Endangered Species Research, 26*, 93–102. https://doi.org/10.3354/esr00620

Nadler, T. (2013). Twenty years Endangered Primate Rescue Center, Vietnam – Retrospect and outlook - report 2012. *Vietnamese Journal of Primatology, 2*(12), 1–12.

Nascimento, R. A., Schiavetti, A., & Montaño, R. A. (2013). An assessment of illegal capuchin monkey trade in Bahia State, Brazil. *Neotropical Biology and Conservation, 8*, 79–87.

Nekaris, K. A.-I., Campbell, N., Coggins, T. G., Rode, E. J., & Nijman, V. (2013). Tickled to death: Analysing public perceptions of 'cute' videos of threatened species (Slow Lorises – *Nycticebus spp.*) on Web 2.0 Sites. *PLoS One, 8*(7), e69215. https://doi.org/10.1371/journal.pone.0069215

Nijman, V., Martinez, C. Y., & Shepherd, C. R. (2009). Saved from trade: Donated and confiscated gibbons in zoos and rescue centres in Indonesia. *Endangered Species Research, 9*, 151–157. https://doi.org/10.3354/esr00218

Nijman, V., & Nekaris, K. A.-I. (2014). Traditions, taboos and trade in slow lorises in Sundanese communities in southern Java, Indonesia. *Endangered Species Research, 25*, 79–88. https://doi.org/10.3354/esr00610

Palmer, A. (2018). Kill, incarcerate, or liberate? Ethics and alternatives to orangutan rehabilitation. *Biological Conservation, 227*(August), 181–188. https://doi.org/10.1016/j.biocon.2018.09.012

Parreñas, J. S. (2018). *Decolonizing extinction: The work of care in orangutan rehabilitation.* Duke University Press.

Pyke, G. H., & Szabo, J. K. (2017). Conservation and the four Rs, which are rescue, rehabilitation, release, and research. *Conservation Biology: the Journal of the Society for Conservation Biology, 32*(1), 5–59. https://doi.org/10.1111/cobi.12937

Romo, V. (2019, Oct 10). *Twenty countries to fight wildlife trafficking as organised crime.* https://dialogochino.net/en/trade-investment/30761-twenty-countries-to-fight-wildlife-trafficking-as-organised-crime/). Accessed 30 Mar 2021.

Ross, S. R., & Leinwand, J. G. (2020). A review of research in primate sanctuaries. *Biology Letters, 16*(4), 20200033. https://doi.org/10.1098/rsbl.2020.0033

Ruiz-Miranda, C. R., De Morais, M. M., Jr., Dietz, L. A., Rocha Alexandre, B., Martins, A. F., Ferraz, L. P., Mickelberg, J., Hankerson, S. J., & Dietz, J. M. (2019). Estimating population sizes to evaluate progress in conservation of endangered golden lion tamarins (*Leontopithecus rosalia*). *PLoS One, 14*(6), 1–18. https://doi.org/10.1371/journal.pone.0216664

Russon, A. E. (2009). Orangutan rehabilitation and reintroduction: Successes, failures, and role in conservation. *Orangutans: Geographic Variation in Behavioral Ecology and Conservation*, 327–350. https://doi.org/10.1093/acprof:oso/9780199213276.003.0023

Schwartz, J. W., Hopkins, M. E., & Hopkins, S. L. (2016). Group Prerelease training yields positive rehabilitation outcomes among Juvenile Mantled Howlers (*Alouatta palliata*). *International Journal of Primatology, 37*(2), 260–280. https://doi.org/10.1007/s10764-016-9900-6

Sherman, J., Ancrenaz, M., & Meijaard, E. (2020). Shifting apes: Conservation and welfare outcomes of Bornean orangutan rescue and release in Kalimantan, Indonesia. *Journal for Nature Conservation, 55*, 125807. https://doi.org/10.1016/j.jnc.2020.125807

Snowdon, C. T. (1989). The criteria for successful captive propagation of endangered primates. *Zoo Biology, 8*(1 S), 149–161. https://doi.org/10.1002/zoo.1430080515

Southwick, C. H., & Lindburg, D. G. (1986). The primates of India: Status, trends, and conservation. In K. Benirschke (Ed.), *Primates. Proceedings in life sciences*. Springer. https://doi.org/10.1007/978-1-4612-4918-4_12

Spehar, S. N., Sheil, D., Harrison, T., Louys, J., Ancrenaz, M., Marshall, A. J., Wich, S. A., Bruford, M. W., & Meijaard, E. (2018). Orangutans venture out of the rainforest and into the anthropocene. *Science Advances, 4*, e1701422.

Speiran, S. I. (2021). Monkey see, monkey do: The work of primates in Costa Rican sanctuaries. In C. Kline & J. M. Rickly (Eds.), *Exploring non-human work in tourism* (pp. 181–206). De Gruyter Oldenbourg. https://doi.org/10.1515/9783110664058-012

Stan, K., & Sanchez-Azofeifa, A. (2019). Deforestation and secondary growth in Costa Rica along the path of development. *Regional Environmental Change, 19*(2), 587–597.

Stokes, R., Tully, G., & Rosati, A. G. (2018). Pan African Sanctuary Alliance: Securing a future for the African great apes. *International Zoo Yearbook, 52*, 173–181.

Suarez, C. E., Gamboa, E. M., Claver, P., & Montoya, F. N. (2001). Survival and reproduction of confiscated capuchin monkeys. *Animal Welfare, 10*, 191–203.

The Indian Express. (2020, Dec 19). *Telangana launches first rescue, rehabilitation centre for monkeys.* https://indianexpress.com/article/cities/hyderabad/telangana-launches-first-rescue-rehabilitation-centre-for-monkeys-7111481/ Accessed 30 Mar 2021.

The MONA Foundation. https://fundacionmona.org/en/nuestros-primates/. Accessed 15 Jan 2021.

Tutin, C. E. G., Ancrenaz, M., Paredes, J., Vacher-Vallas, M., Vidal, C., Goossens, B., Bruford, M. W., & Jamart, A. (2001). Conservation biology framework for the release of wild-born orphaned chimpanzees into the Conkouati Reserve, Congo. *Conservation Biology, 15*(5), 1247–1257. https://doi.org/10.1046/j.1523-1739.2001.00046.x

Uhm, D. (2016). Monkey business: The illegal trade in Barbary macaques. *Journal of Trafficking, Organized Crime and Security © 2016, 2*(1), 36–49.

Vogel, I., Glowing, B., Saint, P. I., Bayart, F., Contamin, H., & de, T. B. (2002). Squirrel monkey rehabilitation in French Guiana: A case study. *Neotropical Primates, 10*, 147–149.

Wallis, J., & Lee, D. R. (1999). Primate conservation: The prevention of disease transmission. *International Journal of Primatology, 20*(6), 803–826.

Wimberger, K., Downs, C. T., & Boyes, R. S. (2010). A survey of wildlife rehabilitation in South Africa: Is there a need for improved management? *Animal Welfare, 19*, 481–499.

Wobber, V., & Hare, B. (2011). Psychological health of orphan bonobos and chimpanzees in African sanctuaries. *PLoS One, 6*(6), e17147. https://doi.org/10.1371/journal.pone.0017147

Yeager, C. P. (1997). Orangutan rehabilitation in Tanjung Puting National Park, Indonesia. *Conservation Biology, 11*, 802–805.

Chapter 16
Through the Looking Glass: Effects of Visitors on Primates in Zoos

Ashley N. Edes and Katie Hall

Contents

Abstract Daily exposure to visitors as well as caretakers makes the zoo environment a unique setting for primates in human care. Understanding how visitors impact zoo-housed primates is key to continually improving their welfare. Herein, we review decades of research on visitor effects in primates, many of which report a combination of effects. The majority of studies suggest visitors are a negative or neutral stimulus, although nearly a third also report some positive effects. Limitations in existing research impede our ability to fully understand how primates perceive visitors. Furthermore, a reliance on negative indicators of welfare and the continued assumption of negative impacts due to early research is likely to bias how results are interpreted. We discuss a need to critically reevaluate our assumption that visitors are inherently negative, especially in modern zoos which have larger, more spacious habitats that allow animals to express greater choice and control and provide potential mitigation strategies if some visitor stimuli are found aversive.

A. N. Edes (✉)
Department of Reproductive and Behavioral Sciences, Saint Louis Zoo, St. Louis, MO, USA
e-mail: aedes@stlzoo.org

K. Hall
Department of Animal Welfare, Sedgwick County Zoo, Wichita, KS, USA

© The Author(s), under exclusive license to Springer Nature Switzerland AG 2023
T. McKinney et al. (eds.), *Primates in Anthropogenic Landscapes*, Developments in Primatology: Progress and Prospects, https://doi.org/10.1007/978-3-031-11736-7_16

Keywords Human-animal relationships · Behavior · Visitor presence · Crowd size · Visitor activity · Crowd composition · Visitor noise levels

16.1   Introduction

The zoo environment is a unique setting for primates in human care, with daily exposure to visitors in addition to their caretakers. As tens of thousands of primates are housed in zoos around the world, understanding the challenges they face and how to mitigate them is essential to positive welfare. The bulk of research on **human-animal relationships (HARs)** in zoos over the past 45 years focuses on how visitors affect animal well-being (Sherwen & Hemsworth, 2019), and the majority are on primates (Hosey, 2000, 2008; Sherwen & Hemsworth, 2019). Negative **visitor effects** presumably include vigilance, hiding, aggression, and increased distance. Neutral effects are characterized by minimal behavior changes. Positive effects may include seeking proximity to humans and low levels of fear, aggression, or vigilance (Smith, 2014). Despite the focus on primates, we know relatively little about each species, and studies on the same species often are contradictory. As responses to visitors may vary based on social structure, body size, and other factors, the substantial phylogenetic variation across primates makes generalizing results difficult. Moreover, researchers have suggested the evolutionary proximity between humans and primates may increase the likelihood they will interpret some human actions as threatening (e.g., staring, yawning; Nimon & Dalziel, 1992; Birke, 2002; Sherwen et al., 2015; Hashmi & Sullivan, 2020). This may be amplified by bidirectional mimicry, which occurs when a primate performs a hostile behavior and humans, likely unaware of what the behavior means, mimic those actions, or vice versa (Nimon & Dalziel, 1992). In this chapter, we review 46 studies on how visitors affect primates in zoos. Unfortunately, relationships between zoo-housed animals and their caretakers have yet to be robustly studied (Sherwen & Hemsworth, 2019). As such, though keeper-animal relationships are a major source of anthropogenic interaction in zoos and are critical to the well-being of primates in human care, we do not discuss them here. Nearly half of the studies we review on visitor effects present a mixture of results. Of all the studies reviewed, approximately 65% report at least one negative, 65% at least one neutral, and 33% at least one positive effect. We recognize that the terms positive, neutral, and negative may be limiting, as the dynamic relationship between primates and zoo guests is complex and nuanced, but we use them here to maintain consistency with the literature. It is beyond the scope of this review to go into specific details about the species from each study. However, given that primates are a large taxonomic group with considerable diversity, we encourage readers to keep in mind that how primates are managed in human care (e.g., group composition, feeding strategies, rearing, enrichment, and training protocols) as well as their natural histories (e.g., terrestrial vs. arboreal, social hierarchies, activity patterns across the 24-hour day) can impact how they respond to visitors. We then discuss how study limitations and researcher bias may influence how results are interpreted, contributing to a potential misconception that visitors are a primarily negative stimulus, and suggest ways visitor effect research may be improved. Finally, we suggest potential strategies for mitigating negative impacts. As animal welfare science moves away from simply reducing negative experiences and toward increasing positive experiences, better understanding how primates respond to and potentially benefit from the presence of and interactions with visitors will allow us to improve their lives in human care.

16.2   Visitor Effects by Variable

16.2.1   Presence of Visitors

Presence or absence is one of the most commonly analyzed visitor effect variables. The mere presence of visitors is associated with increased conspecific-directed aggression and decreased **affiliative behavior** in numerous species (ring-tailed lemurs [*Lemur catta*], Diana monkeys [*Cercopithecus diana*]: Chamove et al., 1988; cotton-top tamarins [*Saguinus oedipus*]: Glatston et al., 1984; Chamove et al., 1988; lion-tailed macaques [*Macaca silenus*]: Mallapur et al., 2005). However, other studies have observed no changes in affiliative or overall social interactions with conspecifics based on visitor presence (ring-tailed lemurs, brown lemurs [*Eulemur fulvus*], black spider monkeys [*Ateles paniscus*], white-fronted capuchins [*Cebus albifrons*], patas monkeys [*Erythrocebus patas*], De Brazza monkeys [*Cercopithecus neglectus*], Sykes monkeys [*Cercopithecus albogularis*], Talapoin monkeys [*Miopithecus talapoin*], Barbary macaques [*Macaca sylvanus*], lion-tailed macaques, black macaques [*Macaca nigra*], Hamadryas baboons [*Papio hamadryas*]: Hosey & Druck, 1987; white-cheeked gibbons [*Nomascus leucogenys*]: Lukas et al., 2002). Increased activity in relation to visitor presence has been reported in multiple species (ring-tailed lemurs, brown lemurs, Talapoin monkeys, white-fronted capuchins, black spider monkeys, lion-tailed macaques, De Brazza monkeys, Sykes monkeys, patas monkeys, black macaques, Hamadryas baboons: Chamove et al., 1988), but not in siamangs (*Symphalangus syndactylus*; Nimon & Dalziel, 1992). Negative behaviors have increased in the presence of visitors (e.g., aggression, pacing; lion-tailed macaques, mandrills [*Mandrillus sphinx*]: Chamove et al., 1988). A recent study comparing behavior in a gorilla (*Gorilla gorilla*) troop during the zoo's closure for the COVID-19 pandemic with data from the previous year saw no significant changes in behavior (Miller et al., 2021). While not significant, the decrease in foraging and activity levels in most of the gorillas during the closure suggests a positive impact of visitors, while the decline in negative behaviors suggests the opposite (Miller et al., 2021), exemplifying how even a single group can show inconsistent effects of visitors.

Studies using visual barriers provide additional examples of how visitor presence can affect primates in human care. Orangutans (*Pongo* spp.) spent similar amounts of time on a platform near the public viewing window and showed little preference in orientation toward the window regardless of whether it was fully open or partially blocked (Bloomfield et al., 2015). Visual barriers did not change the time gorillas spent standing, resting, locomoting, socializing near the window, or banging on the glass (Blaney & Wells, 2004). One-way viewing screens had no effect on the time black-capped capuchins (*Sapajus apella*) spent out of sight or engaged in behaviors such as resting, locomotion, foraging, or playing, nor on their proximity to the viewing window, but levels of aggression, negative behaviors, and glucocorticoids were reduced when the screens were in place (Sherwen et al., 2015). Additionally, ring-tailed lemurs in a **free-ranging**, immersive habitat increased locomotion and more animals came to the ground when visitors were present, suggesting interest in people or a lack of fear (Collins et al., 2017). In contrast, free-ranging white-faced saki (*Pithecia pithecia*) spent more than a third of their time out of sight (Sha Chih Mun et al., 2013).

16.2.2 Crowd Size/Visitor Density

Similarly, primates respond to larger crowd sizes/higher visitor densities with many of the same behavior patterns. For example, some studies have shown increased aggression with larger crowds (golden-bellied mangabeys [*Cercocebus chrysogaster*], Mitchell et al., 1991, 1992; gorillas, Wells, 2005; Kuhar, 2008; Stoinski et al., 2012; Clark et al., 2012; Smith, 2014; Lewis et al., 2020; orangutans, Smith, 2014; mandrills, Chamove et al., 1988), whereas others have not (black macaques, Dancer & Burn, 2019; Japanese macaques [*Macaca fuscata*], Woods et al., 2019; siamangs, white-cheeked gibbons, Smith & Kuhar, 2010; gorillas, Stoinski et al., 2012; Bonnie et al., 2016; chimpanzees, Bonnie et al., 2016). Relatedly, daily wounding rates did not vary based on crowd size in gorillas (Stoinski et al., 2012) or daily total gate counts in ring-tailed lemurs or chimpanzees (*Pan troglodytes*; Hosey et al., 2016). Changes in affiliative behavior also are mixed, with some studies showing reductions (chimpanzees, Wood, 1998; gorillas, Wells, 2005), some showing no change (golden-bellied mangabeys, Mitchell et al., 1991; gorillas, Wells, 2005; Kuhar, 2008; Stoinski et al., 2012; Lewis et al., 2020; siamangs, white-cheeked gibbons, Smith & Kuhar, 2010), and others showing increases (Diana monkeys, Todd et al., 2007). In gorillas, larger crowd sizes have been associated with increases (Wells, 2005) or no change in resting (Kuhar, 2008). In a study on gorillas, pileated gibbons (*Hylobates pileatus*), and orangutans, inactivity levels and locomotion did not vary based on visitor densities (Hashmi & Sullivan, 2020). In response to crowd size, gorillas (Carder & Semple, 2008; Lewis et al., 2020) and white-handed gibbons (*Hylobates lar*; Cooke & Schillaci, 2007) increased scratching, which may indicate anxiety, but crowd size did not impact behaviors such as **stereotypies** in gorillas or orangutans (Hashmi & Sullivan, 2020). In chimpanzees, higher visitor count was associated with an increase in yawning, another potential indicator of anxiety, but not scratching or regurgitation and reingestion (Wallace et al., 2019).

In addition, many species showed increased visual monitoring, often assumed to indicate vigilance, with higher visitor densities (mandrills, Chamove et al., 1988; gorillas, Clark et al., 2012; Lewis et al., 2020; black macaques, Dancer & Burn, 2019; white-handed gibbons, Cooke & Schillaci, 2007). Studies examining food-related behaviors (e.g., searching for, processing, or ingesting food items) have found a combination of negative (ring-tailed lemurs, Goodenough et al., 2019; chimpanzees, Wood, 1998; gorillas, Clark et al., 2012; Lewis et al., 2020), neutral (gorillas, Kuhar, 2008), and positive effects of crowd size (black macaques, Dancer & Burn, 2019; Diana monkeys, Todd et al., 2007; gorillas, orangutans, Hashmi & Sullivan, 2020). Some great apes decrease enrichment use with larger crowds (chimpanzees, Wood, 1998; gorillas, Lewis et al., 2020). There is also evidence of multiple species engaging in avoidance behavior in response to larger crowds. For example, gorillas, white-cheeked gibbons, and siamangs spent more time out of view and decreased their proximity to viewing windows (Kuhar, 2008; Smith & Kuhar, 2010), and gorillas (Collins & Marples, 2016) and orangutans spent less time oriented toward viewing windows when visitor density was high (Hashmi & Sullivan, 2020). However, other studies have shown crowd size did not impact space use patterns or proximity to viewing windows (chimpanzees, Ross & Lukas, 2006; Bonnie et al., 2016; gorillas, Wells, 2005; Ross & Lukas, 2006; Bonnie et al., 2016). Similarly, visitor density did not alter the amount of time gorillas or pileated gibbons spent with their back toward the window (Hashmi & Sullivan, 2020), a behavior that may be avoidant (Collins & Marples, 2016; Hashmi & Sullivan, 2020). Larger crowds were even associated with more time spent at viewing windows in common squirrel monkeys (*Saimiri sciureus*; Polgár et al., 2017).

In other research on the effects of crowd size, Hosey and Druck (1987) observed no relationship with visitor-directed behavior in the 12 species they studied. Crowd size was not associated with behavioral diversity in ring-tailed lemurs (Collins et al., 2017). Other studies reporting a neutral impact of visitor density have used unique measures, such as infants clinging to their mothers (gorillas, pileated gibbons, Hashmi & Sullivan, 2020), distribution of parturition and infant survival across days of the week (gorillas, Kurtycz & Ross, 2015; chimpanzees, Wagner & Ross, 2008), and willingness to participate in cognitive tasks (Japanese macaques, Huskisson et al., 2020; chimpanzees, Hopper et al., 2015).

16.2.3   Visitor Activity

Some have suggested visitor activity is more important than presence or crowd size. In their study on 12 species, Hosey and Druck (1987) reported increased locomotion with more active visitors. Recent studies also described increased locomotion as well as increased vigilance and less resting in ring-tailed lemurs with a higher number of active visitors (Goodenough et al., 2019). In Japanese macaques, "frenetic" crowd activity predicted increased visitor-directed aggression (Woods et al., 2019). Chimpanzees were more likely to scratch when visitors used flash photography and yawn when children screamed, but visitor behavior did not affect regurgitation and reingestion (Wallace et al., 2019).

16.2.4   Crowd Composition

Crowd composition is a rarely investigated visitor variable. While white-handed gibbons showed more scratching and alertness when there were more children, they had similar rates of climbing and hanging behavior regardless of crowd composition (Cooke & Schillaci, 2007). Gorillas also have shown increased scratching and alertness when there were more children, but crowd composition was not associated with affiliative or agonistic social behavior, activity levels, foraging, or engaging with enrichment (Lewis et al., 2020).

16.2.5   Visitor Noise Levels

While many visitor variables studied are visual, auditory stimuli also may play a role in how primates respond to guests. Orangutans showed fewer behavioral responses to quiet crowds compared to louder ones (Birke, 2002). Conversely, no changes in overall behavior despite "the different noise levels due to visitation" were reported for multiple species (brown howler monkeys [*Alouatta guariba*], golden-bellied capuchins [*Sapajus xanthosternos*], golden-bellied lion tamarins [*Leontopithecus chrysomelas*], gorillas, chimpanzees, Quadros et al., 2014). Louder crowds have been associated with increased aggression or threatening behaviors in white-handed gibbons (Cooke & Schillaci, 2007) and gorillas (Lewis et al., 2020). Also in white-handed gibbons, researchers observed a decrease in affiliative behavior with louder crowds (Cooke & Schillaci, 2007), but there was no change in gorillas (Lewis et al., 2020) and play increased in black macaques (Dancer & Burn, 2019). Noise levels were not associated with inactivity in pileated gibbons or orangutans, with locomotion in pileated gibbons, orangutans, and gorillas, or with stereotypic behaviors in gorillas and orangutans (Hashmi & Sullivan, 2020). Multiple species increased alertness or vigilance in response to louder crowds (black macaques, Dancer & Burn, 2019; gorillas, Clark et al., 2012; white-handed gibbons, Cooke & Schillaci, 2007). Reports suggest some species engage in less feeding behavior when louder crowds are present (gorillas, Hashmi & Sullivan, 2020; Lewis et al., 2020), but others have observed neutral (orangutans, Hashmi & Sullivan, 2020) or positive effects (black macaques, Dancer & Burn, 2019). Additionally, Hashmi and Sullivan (2020) reported no association between noise levels and infant clinging in gorillas and pileated gibbons.

16.2.6 Miscellaneous Visitor Effect Research

In addition to traditional variables, other studies have documented visitor effects in various ways. For example, one study combined crowd size, noise level, and activity into a single metric the authors called a "visitor impact score" (Roth & Cords, 2020). Larger visitor impact scores were positively associated with aggression, scratching, and yawning in ebony langurs (*Trachypithecus auratus*). However, positive social behaviors such as mounting and **allogrooming** also were positively associated with visitor impact scores, and there was no change in **autogrooming** (Roth & Cords, 2020). In another study, free-ranging cotton-top tamarins and white-faced saki occasionally directed agonistic behavior toward visitors, such as facial or vocal threats, lunging, chasing, and even biting and scratching (Sha Chih Mun et al., 2013). However, there was no mention of whether any particular visitor actions preceded these few cases of human-directed aggression. Orangutans exhibited less social and play behavior when visitors were less than 10 m away from the viewing window (Choo et al., 2011). Although this effect appears negative, the animals lived in large habitats with plenty of space to control their proximity to guests regardless of visitor position, so habitat design must be considered when interpreting effects of human proximity on animal behavior. Begging/soliciting food has been observed in ring-tailed lemurs (Collins et al., 2017), cotton-top tamarins (Sha Chih Mun et al., 2013), lion-tailed macaques (Mallapur et al., 2005), chimpanzees (Cook & Hosey, 1995; Wood, 1998), and orangutans (Birke, 2002; Choo et al., 2011). Relatedly, crowned lemurs (*Eulemur coronatus*) increased their interactions with visitors over time during visitor feeding experiences (Jones et al., 2016). Regardless of whether feeding is permitted, active solicitation has been considered a positive visitor effect because it shows a lack of fear of humans. Likewise, juveniles accounted for 84% of all nonaggressive visitor-directed interactions in Japanese macaques, which the authors suggested may indicate curiosity (Woods et al., 2019).

16.3 Limitations of Research to Date and Alternative Interpretations

Though visitor effects studies form the foundation of our knowledge on HARs in primates, several limitations should be considered when contextualizing the literature. First, visitor variables suffer from inconsistent terminology (e.g., crowd size vs. visitor density) and measurement. Methodological inconsistencies, and subjectively or poorly defined variables such as noise or activity, make replication and interpretation difficult. For example, some studies measure visitor density using instantaneous crowd size counts, where the observer counts the number of guests, while others use total daily gate counts as proxies. Although the two variables were correlated, behavior changes in gorillas were explained only by instantaneous crowd size counts (Lewis et al., 2020). Additionally, any one group of visitors may not be stressful, but there could be a cumulative effect throughout the day (Kuhar, 2008; Collins et al., 2017).

Visitor effects research also often ignores animal-based factors that may confound results and interpretations. Even within the same species, primates show substantial individual variation in their responses to humans. Exemplifying this, a recent study on yellow-breasted capuchins (*Sapajus xanthosternos*) showed such marked individual variation in how visitors affected their behavior that no general pattern could be discerned (Rodrigues & Azevedo, 2017). Therefore, studies should incorporate individual characteristics whenever possible, such as age, sex, rearing, personality, and social rank (Hosey, 2008; Stoinski et al., 2012; Bonnie et al., 2016; Polgár et al., 2017; Woods et al., 2019; Hashmi & Sullivan, 2020; Lewis et al., 2020). For example, male gorillas increased aggression with larger crowds while females did not (Stoinski et al., 2012), and personality factors moderate effects of crowd size on behavior in gorillas (Stoinski et al., 2012) and squirrel monkeys (Polgár et al., 2017). The impact of individual variation on group dynamics may also help explain the inconsistent results observed between studies of the same species.

Additionally, few studies incorporate **physiological indicators** of welfare alongside behavior and those that do have measured only glucocorticoids (e.g., cortisol; Davis et al., 2005; Clark et al., 2012; Sherwen et al., 2015). While glucocorticoids can be informative, they have functions beyond stress responses, and accurate interpretation can require considerable context (Cockrem, 2013; McEwen, 2019). Moreover, through habituation (Rault et al., 2020), individuals may learn that experiences they initially found stressful are not a concern, meaning physiological responses may decrease over time and disappear altogether. To better understand if there is a biological cost associated with behavioral changes, more studies need to include physiological data and, whenever possible, biomarkers from multiple systems (e.g., immune, reproductive) alongside glucocorticoids (Brown et al., 2001; Linklater et al., 2010). Behavioral changes in some individuals may successfully moderate physiological changes, indicating successful coping and resilience. Outside of situations known to be harmful (e.g., increased wounding), this would raise the question of whether behavioral changes in response to visitors are indeed negative. One example of behavior potentially attenuating physiological responses is from a recent study on how gorillas responded to a recurring evening event over multiple weeks in which negative behavioral changes were observed (e.g., more contact aggression), but there were no differences in glucocorticoids (Bastian et al., 2020).

It also is uncommon to include environmental variables such as time of day, temperature, or precipitation. These are correlated with visitor variables, for example, more visitors on days with good weather and larger crowds mid-day (Goodenough et al., 2019), but the common assumption is that changes in behavior are driven primarily by guests. However, as shown in two recent studies on ring-tailed lemurs, ignoring environmental variables may overestimate visitor effects. Collins et al. (2017) showed behavioral changes were driven by time of day, season, and weather, with few effects of visitors. More specifically, Goodenough et al.

(2019) reported 10–37% of the behavioral changes observed were explained by time of day and weather, whereas effects of visitors explained only 3–8%.

Furthermore, visitor effect studies often lack **internal**, **external**, and **construct validity**. Likely due to difficulty manipulating conditions, we have yet to determine the cause-and-effect relationship between visitors and changes in behavior. There are two non-mutually exclusive hypotheses for the direction of events (Hosey, 2000, 2008): the **visitor effect hypothesis** suggests visitors induce changes in behavior, and the **visitor attraction hypothesis** suggests active animals attract more attention. The visitor effect hypothesis is the lens through which most studies are viewed (we even refer to them as *visitor effect* studies). In an early influential review on the topic, Hosey (2000) shaped the field by stating the visitor attraction hypothesis "cannot account for most of the observed effects." Although active behavior in animals has the strongest impact on how long visitors spend at exhibits (Johnston, 1998), Choo et al. (2011) wrote that visitors looking at or taking photographs "significantly increased the chances" orangutans would be feeding, playing or socializing, and moving. This framing seems to suggest orangutans *perform* behaviors in response to being watched and photographed, despite these behaviors not being affected by crowd size or activity (Choo et al., 2011). This example demonstrates how ingrained the idea of visitors causing behavioral changes has become. On the other hand, the cause-and-effect direction of both hypotheses is limiting: the relationship between primate and guest is likely to be interactive, dynamic, reciprocal, and involve a degree of turn-taking. Though the study of visitor effects on zoo animals should remain its own body of research, especially considering mitigations implemented that improve animal welfare, perhaps future research should also consider "guest-animal interaction effects" above and beyond visitor effects.

Although the difficulty and expense of multi-institution studies are recognized and understood, because research on visitor effects is primarily conducted at single zoos with small sample sizes, generalizing results across institutions is not possible and makes it difficult to understand the effect visitors have on primates or establish best practice guidelines. These limitations are compounded further by the short duration of many studies, which usually capture only a few weeks or months of time for animals who live for decades. The field would benefit from a meta-analysis quantifying the strength of behavioral changes that have been observed. However, many studies do not report sufficient statistical information to allow for meta-analysis, such as effect sizes and confidence intervals (Cooper, 2017), and even significance can be vague (e.g., $p < 0.05$ vs. $p = 0.021$). More robust statistical reporting would help overcome this issue with external validity, allowing meta-analyses even when multi-institution studies remain infeasible.

Perhaps most importantly, researchers who study visitor effects need to critically examine whether behaviors assumed to be negative truly are, and if they are, whether the impacts are substantial. Though the hope may be for primates in human care to go about their lives as if we are not here, **behavioral flexibility** is an adaptive coping strategy (Hill & Broom, 2009) and it is unrealistic for animals not to show any response to changes in their environment. Akin to how humans behave differently when at work around colleagues than at home with family, such changes are not inherently negative and instead may be neutral or even positive. Behaviors inter-preted as negative are sometimes instead appropriate, species-typical reactions. Though wild-type behavior does not equal good welfare (Learmonth, 2019), the standard in zoos is usually to encourage species-typical responses, yet in some cases, these are then interpreted negatively. For example, silverbacks displaying at crowds while females more passively avoid them (Stoinski et al., 2012; Clark et al., 2012; Lewis et al., 2020) is expected based on gorilla socioecology (Harcourt & Stewart, 2007), but aggression and hiding are predominantly characterized as nega-tive reactions. While that could certainly be the case if they resulted in injury or prevented positive behaviors from occurring, more evidence is needed. If less feed-ing is observed in response to visitors, consistently weighing animals, a standard practice in zoos for many species, will determine whether there is cause for concern. Measures of behavioral diversity across 24-hour periods can be used to know if desired behaviors, such as play or grooming, are negatively impacted by visitors. Recently, the possibility has been raised that visitor-directed aggression may even be beneficial for some individuals. For example, Woods et al. (2019) suggested low-ranking macaques may be the ones who are performing most of the aggression, in which case the public may enhance their welfare by providing a safe outlet for expressing species-typical displacement behaviors. Comparably, some ambiguous behaviors are frequently assumed to be negative. Increases in activity and decreases in inactivity have been interpreted as agitation or interfering with rest (Chamove et al., 1988), yet zoos are often concerned about lower activity levels in captive col-lections relative to wild populations (e.g., Ross & Shender, 2016). Instead, changes in activity levels may indicate interest and stimulation; unless there are data animals are not getting sufficient rest or are losing weight, changes in activity could be con-sidered neutral or positive effects. Visual attention also is an ambiguous behavior that is frequently interpreted negatively, as indicated by the practice of calling it vigilance. However, absent of other evidence, visual attention simply indicates awareness of their environment, and may even be a sign of interest or curiosity. For example, gorillas and orangutans spend more time watching familiar humans, with whom they have positive HARs, than unfamiliar (Smith, 2014). The assumption that visual attention equates to vigilance and feeling threatened needs explicitly investigated to determine if this normal, appropriate response to a changing envi-ronment truly indicates a welfare concern. In modern zookeeping techniques, primate-keeper interactions are essential for their husbandry and health care. Should we instead reevaluate the goal that primates should behave more "wild" when in view of visitors? In order to give primates the best opportunities to exhibit species-typical behaviors and to limit the impact of visitor effects, perhaps we should instead reconceptualize the visitor experience: how visitors should behave, education on what behaviors to look for and expect from a variety of species, and what lessons about conservation they should take home.

The bias some researchers seem to have toward negative conclusions further emphasizes these issues with construct validity. For example, despite reporting no changes in behavior in 14 different primate and non-primate species, Quadros et al. (2014) wrote visitor noise was "almost certainly having a negative impact on the welfare of these species." Similarly, after observing evidence of mostly neutral and even some positive visitor effects in gorillas, orangutans, and pileated gibbons, Hashmi and Sullivan (2020) concluded the visitor effect "ranged from no effect to detrimental." Part of this bias may stem from a historical emphasis on negative indicators, but an absence of negative indicators does not always translate to a positive welfare state. Modern animal welfare science recognizes welfare exists on a spectrum ranging from poor to good and actively promotes positive welfare states (Yeates & Main, 2008; Mellor & Beausoleil, 2015). While more research is needed to determine indicators of positive welfare (Yeates & Main, 2008), there should be equal emphasis between these and negative indicators. Researchers (and readers) should be aware of how an emphasis on negative indicators and a prior belief that visitors are stressful is likely to shape everything from how behaviors are perceived (e.g., the often-subtle distinction between play and aggression) through to how results are interpreted and reported.

Finally, results from early research may lead to erroneous conclusions about impacts today. Modern zoo habitats are a far cry from the cages of decades ago (Fig. 16.1). Major changes in habitat design, some directly driven by studies on visitor effects, have resulted in larger, more **naturalistic** habitats that provide greater opportunities for choice and control. Such changes are likely to mitigate many negative impacts of visitors (Hosey, 2005; Ross et al., 2009). Moreover, early enclosure design often permitted visitor-primate interactions that are no longer possible. Unfortunately, multiple studies described harassment and teasing by visitors, for instance, by climbing or kicking fences (Mitchell et al., 1992), throwing rocks (Wood, 1998), or even physical harm (Mallapur et al., 2005). Considering this, aversive reactions to visitors in some individuals are unsurprising, and given the long memories and life spans of these animals, may persist even after moving into updated habitats. Both shifting values of the public toward increased animal welfare, which impacts their behavior and expectations when visiting zoos, and the distance or barriers preventing public interaction that are common in newer habitats may improve animals' perceptions of visitors. Several recently published studies demonstrate visitor effects are less negative than predicted based on reports from earlier decades (Wagner & Ross, 2008; Choo et al., 2011; Stoinski et al., 2012; Clark et al., 2012; Kurtycz & Ross, 2015). These shifting perspectives are perhaps best encapsulated by a statement in a recent paper from Hosey et al. (2016, p. 209): "We can only agree… that the effects of zoo visitors on captive animals may be less profound than previous studies suggested."

16.4 Potential Strategies for Mitigating Negative Impacts

Although changes in enclosure design and management strategies over time have mediated some concerns, there are still cases where visitors may be detrimental to the welfare of zoo-housed primates. Successfully mitigating negative impacts requires first identifying which visitor variables are responsible, and we recommend

Fig. 16.1 A comparison of a historic primate exhibit (**a**) with the larger, naturalistic habitats in many modern zoos (**b–d**). Panel (**a**) Monkey House interior (Courtesy of the Saint Louis Zoo). Panel (**b**) Donn and Marilyn Lipton Fragile Forest (Courtesy of the Saint Louis Zoo). Panel (**c**) Treetop path, Masoala Halle (Copyright: Zoo Zürich, Corinne Invernizzi). Panel (**d**) Gondwanaland. (Copyright: Zoo Leipzig, Maria Saegebarth)

comprehensive studies before implementing costly modifications that may have limited benefit. For example, a study on black-capped capuchins showed no evidence that visitor behaviors (e.g., banging on windows) or noise levels decreased when a one-way viewing screen was put in place, allowing the authors to conclude the behavioral changes they observed after its implementation was likely due to reduced visual stimuli (Sherwen et al., 2015). Given that many visitor variables are visual (e.g., crowd size, activity), multiple studies have implemented visual barriers with varying degrees of success. The study on black-capped capuchins reported a 68% reduction in conspecific-directed aggression and a further reduction in already low rates of visitor-directed aggression, less time visually oriented toward visitor areas, and a 38% reduction in abnormal behaviors in the two individuals who performed them (Sherwen et al., 2015). As described in Sect. 16.2.1, implementing visual barriers may also result in no changes to a variety of behaviors (Blaney & Wells, 2004; Bloomfield et al., 2015; Sherwen et al., 2015). A male drill (*Mandrillus leucophaeus*) even sought higher ground to continue engaging in visitor-directed aggression when a barrier was put in place (Martín et al., 2016). Bloomfield et al.

(2015) postulated people themselves may not be enriching but the changing scenery in the visitor viewing area may be. Animals also may be more responsive to visitors when they have nothing else to do. As such, increasing or changing **enrichment** may decrease boredom. At one zoo, gorillas engaged in more self-scratching and visual monitoring with higher visitor numbers, but these behaviors disappeared when food enrichment was provided (Carder & Semple, 2008). However, other studies indicate visitors reduce the likelihood of chimpanzees (Wood, 1998) and gorillas (Lewis et al., 2020) engaging with enrichment. Although a feeding enrichment program successfully modified behaviors such as stereotypies in a male drill, it was not successful at decreasing visitor-directed aggression (Martín et al., 2016).

Studies on primates in large, naturalistic enclosures such as those described in Choo et al. (2011) show little evidence that visitors are aversive, likely because they can maintain comfortable distances and have control over their level of visibility. Given that some primates could interpret human actions as threatening, providing spaces where individuals can retreat is important. This could be accomplished by allowing access to off-exhibit holding areas even during open hours, ensuring exhibits are outfitted with on-exhibit hiding areas, and designing exhibits that offer sufficient retreat distances from the public without having to sacrifice access to important resources (e.g., shade, water sources). Such provisions give individuals more choice and control over their experiences, which is essential to positive welfare (Broom, 1988; Hill & Broom, 2009). As multiple studies have documented no change in time spent out of sight, proximity to viewing windows, or space use based on different visitor variables (Blaney & Wells, 2004; Wells, 2005; Ross & Lukas, 2006; Quadros et al., 2014; Bloomfield et al., 2015; Sherwen et al., 2015; Bonnie et al., 2016), it is unlikely many primates would choose to spend the majority of their time in off-display areas compared to their larger, more enriched habitats. Providing higher vantage points or other ways of elevating primates relative to guests also may improve visitor effects, especially for arboreal species.

Finally, providing increased opportunities for positive interactions may help strengthen and improve visitor-primate relationships. For example, there is evidence showing primates form positive bonds with familiar humans such as their caretakers (Herrelko et al., 2012; Smith, 2014), who may even be an integral part of their social networks (Funkhouser et al., 2020), as well as evidence demonstrating the enriching effects of training (Savastano et al., 2003) and participating in cognitive tasks (Herrelko et al., 2012). Implementing **positive reinforcement training** in Abyssinian colobus (*Colobus guereza*) resulted in an elimination of interactions with visitors (Melfi & Thomas, 2005). Conducting training sessions or even cognitive research in full view of the public improves visitor understanding of the work done in zoos and appreciation for the animals themselves (Hopper, 2017). It also helps primates form positive associations between activities they enjoy and the presence of guests, indirectly improving their perception of unfamiliar humans (see also Polgár et al., 2017). Depending on the type of cognitive research being conducted, it may even be possible for visitors to interact with primates through touchscreen activities. For primates housed in zoos that will not be released into the wild, we should consider potential avenues to further improve their experiences in human care even if the behaviors are not "natural" (see also Brando & Herrelko, 2021), especially if those experiences could mimic the cognitive challenges wild primates experience but are often lacking in captive environments. However, given the consequences associated with viewing images and videos of primates in inappropriate contexts with humans, such as decreased perception of their conservation status or increased perception of their suitability as pets (Ross et al., 2011; Nekaris et al., 2013), such nonnatural interactions would need to be carefully moderated to provide the necessary context to visitors.

16.5   Conclusion

Early research in primates indicated visitors had primarily negative impacts on behavior. However, primates can form positive relationships with humans, and accordingly there also is evidence to support neutral or even positive visitor effects. Given the limitations inherent to research on visitor effects and the potential for alternative interpretations of some results, as well as the considerable changes in zoo habitat design in the 45-year history of these studies, we suggest critical reevaluation of the common assumption that visitors are a negative stimulus. Incorporating consistent methods for measuring visitor variables as well as additional animal- and environment-based variables would greatly improve our understanding of how visitors impact primates in human care. It also is necessary to improve construct validity across studies, ensuring that we accurately categorize behaviors as indicating positive or negative effects. Whenever possible, we should design studies with high internal validity that can determine the cause-and-effect relationship between visitors and behavioral changes. Furthermore, as visitor effect studies are essentially a compendium of case studies, researchers should report data that are amenable to meta-analysis, such as effect sizes, which would allow us to generalize effects across zoos, strengthening external validity. Finally, we encourage researchers to be cautious in their interpretations, provide detailed alternatives, and explicitly acknowledge limitations. In cases where visitor effects may be negative, strategies such as providing hiding areas or increasing enrichment and training may successfully mitigate these effects. Evaluating such changes systematically will help determine best practices for care and management, and we encourage publishing both successful and unsuccessful attempts. Promoting positive visitor-primate relationships will continue building on the vast improvements already made in animal welfare over the past several decades, allowing species in human care to thrive.

Acknowledgments We are grateful to David Powell for thorough suggestions on chapter organization, to an anonymous peer reviewer for careful comments, and to the editors for their thought-provoking questions on a draft of the manuscript. We are also thankful for the invitation to contribute to this volume.

References

Bastian, M. L., Glendinning, D. R., Brown, J. L., et al. (2020). Effects of a recurring late-night event on the behavior and welfare of a population of zoo-housed gorillas. *Zoo Biology, 39*(4), 217–229. https://doi.org/10.1002/zoo.21553

Birke, L. (2002). Effects of browse, human visitors and noise on the behaviour of captive orangutans. *Animal Welfare, 11*, 189–202.

Blaney, E. C., & Wells, D. L. (2004). The influence of a camouflage net barrier on the behavior, welfare and public perceptions of zoo-housed gorillas. *Animal Welfare, 13*, 111–118.

Bloomfield, R. C., Gillespie, G. R., Kerswell, K. J., et al. (2015). Effect of partial covering of the visitor viewing area window on positioning and orientation of zoo orangutans: A preference test. *Zoo Biology, 34*, 223–229.

Bonnie, K. E., Ang, M. Y. L., & Ross, S. R. (2016). Effects of crowd size on exhibit use by and behavior of chimpanzees (*Pan troglodytes*) and Western lowland gorillas (*Gorilla gorilla*) at a zoo. *Applied Animal Behaviour Science, 178*, 102–110.

Brando, S., & Herrelko, E. S. (2021). Wild animals in the city: Considering and connecting with animals in zoos and aquariums. In B. Bovenkerk & J. Keulartz (Eds.), *Animals in our midst: The challenges of co-existing with animals in the Anthropocene* (pp. 341–360). Springer International Publishing.

Broom, D. R. (1988). The scientific assessment of animal welfare. *Applied Animal Behaviour Science, 20*, 5–19.

Brown, J. L., Bellem, A. C., Fouraker, M., et al. (2001). Comparative analysis of gonadal and adrenal activity in the black and white rhinoceros in North America by noninvasive endocrine monitoring. *Zoo Biology, 20*, 463–486.

Carder, G., & Semple, S. (2008). Visitor effects on anxiety in two captive groups of western lowland gorillas. *Applied Animal Behaviour Science, 115*, 211–220.

Chamove, A. S., Hosey, G. R., & Schaetzel, P. (1988). Visitors excite primates in zoos. *Zoo Biology, 7*, 359–369.

Choo, Y., Todd, P. A., & Li, D. (2011). Visitor effects on zoo orangutans in two novel, naturalistic enclosures. *Applied Animal Behaviour Science, 133*, 78–86.

Clark, F. E., Fitzpatrick, M., Hartley, A., et al. (2012). Relationship between behavior, adrenal activity, and environment in zoo-housed western lowland gorillas (*Gorilla gorilla gorilla*). *Zoo Biology, 31*, 306–321.

Cockrem, J. F. (2013). Individual variation in glucocorticoid stress responses in animals. *General and Comparative Endocrinology, 181*, 45–58.

Collins, C. K., & Marples, N. M. (2016). The effects of zoo visitors on a group of Western lowland gorillas Gorilla gorilla gorilla before and after the birth of an infant at Dublin Zoo. *International Zoo Yearbook, 50*, 183–192.

Collins, C., Corkery, I., Haigh, A., et al. (2017). The effects of environmental and visitor variables on the behavior of free-ranging ring-tailed lemurs (*Lemur catta*) in captivity. *Zoo Biology, 36*, 250–260.

Cook, S., & Hosey, G. R. (1995). Interaction sequences between chimpanzees and human visitors at the zoo. *Zoo Biology, 14*, 431–440.

Cooke, C. M., & Schillaci, M. A. (2007). Behavioral responses to the zoo environment by white handed gibbons. *Applied Animal Behaviour Science, 106*, 125–133.

Cooper, H. (2017). *Research synthesis and meta-analysis: A step-by-step approach* (5th ed.). Sage Publications, Inc.

Dancer, A. M. M., & Burn, C. C. (2019). Visitor effects on zoo-housed Sulawesi crested macaque (*Macaca nigra*) behaviour: Can signs with 'watching eyes' requesting quietness help? *Applied Animal Behaviour Science, 211*, 88–94.

Davis, N., Schaffner, C. M., & Smith, T. E. (2005). Evidence that zoo visitors influence HPA activity in spider monkeys (*Ateles geoffroyii rufiventris*). *Applied Animal Behaviour Science, 90*, 131–141.

Funkhouser, J. A., Mayhew, J. A., Mulcahy, J. B., & Sheeran, L. K. (2020). Human caregivers are integrated social partners for captive chimpanzees. *Primates, 62*(2), 297–309. https://doi.org/10.1007/s10329-020-00867-6

Glatston, A. R., Geilvoet-Soeteman, E., Hora-Pecek, E., & van Hooff, J. A. R. A. M. (1984). The influence of the zoo environment on social behavior of groups of cotton-topped tamarins, *Saguinus oedipus oedipus. Zoo Biology, 3*, 241–253.

Goodenough, A. E., McDonald, K., Moody, K., & Wheeler, C. (2019). Are "visitor effects" over-estimated? Behaviour in captive lemurs is mainly driven by co-variation with time and weather. *Journal of Zoo and Aquarium Research, 7*, 59–66.

Harcourt, A. H., & Stewart, K. J. (2007). Gorilla society: What we know and don't know. *Evolutionary Anthropology, 16*, 147–158.

Hashmi, A., & Sullivan, M. (2020). The visitor effect in zoo-housed apes: The variable effect on behaviour of visitor number and noise. *Journal of Zoo and Aquarium Research, 8*, 268–282.

Herrelko, E. S., Vick, S.-J., & Buchanan-Smith, H. M. (2012). Cognitive research in zoo-housed chimpanzees: Influence of personality and impact on welfare. *American Journal of Primatology, 74*, 828–840.

Hill, S. P., & Broom, D. M. (2009). Measuring zoo animal welfare: Theory and practice. *Zoo Biology, 28*, 531–544.

Hopper, L. M. (2017). Cognitive research in zoos. *Current Opinion in Behavioral Sciences, 16*, 100–110.

Hopper, L. M., Kurtycz, L. M., Ross, S. R., & Bonnie, K. E. (2015). Captive chimpanzee foraging in a social setting: A test of problem solving, flexibility, and spatial discounting. *PeerJ, 3*, e833.

Hosey, G. R. (2000). Zoo animals and their human audiences: What is the visitor effect? *Animal Welfare, 9*, 343–357.

Hosey, G. R. (2005). How does the zoo environment affect the behaviour of captive primates? *Applied Animal Behaviour Science, 90*, 107–129.

Hosey, G. R. (2008). A preliminary model of human–animal relationships in the zoo. *Applied Animal Behaviour Science, 109*, 105–127.

Hosey, G. R., & Druck, P. L. (1987). The influence of zoo visitors on the behaviour of captive primates. *Applied Animal Behaviour Science, 18*, 19–29.

Hosey, G. R., Melfi, V., Formella, I., et al. (2016). Is wounding aggression in zoo-housed chimpanzees and ring-tailed lemurs related to zoo visitor numbers? *Zoo Biology, 35*, 205–209.

Huskisson, S. M., Ross, S. R., & Hopper, L. M. (2020). Do zoo visitors induce attentional bias effects in primates completing cognitive tasks? *Animal Cognition, 24*(3), 645–653. https://doi.org/10.1007/s10071-020-01445-5

Johnston, R. J. (1998). Exogenous factors and visitor behavior: A regression analysis of exhibit viewing time. *Environment and Behavior, 30*, 322–347.

Jones, H., McGregor, P. K., Farmer, H. L. A., & Baker, K. R. (2016). The influence of visitor interaction on the behavior of captive crowned lemurs (*Eulemur coronatus*) and implications for welfare. *Zoo Biology, 35*, 222–227.

Kuhar, C. W. (2008). Group differences in captive gorillas' reaction to large crowds. *Applied Animal Behaviour Science, 110*, 377–385.

Kurtycz, L. M. B., & Ross, S. R. (2015). Western lowland gorilla (*Gorilla gorilla gorilla*) birth patterns and human presence in zoological settings. *Zoo Biology, 34*(6), 518–521. https://doi.org/10.1002/zoo.21243

Learmonth, M. J. (2019). Dilemmas for natural living concepts of zoo animal welfare. *Animals (Basel), 9*, 318. https://doi.org/10.3390/ani9060318

Lewis, R. N., Chang, Y.-M., Ferguson, A., et al. (2020). The effect of visitors on the behaviour of zoo-housed western lowland gorillas (*Gorilla gorilla gorilla*). *Zoo Biology, 39*(5), 283–296. https://doi.org/10.1002/zoo.21552

Linklater, W. L., MacDonald, E. A., Flamand, J. R. B., & Czekala, N. M. (2010). Declining and low fecal corticoids are associated with distress, not acclimation to stress, during the translocation of African rhinoceros. *Animal Conservation, 13*, 104–111.

Lukas, K. E., Barkauskas, R. T., Maher, S. A., et al. (2002). Longitudinal study of delayed reproductive success in a pair of white-cheeked gibbons (*Hylobates leucogenys*). *Zoo Biology, 21*, 413–434.

Mallapur, A., Sinha, A., & Waran, N. (2005). Influence of visitor presence on the behaviour of captive lion-tailed macaques (*Macaca silenus*) housed in Indian zoos. *Applied Animal Behaviour Science, 94*, 341–352.

Martín, O., Vinyoles, D., García-Galea, E., & Maté, C. (2016). Improving the welfare of a zoo-housed male drill (*Mandrillus leucophaeus poensis*) aggressive toward visitors. *Journal of Applied Animal Welfare Science, 19*, 323–334.

McEwen, B. S. (2019). What is the confusion with cortisol? *Chronic Stress, 3*, 2470547019833647.

Melfi, V. A., & Thomas, S. (2005). Can training zoo-housed primates compromise their conservation? A case study using Abyssinian colobus monkeys (*Colobus guereza*). *Anthrozoös, 18*, 304–317.

Mellor, D. J., & Beausoleil, N. J. (2015). Extending the "five domains" model for animal welfare assessment to incorporate positive welfare state. *Animal Welfare, 24*, 241–253.

Miller, M. E., Robinson, C. M., & Margulis, S. W. (2021). Behavioral implications of the complete absence of guests on a zoo-housed gorilla troop. *Animals (Basel), 11*, 1346. https://doi.org/10.3390/ani11051346

Mitchell, G., Herring, F., Obradovich, S., et al. (1991). Effects of visitors and cage changes on the behaviors of mangabeys. *Zoo Biology, 10*, 417–423.

Mitchell, G., Herring, F., & Obradovich, S. (1992). Like threaten like in mangabeys and people? *Anthrozoös, 5*, 106–112.

Nekaris, K. A.-I., Campbell, N., Coggins, T. G., et al. (2013). Tickled to death: Analysing public perceptions of "cute" videos of threatened species (slow lorises – *Nycticebus spp.*) on web 2.0 sites. *PLoS One, 8*, e69215.

Nimon, A. J., & Dalziel, F. R. (1992). Cross-species interaction and communication: A study method applied to captive siamang (*Hylobates syndactylus*) and long-billed corella (*Cacatua tenuirostris*) contacts with humans. *Applied Animal Behaviour Science, 33*, 261–272.

Polgár, Z., Wood, L., & Haskell, M. J. (2017). Individual differences in zoo-housed squirrel monkeys' (*Saimiri sciureus*) reactions to visitors, research participation, and personality ratings. *American Journal of Primatology, 79*, 22639. https://doi.org/10.1002/ajp.22639

Quadros, S., Goulart, V. D. L., Passos, L., et al. (2014). Zoo visitor effect on mammal behaviour: Does noise matter? *Applied Animal Behaviour Science, 156*, 78–84.

Rault, J.-L., Waiblinger, S., Boivin, X., & Hemsworth, P. (2020). The power of a positive human–animal relationship for animal welfare. *Frontiers in Veterinary Science, 7*, 590867.

Rodrigues, N. S. S. O., & Azevedo, C. S. (2017). Influence of visitors on the behaviour of Yellow-breasted capuchins *Sapajus xanthosternos* at Belo Horizonte Zoo (BH Zoo), Brazil. *International Zoo Yearbook, 51*, 215–224.

Ross, S. R., & Lukas, K. E. (2006). Use of space in a non-naturalistic environment by chimpanzees (*Pan troglodytes*) and lowland gorillas (*Gorilla gorilla gorilla*). *Applied Animal Behaviour Science, 96*, 143–152.

Ross, S. R., & Shender, M. A. (2016). Daily travel distances of zoo-housed chimpanzees and gorillas: Implications for welfare assessments and space requirements. *Primates, 57*(3), 395–401. https://doi.org/10.1007/s10329-016-0530-6

Ross, S. R., Schapiro, S. J., Hau, J., & Lukas, K. E. (2009). Space use as an indicator of enclosure appropriateness: A novel measure of captive animal welfare. *Applied Animal Behaviour Science, 121*, 42–50.

Ross, S. R., Vreeman, V. M., & Lonsdorf, E. V. (2011). Specific image characteristics influence attitudes about chimpanzee conservation and use as pets. *PLoS One, 6*, e22050.

Roth, A. M., & Cords, M. (2020). Zoo visitors affect sleep, displacement activities, and affiliative and aggressive behaviors in captive ebony langurs (Trachypithecus auratus). *Acta Ethologica, 23*, 61–68.

Savastano, G., Hanson, A., & McCann, C. (2003). The development of an operant conditioning training program for new world primates at the Bronx Zoo. *Journal of Applied Animal Welfare Science, 6*, 247–261.

Sha Chih Mun, J., Kabilan, B., Alagappasamy, S., & Guha, B. (2013). Benefits of naturalistic free-ranging primate displays and implications for increased human–primate interactions. *Anthrozoös, 26*, 13–26.

Sherwen, S. L., & Hemsworth, P. H. (2019). The visitor effect on zoo animals: Implications and opportunities for zoo animal welfare. *Animals (Basel), 9*, 366. https://doi.org/10.3390/ani9060366

Sherwen, S. L., Harvey, T. J., Magrath, M. J. L., et al. (2015). Effects of visual contact with zoo visitors on black-capped capuchin welfare. *Applied Animal Behaviour Science, 167*, 65–73.

Smith, J. J. (2014). Human-animal relationships in zoo-housed orangutans (*P. abelii*) and gorillas (*G. g. gorilla*): The effects of familiarity. *American Journal of Primatology, 76*, 942–955.

Smith, K. N., & Kuhar, C. W. (2010). Siamangs (*Hylobates syndactylus*) and white-cheeked gibbons (*Hylobates leucogenys*) show few behavioral differences related to zoo attendance. *Journal of Applied Animal Welfare Science, 13*, 154–163.

Stoinski, T. S., Jaicks, H. F., & Drayton, L. A. (2012). Visitor effects on the behavior of captive western lowland gorillas: The importance of individual differences in examining welfare. *Zoo Biology, 31*, 586–599.

Todd, P. A., Macdonald, C., & Coleman, D. (2007). Visitor-associated variation in captive Diana monkey (*Cercopithecus diana diana*) behaviour. *Applied Animal Behaviour Science, 107*, 162–165.

Wagner, K. E., & Ross, S. R. (2008). Chimpanzee (*Pan troglodytes*) birth patterns and human presence in zoological settings. *American Journal of Primatology, 70*, 703–706.

Wallace, E. K., Herrelko, E. S., Koski, S. E., et al. (2019). Exploration of potential triggers for self-directed behaviours and regurgitation and reingestion in zoo-housed chimpanzees. *Applied Animal Behaviour Science, 221*, 104878.

Wells, D. L. (2005). A note on the influence of visitors on the behaviour and welfare of zoo-housed gorillas. *Applied Animal Behaviour Science, 93*, 13–17.

Wood, W. (1998). Interactions among environmental enrichment, viewing crowds, and zoo chimpanzees (*Pan troglodytes*). *Zoo Biology, 17*, 211–230.

Woods, J. M., Ross, S. R., & Cronin, K. A. (2019). The social rank of zoo-housed Japanese macaques is a predictor of visitor-directed aggression. *Animals (Basel), 9*, 316. https://doi.org/10.3390/ani9060316

Yeates, J. W., & Main, D. C. J. (2008). Assessment of positive welfare: A review. *Veterinary Journal, 175*, 293–300.

Chapter 17
Primate Portrayals: Narratives and Perceptions of Primates in Entertainment

Brooke C. Aldrich, Kim Feddema, Anna Fourage, K. A. I. Nekaris, and Sam Shanee

Contents

Abstract Across the globe and across time, primates have been used in live performances and depicted through imagery to entertain audiences and tell stories. Technological advances have led to a proliferation of ways in which we consume media and with that, audiences for primates in entertainment have flourished. Here we review some of the ways primates are used as entertainers and examine representations of primates in contemporary media. We provide an overview of the role of primates in the entertainment industry and discuss issues of animal welfare and conservation. An understanding of the history primates in media and entertainment

B. C. Aldrich (✉)
Neotropical Primate Conservation, Seaton, Cornwall, UK

Asia for Animals Coalition, Hong Kong, China
e-mail: brooke@neoprimate.org

K. Feddema
Business School, University of Western Australia, Perth, WA, Australia

A. Fourage · K. A. I. Nekaris
Noctural Primates Research Group, Oxford Brookes University, Oxford, UK

S. Shanee
Neotropical Primate Conservation, Seaton, Cornwall, UK

© The Author(s), under exclusive license to Springer Nature Switzerland AG 2023
T. McKinney et al. (eds.), *Primates in Anthropogenic Landscapes*, Developments in Primatology: Progress and Prospects, https://doi.org/10.1007/978-3-031-11736-7_17

is critical to regulating these practices and ensuring the health and welfare of both humans and animals.

Keywords Actors · Ethics · Film · Gaming · Selfies · Social media · Welfare · Photo props

17.1   Introduction

To entertain is 'to provide (someone) with amusement or enjoyment' (Oxford Dictionary, 2021). For millennia, people have used animals, including primates, as sources of entertainment in a myriad of forms. As early as the first century BC, apes and monkeys were hunted in *Venatio* blood sports in the arenas of the Roman Empire (Mechikoff, 2014 in Ćurko, 2020). In nineteenth-century England, primates, including an individual called Jacko Macauco ('the champion of monkeys'), were pitted against fighting dogs (Lennox, 1860). Today, the use of primates in entertainment remains prevalent. Primates are used for live entertainment across the globe as photo props (LaFleur et al., 2019; Stazaker & Mackinnon, 2018), circus performers (Agoramoorthy & Hsu, 2005; Gotsis, 2018; Shanee et al., 2017), objects of novel experience (Romano, 2019) and as street performers (Fragaszy & Visalberghi, 2004). Primate entertainment is not only performed by live actors but also delivered to audiences across numerous media platforms. Primates are used as film and television actors (Aldrich, 2018; Wallace, 2012) and in the creation of social media content (Nekaris et al., 2013, 2015).

The International Primatological Society developed a statement opposing such use (IPS, 2008). The Society objects to the caricaturisation of primates (including dressing them and training them in ways that compromise welfare), failure to keep primate performers in species-appropriate social groups, the use of aversive training, inaccurate portrayal of primate biology and conservation status and the risk of disease transfer between people and primates. All are common aspects of the use of primates as performers. The Society's statement is closely linked with the Five Freedoms of animal welfare: freedom from hunger or thirst; freedom from discomfort; freedom from pain, injury or disease; freedom from fear and distress; and freedom to express normal behaviours (Webster, 2016). We recognise that we cannot include in this chapter all examples of primates in entertainment nor detail all the complex ethical and welfare issues that arise. Instead, we focus on those forms of entertainment that deprive primates of these freedoms or compromise their conservation. Further, we discuss primates in film, television, social media and gaming, because images of primates in these mediums have implications for public understanding of primates. Pet keeping, ecotourism, zoos and roadside attractions, examined in other chapters of this book and elsewhere, could also be viewed as forms of entertainment (Russon & Wallis, 2014; Chaps. 11, 14 and 16, this volume).

17.2 Live Entertainment

17.2.1 Circuses and Live Performances

Travelling circuses and similar acts featuring performing primates still draw crowds, whether legally permitted or not (Born Free Foundation, 2014; Shanee & Shanee, 2021; Thomas, 2018). The use of wild animals in circuses has been banned by 31 countries, while other nations have implemented regional bans and restrictions (Dykes, 2020). Most bans are implemented on welfare grounds (Tyson, 2020). Yet, in much of the world, the use of primates as performers on stage, in circuses or in other public venues, is permitted, and the practice continues illegally in many other countries. In Vietnam, 22 circuses that feature performing wild animals have been identified, 18 of which include primates. Most often, the primates involved are macaques. As of October 2020, 1 Vietnamese circus alone owned 24 macaques (Animals Asia Foundation, unpublished data).

In the USA, wild animal acts regularly travel the state and county fair circuit. The state of Minnesota alone hosts over 90 such annual events (Qian et al., 2018). During the summer of 2014, one of these events was observed by Aldrich (2014). Several times a day, two ring-tailed lemurs (*Lemur catta*) were removed from a small display in a trailer and placed in carrying containers. These containers were then connected to an enclosed metal 'racetrack' with another carrying container at the other end (Fig. 17.1a). Children were selected from the audience to open hatches, allowing the lemurs to 'race' to the other end (Aldrich, 2014). Similarly, an act called 'The Banana Derby' has appeared regularly at fairs across the country, featuring clothed capuchin monkeys strapped to the backs of dogs, 'racing' around a small course (Space Coast Daily, 2020). In South America, primates, including capuchins (*Cebus* and *Sapajus* spp.), spider monkeys (*Ateles* spp.) and woolly monkeys (*Lagothrix* spp.), are regularly used as performers in travelling circuses and fairs (S. Shanee, personal observation; Fig. 17.1b).

In Indonesia, long-tailed macaques (*Macaca fascicularis*) are trained to perform in masked performances (Fig. 17.1c). The animals are referred to as 'masked monkeys' – *topeng monyet* in Bahasa Indonesia or *ledhek kethek* and *tandhak bedhes* in Javanese (Hansen et al., 2021; Kencana, 2017). The practice draws on traditional Indonesian dance dramas in which human dancers wear masks to embody nature, ancestors, mythical beings or characters from traditional fables (Emigh, 1996). Originating in the 1890s, *topeng monyet* replaced the human dancers with trained macaques. Over time, the practice became popular and evolved to include young wild-caught macaques dancing, praying, riding bicycles or performing other tricks. Aversive handling and training methods are often used. Trainers may use pliers to remove canine teeth without anaesthesia or restrain animals on short metal chains to force them to stand bipedally in cramped cages for long periods. Often these chains are put on the animals when they are young and become embedded into the skin as they grow. These practices can result in painful infections, injuries and behavioural problems (JAAN, 2015). As a result of pressure from animal welfare groups, the

Fig. 17.1 (**a**) Ring-tailed lemur (*Lemur catta*) at a racetrack attraction at a county fair in Massachusetts, USA. (Photo B. Aldrich). (**b**) Black-faced spider monkeys (*Ateles chamek*) in a travelling circus in Bolivia. (Photo N. Shanee/Neotropical Primate Conservation). (**c**) Masked monkey (*Macaca fascicularis*) in Tasikmalaya, Indonesia. (Photo A. Walmsley/Little Fireface Project)

practice was officially outlawed across the island of Java in 2013. However, public sentiment has changed but little, and the practice still occurs across Indonesia (K. Feddema and K.A. Nekaris, personal observation). The practice of *topeng monyet* exemplifies the complex interaction between sociocultural, economic and animal welfare factors that shape entertainment practices.

17.2.2 Paid Interactions

Paid interactions between people and primates can occur alongside live performances or separately from them. These interactions tend to fall into two overlapping categories. Animal experiences involve payment to a handler in exchange for the opportunity to physically interact with a primate, including feeding it, touching it or engaging in a planned activity (e.g. lemur yoga; Romano, 2019). Photo opportunities involve payment for a photo with the primate, which often entails physical contact between the person and the animal. In some cases, where these interactions are not regulated and are illegal, vendors may also allow clients to purchase the animal as a pet (Osterberg & Nekaris, 2015). The industry built around this sort of interaction is known as the photo prop trade.

Box 17.1 Thailand's Wildlife Tourist Attractions

Thailand is a leading global tourist destination, which attracted 39.8 million visitors in 2019 (Ministry of Tourism and Sports, 2021). Many tourists visit zoos and wildlife tourist attractions (WTAs), which range from local zoos and single species attractions like elephant (*Elephas maximus*) camps to safari parks or even a zoo on top of a department store. WTAs offer a wide selection of animal interactions, including circus-style shows, photos and feeding opportunities. These shows routinely feature primates. Of the primates, macaques (*Macaca* spp.), mainly pig-tailed macaques (*M. nemestrina*), are the most widely used. Schmidt-Burbach et al. (2015) recorded 371 individual macaques at 21 venues offering monkey shows. Often these macaques are forced to ride bicycles, play basketball, do push-ups, pose for photos and mingle with the crowd to sell souvenirs and beg for tips. Apes are exploited for the same purposes, with gibbons (Hylobatidae), orangutans (*Pongo* spp.) and chimpanzees (*Pan troglodytes*) common as photo props (Kerr, 2017). One WTA in Bangkok is well-known to animal advocates, who have repeatedly expressed concern over its treatment of animals, including the great apes (PETA, 2021; WFFT, 2019). Upon arrival at this venue, tourists see infant chimpanzees, orangutans and gibbons dressed like human babies and surrounded by plush toys (Fig. 17.2; Cohen, 2009). Keepers charge tourists to bottle-feed the apes or to have their photo taken with them (A. Fourage, personal observation).

Given the perceived legitimacy of venues, tourists may assume that the animals are well cared for and fail to realise the welfare harms caused by their involvement. Experiences such as these are known drivers of the wildlife trade, encouraging demand for primate pets (Norconk et al., 2020). The acquisition of non-native species is frequently connected to illegal activity, and in many cases, their origins cannot be determined (Beastall et al., 2016). If primates are wild-caught, their mothers or group mates may be killed during capture (Norconk et al., 2020; Peres, 1991; Stiles et al., 2013), which can cause psychological trauma from capture or from premature maternal separation (Bradshaw et al., 2005; Mallonee & Joslin, 2004). Captive environments for these individuals are frequently substandard, with movement severely restricted. Without the ability to express natural behaviours, animals may develop abnormal behaviours, including coprophagy, regurgitation and reingestion, hair plucking, stereotypies (e.g. rocking back and forth) and auto-aggression (e.g. self-biting and wounding) (Jacobson et al., 2016; Lopresti-Goodman et al., 2013). Current legislation does not prohibit the use of wild animals for entertainment in Thailand nor does it sufficiently govern husbandry practices (Dorloh, 2017; Schmidt-Burbach et al., 2015). Therefore, welfare harms will continue as long as demand for interactions persists.

(continued)

Fig. 17.2 Two infant chimpanzees at a Thai WTA wearing diapers and chained at the wrist. (Photo A. Fourage)

In Madagascar, where the photo prop trade overlaps with pet keeping (Chap. 14, this volume), lemurs are often kept in cages in front of homes, hotels or businesses. Nearly half of owners said that these 'pets' were used to generate income from tourists (LaFleur et al., 2019). A 2016 analysis of the welfare of captive lemurs in Madagascar found that the majority were fed unnatural diets and were in bad health (53%), kept in small cages (72%) and restrained by leashes and chains (67%) (Reuter & Schaefer, 2016). In the Colombian and Peruvian Amazon, photo opportunities with primates are often offered by hotels, and primates are kept as attractions for viewing, petting or feeding in bars, where tourists or staff may offer cigarettes or alcoholic drinks to the animals (S. Shanee, personal observation). The photo prop and pet trade industries often overlap, with individual owners and wildlife vendors offering their monkeys as photo props to generate additional income (Shanee & Shanee, 2021).

In popular tourist destinations such as beaches, bars and street markets, primates are illegally touted as photo props or offered for sale. Animals must often work long hours in crowded, hot conditions with no access to food or water (Stazaker & Mackinnon, 2018). Some are drugged and others have teeth removed to prevent injury to tourists, handlers or cage mates. Osterberg and Nekaris (2015) surveyed the illegal use of slow lorises (*Nycticebus* spp.) as photo props at PaTong, a resort town in Puket, Thailand. Tourists paid between USD $3.15 and $15.70 to have their photographs taken with slow lorises. Gibbons are used for similar purposes in Thai tourist resorts (Osterberg & Nekaris, 2015). Grey (2012) documented their presence in tourist bars, restaurants and beaches, where **touts** charged an average of USD $6.34 per photo. Osterberg and Nekaris suggested that an increase in the use of lorises may have been due to difficulty in obtaining baby gibbons, as lorises had not been a well-established part of the photo prop trade prior to the study (Osterberg & Nekaris, 2015). In Turkey, tourists could take photos with slow lorises while playing with them and feeding them fruit (Kitson & Nekaris, 2017). In Morocco, people could purchase photo opportunities with Barbary macaques (*M. sylvanus*). Stazaker and MacKinnon (2018) reported that young macaques were used for as many as 18 interactions per hour in Marrakech's Djma el Fnaa square. Foreign tourists paid up to USD $12 per transaction, though Moroccan tourists paid less. It is illegal to keep Barbary macaques in Morocco, but the authorities grant special dispensations to photo touts (S. Waters, personal communication). Tourist guides on jungle tours in South America and also in India may temporarily capture animals so that tourists can pose for 'wild selfies' (Kanagavel et al., 2013; WAP, 2020).

Myrtle Beach Safari in South Carolina, USA, offers encounters and photo sessions with gibbons and infant chimpanzees (*Pan troglodytes*), at prices ranging from USD $100 to $789 (Myrtle Beach Safari, 2021). The Three Monkey's [sic] Photo Emporium in Tennessee describes itself as 'an award winning, one-of-a-kind experiential photo studio … Our three Capuchin Monkeys, Sasha, Cody & Lola look forward to interacting with you during a fun photo session' (Three Monkey's Photo Emporium, 2021). The owner of the business previously ran a cafe in which visitors could interact with monkeys but was forced to shut after a child was bitten in 2014 (Hlavaty, 2015). In 2019, at Armathwaite Hall Hotel and Spa in the UK's Lake District, tourists were invited to do yoga with ring-tailed lemurs (*Lemur catta*) as part of a wellness programme (Romano, 2019). In March 2021, hotel staff confirmed by telephone that the 'lemoga' programme is no longer running but that it may be revisited in the future.

Between 1926 and 1972, London Zoo regularly held 'chimpanzee tea parties' in which resident chimpanzees were dressed in clothes and fed tea and cakes (Allen et al., 1994). Such events were common in western zoos until the 1960s and 1970s, and similar events still occur in other countries, such as Indonesia (Klook Travel, 2021), China and Vietnam (T. Nguyen and H. Wang, personal communication). In Japan, the NOAH Inner City Zoo in Yokohama allows tourists, for a fee, to take photos with the animals (Musing et al., 2015). Even in modern zoos that ascribe to ethical standards, D'Cruze et al. (2019) reported that opportunities to hand feed or pet wild animals, including primates, and walk-through enclosures were still prevalent. Accredited facilities, too, offer such opportunities. For example, both VinPearl Safari Park in Vietnam (a member of the World Association of Zoos and Aquariums (WAZA)) and Howlett's Zoo in the UK (a member of the European Association of Zoos and Aquaria (EAZA)) offer opportunities to interact with and hand-feed lemurs (Hoàng, 2021; Howlett's Zoo, 2021). The authors suggest that walk-through exhibits may increase the public's likelihood of wanting wild animals as pets and note that in many of these institutions, appropriate accompanying information on species welfare and conservation issues is lacking.

17.3 Primates on Screen

17.3.1 Film

Primates have performed on screen since the early twentieth century. A capuchin monkey appeared in the 1928 Buster Keaton film *The Cameraman* (Fragaszy & Visalberghi, 2004). The Tarzan movies of the 1930s featured a chimpanzee named Cheeta, most likely played by several different chimpanzees (Child, 2011). In 1951, a chimpanzee named Bonzo appeared in *Bedtime for Bonzo*. In fact, following her import to the USA from Liberia in 1948, she appeared in film and television under at least four different names (Molloy, 2011). The use of great apes has not been restricted to chimpanzees. *Every Which Way but Loose* (1978) and its sequel *Any Which Way You Can* (1980) featured an orangutan character named Clyde, played by three separate orangutans (IMDB, 2021a, b). The film *Gorillas in the Mist* included footage of wild gorillas in Rwanda, and primatologists were hired to oversee the lead actor's close proximity to the animals (Rule, 1988). The Walt Disney Company introduced a policy in 2012 prohibiting the use of great ape actors, additionally stating that 'other large primates' will not be used, specifying baboons and macaques (Leighty et al., 2015; Walt Disney Company, 2012). The company has, however, used other primate actors, such as capuchins, in their productions since this policy was introduced (Aldrich, 2018).

A study of primate actors in wide-release English language films found that from 1990 to 2013, chimpanzees (*Pan troglodytes*), capuchins (*Cebus* and *Sapajus* spp.) and cercopithecines (*Macaca* and *Papio* spp.) were frequently used as actors (Aldrich, 2018). Comparison to an earlier, unpublished study (Gil Vazquez & Nekaris, unpublished) indicated that the use of orangutans (*Pongo* ssp.) had decreased over time. Aldrich (2018) found no significant increase or decrease in the frequency with which primate actors in general were used during the study period. Aldrich also analysed content in 38 film trailers that included primate actors. The actors were shown 'grinning' for 19% of the time they were on-screen. Bare-teeth displays often signal fear or submission in chimpanzees, capuchins, macaques and other primates (Preuschoft & Van Hooff, 1995), but in the context of the trailers, always signified pleasure or humour. The primate actors were clothed for 50% of their time on-screen and shown in anthropogenic environments 87% of the time. The presence of such behaviours is of concern as such portrayal can foster false understanding about the suitability of primates as human companions and about their conservation status (Leighty et al., 2015; Ross et al., 2011; Schroepfer et al., 2011). Other inaccuracies, for example, about behaviour, diet or distribution, may also lead to misunderstandings.

Primate actors have appeared in cinema around the world. In Vietnam, a film entitled *Con Cu Li Bé Nhỏ (The Little Slow Loris)* (2014) featured a live loris and touched upon issues of illegal wildlife trade and the use of lorises in folk medicine. The film ended with the loris escaping and returning to the wild – into a potentially inappropriate habitat, and at risk of disease (Thạch et al., 2018). India's highest grossing film of 1993, *Aankhen*, featured a performing macaque (Dhawan, 1993; Newsable, 2018). In 2019, The Animal Welfare Board of India began regulating the use of animals in Indian cinema (Jha, 2019), but it is unclear whether this will reduce the country's use of primate actors. Primates continue to be used in harmful ways in US film and television, despite the American Humane Association's long-standing monitoring of the use of animals in film and television (AHA, 2021; Aldrich, 2018; Geary, 2019).

17.3.2 Television

Primate actors are also regularly used in live-action television. *BJ and the Bear* (1978–1981) featured a chimpanzee (Bear) who was played by several different chimpanzees, one of whom later attacked and injured his owner (ABC News, 2006). The successful sitcom *Friends* (1994–2004) featured a white-fronted capuchin monkey (*Cebus capucinus*) in eight episodes. More recently, the US television miniseries *I Know This Much is True* (2020) featured a character played by Crystal, a capuchin monkey who has appeared on-screen in many other productions (Clarke, 2020). The 2020 Netflix show *Ratched* featured a clothed white-fronted capuchin in constant contact with the main human character. At the time of writing, two BBC drama series feature monkeys in close contact with people: *The Serpent* (featuring a collared macaque) and *The Terror* (featuring a clothed capuchin monkey).

Television advertisements have also featured primates. Inspired by the London Zoo tea parties, the PG Tips advertisements, a staple feature of British television between 1957 and 2002, featured Louis and Choppers, residents of Twycross Zoo. They were dressed in human clothing and voiced over for the advertisements. In 2014, Twycross Zoo admitted that it had been ethically problematic to use the chimpanzees in such a way and that the animals had difficulty interacting with other chimpanzees or otherwise behaving naturally (Heath, 2014).

17.3.3 Animated Representations and Anthropomorphism

Animated representations of primate characters are ubiquitous and have long been a media staple for children, including popular cartoons, puppets and video games. Animated primates are typically anthropomorphised and characterised as 'mischievous' (such as Abu from *Aladdin* (1992)) or 'eccentric' (such as King Louie of the *Jungle Book* (1967, 1995) or *The Lion King*'s Rafiki (1994)). Although some characters are drawn in broad generalisations that do not correspond to actual species, animations have included species from across the primate taxa, including platyrrhines, catarrhines and strepsirrhines (e.g. Abu the capuchin monkey, in Disney's *Aladdin* (1992), lemurs from the film *Madagascar* (2005), the sifaka from the television show *Zoboomafoo*, and galagos from the TV series *The Bush Baby* (1992)). Even in 2021, a pet slow loris holding a tiny umbrella is the only 'exotic' pet amongst a series of domestic pets in Fox's cartoon series HouseBroken. More recently, computer-generated imagery (CGI) has been used to create more detailed, realistic renderings of primates in films such as *War for the Planet of the Apes* (2017) and *Godzilla* vs *Kong* (2021). This technology allows filmmakers to include primate characters in live-action productions without using live primate actors and has therefore been suggested as a cruelty-free alternative for the industry (Craig, 2018). The popularity of such animated characters has also extended to other media, including video games and augmented reality.

Video games have produced arguably one of the most famous and influential animated primate characters – Donkey Kong. In 1980, *PacMan* became the first video game to feature named characters, paving the way for the success of *Donkey Kong* in 1981 (DeMaria, 2018). Forty years later, *Donkey Kong* remains one of the most recognisable game characters. *Donkey Kong* was designed to be 'comical and entertaining' such that 'the farcical, childlike and nonsexual *Donkey Kong* creates a humorous impression' while in contrast, King Kong is 'a ferocious gorilla in search of a beautiful woman who goes on rampages…culminating in his tragic and bloody death' (SDNY, 1983).

By comparison, the monkey characters in the critically acclaimed game *Ape Escape* were described as 'adorable-as-can-be', with the reviewer stating 'there is something extremely cute about those damn monkeys. With their goofy little helmets, their puny little screeches' (Perry, 1999). Finally, a review of a mobile phone game featuring a troupe of small monkeys stated, 'the beautifully animated monkeys are everything cartoon monkeys should be – nimble, mischievous and charming' (Mundy, 2009). These characteristics are congruent with the anthropomorphic traits seen in cartoons and animations and represent the most common perception of primate behaviour. While it may seem irrelevant to discuss fictional primate characters, the personality traits attributed to both the animated representations discussed above and the game versions provide key insights into public perceptions of the primate behaviour, and indeed what to expect from an interaction with a live primate, particularly given that many of these characters regularly interact with humans within their storylines.

17.4 Primates Online

17.4.1 Social Media

Since the age of social media began in 2005–2006, sites like MySpace, Twitter, Facebook and YouTube have created a social landscape that allows individuals to upload, share and discuss text, image and video content. The availability of these forms of media has significantly altered how people access and process information. In 2020, over half of the global population were active Internet users and approximately 3.6 billion were users of social networking sites (Statista, 2021). The rise of social media provided the entertainment industry with a platform to offer free, widespread access to previously difficult-to-access content. Films and television shows like *Tarzan* could be uploaded to YouTube, for example, and reach a much wider audience, or images of *topeng monyet* could be shared, spreading awareness about different cultural practices. One of the major opportunities that social media provided was the ability for individuals to upload content that they created themselves, without requiring specialised equipment or large budgets.

Engagement behaviours such as liking, sharing and commenting are the currency of social media, as these promote channels and increase audiences. These behaviours can provide or increase advertising revenue for monetised channels (Hollebeek et al., 2014). Social media profiles generally promote the most socially desirable aspects of a user's life, which has created a demand for ever more elaborate 'selfies', photographs taken by the subject (Kurniawan et al., 2013). This drive has led to the proliferation of 'wildlife selfies' in which the subject is in close proximity to or in contact with wildlife (Lenzi et al., 2020). Resulting from interactions like those discussed above, these photos document animal encounters in facilities like tourist centres, zoos or rescue centres, or wild animal encounters and photo prop experiences. Photos of users holding slow lorises, for example, became popular on the site Instagram after tourists in Thailand were encouraged to take them by touts at beaches, nightlife hotspots and hotels (Osterberg & Nekaris, 2015). The popularity of such photos, including those taken and posted on social media by celebrities, further promoted the photo prop trade and encouraged the behaviour to be repeated in other tourist destinations such as Turkey (Kitson & Nekaris, 2017). Many social media channels offer no means to report illegal ownership or trading of primates, and limited means for reporting animal abuse. In 2017, Instagram became the first social media platform to provide an alert system warning of images promoting animal cruelty when certain animal hashtags are searched for (Instagram, 2021). To date, it does not appear that any further action has been taken, and cruelty content is still easy to find on Instagram.

17.4.2   YouTube Channels

Videos of pet primates are very popular on social media, and several have gone viral, including a video of a pygmy slow loris (*N. pygmaeus*) which yielded millions of views across multiple platforms (Nekaris et al., 2013). Many celebrities shared and commented positively on the video. Videos of pet primates often feature animals that are owned illegally and frequently show the poor conditions in which the animals are kept. Nekaris et al. (2015) assessed 100 videos of pet slow lorises on video sharing sites and found that all showed at least one instance of compromised animal welfare, with one third of videos violating all five freedoms (see Sect. 17.1). Perhaps more troubling is the existence of channels that profit from the deliberate abuse of animals, showing content such as scaring animals with firecrackers, forcing them to walk bipedally, or in severe cases physically assaulting or killing them (Knox, 2021). Account owners profit from such content both through the revenue collected from advertisements and by soliciting viewers to send money as 'donations'. Via commenting, viewers can make suggestions for new ways to 'discipline' or otherwise hurt the primates. While some platforms have community standards that prohibit such content, they rely on the videos being reported by viewers and often employ a 'three strikes policy' that allow channels to continue posting content even after being flagged (Google, 2021).

17.4.3   Fan Accounts

Through social media, we see the rise of the 'primate celebrities' as personalities with their own fanbases. Primate actors like Crystal the Monkey, discussed in Sect. 17.3.2, have accounts with which their keepers and trainers can share behind-the-scenes content. Pet owners can create accounts for their animals, creating dedicated fan bases that can be monetised through advertising and merchandise. Social media platform TikTok allows users to upload short video content to the site. The most popular primate accounts on the site have millions of followers. As of February 2021, channel *Heresyourmonkeycontent*, which features a pet capuchin, has 11.2 million followers; *Mojothemonkey*, featuring a pet vervet monkey has 2.4 million; and *Monkeykeer*, featuring an infant macaque, has 1.1 million. Their content typically includes videos of the monkeys playing, performing anthropomorphic behaviours or interacting with humans and other animals. In contrast to the deliberate abuse channels discussed in Sect. 17.4.2, cruelty may not be intended by the creators of such content. However, the welfare issues inherent to the primate pet trade are often apparent in the animals' behaviours or condition, or the visibly inappropriate social or physical environments in which they are kept (Chap. 14, this volume). In most cases, the animals are attributed positive emotions such as 'loving' the activities or 'having fun'.

Fig. 17.3 The *Caring Monkey* Snapchat filter places a virtual macaque on the shoulder of the user that performs grooming behaviours. (Photo K. Feddema)

17.4.4 Augmented Reality

Augmented reality technology such as Snapchat and Instagram filters and virtual backgrounds on video conferencing software have created opportunities for simulated primate encounters. The *Caring Monkey* lens for Snapchat, for example, places a realistic virtual macaque on the shoulder of the user and then simulates grooming behaviours (Fig. 17.3). Video conferencing software allows users to select backgrounds that place them in contact with primates, with one example showing the user next to a capuchin monkey 'waiter' tending a bar (Gray Malin Fine Art Photography, 2021).

These augmented reality options raise some ethical questions. On the one hand, like CGI and animations, virtual primates provide an alternative that replaces contact with animals and reduces health and welfare concerns. On the other hand, the continued use of primate imagery in this way may normalise contact with primates and therefore drive consumer demand for primate experiences 'in real life' (Ross et al., 2011). At present, there is no evidence as to what impact virtual reality will play in the future of primates in entertainment; however, if their virtual counterparts can become indistinguishable from the 'real deal', it may obviate the use of primate photo prop animals.

17.5 Discussion

The use of primate performers and the sharing of primate imagery online can result in negative animal welfare, conservation and human health outcomes (Aldrich, 2018; Feddema & Nekaris, 2020; Waters et al., 2021). In order to be used as sources of entertainment, primates are often isolated from others of their own kind and may be taken from the wild and/or smuggled out of their range countries. When we consider the tenets of the Five Freedoms of animal welfare and the guidelines of the International Society of Primatologists, doubtless many primates used in entertainment are subject to some form of mistreatment.

It is sometimes possible to determine from images and videos whether welfare is explicitly compromised during filming (Nekaris et al., 2015). In order to create content, animals are often placed in human contexts or portrayed exhibiting human behaviours. This impinges on the animal's freedoms in accordance with the Five Freedoms model of animal welfare (Webster, 2016). Forcing animals to perform unnatural behaviour disregards their freedom to express natural behaviour. Animals may be trained using aversive techniques which can violate the freedom from pain and injury and the freedom from fear and distress. In addition, animals may be housed or handled in an inappropriate manner, such as in isolation from conspecifics or in barren environments without appropriate nutrition. This neglects an animal's freedom from discomfort and freedom from hunger and thirst (IPS, 2008).

Seeing primates in a human context can increase people's desire for one as a pet and encourage the belief that primates make suitable companions (Leighty et al., 2015; Ross et al., 2008; Schroepfer et al., 2011). Additionally, it may reflect the Anthropogenic Allee Effect, which describes when both price and interest increase with species rarity. This phenomenon has been shown to occur within the illicit pet trade and can be made worse when images on social media associate ownership with social status and increase desirability for the species (Morgan & Chng, 2018; Siriwat et al., 2019). This creates both an opportunity and a cost for conservationists. On the one hand, the desire to show off rare and charismatic species may provide law enforcement and species protection regulators the ability to detect and prosecute the illegal ownership and trafficking of animals (Nekaris et al., 2013). On the other hand, desire for internet fame and revenue through photos with rare species may increase demand for exotic species in the pet trade or support the lucrative photo prop industry (Kitson & Nekaris, 2017).

Social media provides the ability to engage with content through uploading comments, expressing reactions and sharing content (Hollebeek et al., 2017). Discussions around primate content offer insight into the perception of primates in entertainment. Although limited, research suggests that responses are contextual and influenced by the users' social networks, personal beliefs and level of knowledge (Feddema et al., 2021; Riddle & MacKay, 2020). Online platforms therefore present potential sources of misinformation on primate welfare and conservation if not corrected by the platform or other parties (Thaler & Shiffman, 2015). For example, primate 'influencers' may mislead potential owners about appropriate husbandry by promoting inappropriate diet, housing or enrichment.

The desire for wildlife selfies and videos has been shown to increase risk-taking behaviours (Cherry et al., 2018; Pagel et al., 2020). The drive for 'like currency' may result in people getting too close to wild animals, interacting with dangerous species or provoking aggression. An increase in encounters with wild animals increases the risk of disease transmission, injury or even death (Carne et al., 2017; Gautret et al., 2007; Muehlenbein, 2017). The use of primates as entertainers therefore not only poses significant risk to the animals involved but also to human health through increased opportunities for zoonotic and anthropozoonotic disease transfer. This process can have similar effects to disease transfer in the illegal wildlife trade and in primate-based wildlife tourism (Bezerra-Santos et al., 2021; Devaux et al., 2019; Muehlenbein & Wallis, 2014).

The use of primates in entertainment, whether in public spaces or on the small or large screen, stems from a historically deep-rooted global phenomenon. Its consequences for welfare and conservation are largely negative. Understanding the history of primates within the media landscape from a sociocultural standpoint is critical to regulating these practices and ensuring the health and welfare of both humans and animals. While approximately 60% of the world's primate species are threatened with extinction, and closer to 75% of primate populations are in decline (Estrada et al., 2017), viewing primates in anthropomorphic settings has been shown to reduce the public's perception of this threat (Leighty et al., 2015; Ross et al., 2011). Conservationists should consider the role of the media and other forms of entertainment in shaping the narrative around primates and their conservation status in order to raise public interest and concern.

Acknowledgements The authors would like to thank the editors of this volume, Tracie McKinney, Siân Waters and Michelle Rodrigues, for the opportunity to contribute. We are also grateful to the anonymous reviewer whose input has been valuable. SS thanks his teammates at Neotropical Primate Conservation, as well as the wildlife authorities and rescue centre owners of Peru, Colombia, Bolivia and Ecuador, for their work in confronting the illegal use of wildlife across South America. He is also grateful to the many institutions who have funded his work. KF would like to acknowledge the Whadjuk Noongar nation, the traditional owners of the land and waterways on which she lives and works – land whose sovereignty was never ceded.

References

ABC News. (2006). *Chimp-loving duo sees their darker side.* ABC News. https://abcnews.go.com/Primetime/Science/story?id=691497&page=1. Accessed 6 Mar 2021.

Agoramoorthy, G., & Hsu, M. J. (2005). Use of nonhuman primates in entertainment in Southeast Asia. *Journal of Applied Animal Welfare Science, 8*, 141–149.

AHA. (2021). *No animals were harmed.* American Humane Society. https://www.americanhumane.org/initiative/no-animals-were-harmed/. Accessed 22 Feb 2021.

Aldrich, B. (2014). *All fair video footage.*

Aldrich, B. C. (2018). The use of primate "actors" in feature films 1990–2013. *Anthrozoös, 31*, 5–21. https://doi.org/10.1080/08927936.2018.1406197

Allen, J. S., Park, J., & Watt, S. L. (1994). The chimpanzee tea party: Anthropomorphism, orientalism, and colonialism. *Visual Anthropology Review, 10*, 45–54.

Beastall, C. A., Bouhuys, J., & Ezekiel, A. (2016). *Apes in demand for zoo and wildlife attractions in Peninsular Malaysia and Thailand*. TRAFFIC.

Bezerra-Santos, M. A., Mendoza-Roldan, J. A., Thompson, R. C. A., et al. (2021). Illegal wildlife trade: A gateway to zoonotic infectious diseases. *Trends in Parasitology, 37*, 181–184. https://doi.org/10.1016/j.pt.2020.12.005

Born Free Foundation. (2014). *Circuses and performing animals*. Born Free Foundation. http://www.bornfree.org.uk/campaigns/zoo-check/circuses-performing-animals/. Accessed 9 Jun 2014.

Bradshaw, G. A., Schore, A. N., Brown, J. L., et al. (2005). Elephant breakdown. *Nature, 433*, 807–807. https://doi.org/10.1038/433807a

Carne, C., Semple, S., MacLarnon, A., et al. (2017). Implications of tourist–macaque interactions for disease transmission. *EcoHealth, 14*, 704–717.

Cherry, C., Leong, K. M., Wallen, R., & Buttke, D. (2018). Risk-enhancing behaviors associated with human injuries from bison encounters at Yellowstone National Park, 2000–2015. *One Health, 6*, 1–6.

Child, B. (2011). *Chimp claimed as Cheetah from the Tarzan films dies*. The Guardian.

Clarke, S. (2020). *The monkey from "Community" and "The Hangover Part II" makes an insane $12,000 per episode*. Showbiz Cheat Sheet. https://www.cheatsheet.com/entertainment/the-monkey-from-community-and-the-hangover-part-ii-makes-an-insane-12000-per-episode.html/. Accessed 22 Feb 2021.

Cohen, E. (2009). The wild and the humanized: Animals in Thai tourism. *Anatolia, 20*, 100–118. https://doi.org/10.1080/13032917.2009.10518898

Craig, L. E. (2018). *Putting primates on screen is fuelling the illegal pet trade*. The Conversation. http://theconversation.com/putting-primates-on-screen-is-fuelling-the-illegal-pet-trade-91995. Accessed 28 Mar 2021.

Ćurko, B. (2020). Bioethical questions of animals in sport. *Pannoniana, 4*, 143–153.

D'Cruze, N., Khan, S., Carder, G., et al. (2019). A global review of animal–visitor interactions in modern zoos and aquariums and their implications for wild animal welfare. *Animals, 9*, 332.

DeMaria, R. (2018). *High score! Expanded: The illustrated history of electronic games* (3rd ed.). Routledge.

Devaux, C. A., Mediannikov, O., Medkour, H., & Raoult, D. (2019). Infectious disease risk across the growing human-non human primate interface: A review of the evidence. *Frontiers in Public Health, 7*, 305. https://doi.org/10.3389/fpubh.2019.00305

Dhawan, D. (1993). *Aankhen*. Chiragdeep International.

Dorloh, S. (2017). The protection of animals in Thailand–An insight into animal protection legislation. *International Journal of Humanities and Cultural Studies, 4*, 58–63.

Dykes, J. (2020). *The show can't go on: The fight for an EU-wide ban on wild animals in circuses*. Geographical Magazine. https://geographical.co.uk/nature/wildlife/item/3734-the-show-cant-go-on-can-we-bring-in-an-eu-wide-ban-on-wild-animals-in-circuses. Accessed 6 Mar 2021.

Emigh, J. (1996). *Masked performance: The play of self and other in ritual and theatre*. University of Pennsylvania Press.

Estrada, A., Garber, P. A., Rylands, A. B., et al. (2017). Impending extinction crisis of the world's primates: Why primates matter. *Science Advances, 3*, e1600946. https://doi.org/10.1126/sciadv.1600946

Feddema, K., & Nekaris, K. A. I. (2020). Online imagery and loris conservation. In A. M. Burrows & K. A. I. Nekaris (Eds.), *Evolution, ecology and conservation of lorises and pottos* (pp. 362–373). Cambridge University Press.

Feddema, K., Harrigan, P., & Wang, S. (2021). The dark side of social media engagement: An analysis of user-generated content in online wildlife trade communities. *Australasian Journal of Information Systems, 25*, 1–35. https://doi.org/10.3127/ajis.v25i0.2987

Fragaszy, D. M., & Visalberghi, E. (2004). *The complete capuchin: The biology of the genus Cebus*. Cambridge University Press.

Gautret, P., Schwartz, E., Shaw, M., et al. (2007). Animal-associated injuries and related diseases among returned travellers: A review of the GeoSentinel Surveillance Network. *Vaccine, 25*, 2656–2663.

Geary, D. (2019). Only a couple animals died during the production of this film. *Cinesthesia, 10*, 1–11.

Gil Vazquez, A., & Nekaris, K. A. I. (unpublished). *Primates in movies: Every which way but wild*.

Google. (2021). *Violent or graphic content policies*. YouTube Terms Service. https://support.google.com/youtube/answer/2802008. Accessed 11 Mar 2021.

Gotsis, T. (2018). *Exotic animals in circuses*. NSW Parliamentary Research Service.

Gray Malin Fine Art Photography. (2021). *Brand new: Travel from home with our new zoom conference backgrounds*. Gray Malin. https://www.graymalin.com/lifestyle/travel-from-home-with-our-new-zoom-conference-backgrounds. Accessed 28 Mar 2021.

Grey, S. (2012). Conservation difficulties for *Hylobates lar*: White-handed gibbons and Thailand's illegal pet trade. *Consortium: A Journal of Crossdisciplinary Inquiry, 32*, 45–59.

Hansen, M. F., Gill, M., Nawangsari, V. A., et al. (2021). Conservation of long-tailed macaques: Implications of the updated IUCN status and the CoVID-19 pandemic. *Primate Conservation, 35*, 1–11.

Heath, N. (2014). *PG tips chimps: The last of the tea-advertising apes*. BBC News.

Hlavaty, C. (2015). *Popcorn shop monkeys quarantined after biting child, but whose fault was it?* Chron. https://www.chron.com/neighborhood/bayarea/news/article/Popcorn-shop-monkeys-quarantined-after-biting-6511594.php. Accessed 22 Feb 2021.

Hoàng, H. (2021). *Mê mẩn ngắm sư tử con, hổ… và những góc ảnh dễ thương tại vườn thú bán hoang dã Phú Quốc*. danviet.vn. https://danviet.vn/me-man-ngam-su-tu-con-ho-va-nhung-goc-anh-de-thuong-tai-vuon-thu-ban-hoang-da-phu-quoc-20210208085053959.htm. Accessed 28 Aug 2021.

Hollebeek, L. D., Glynn, M. S., & Brodie, R. J. (2014). Consumer brand engagement in social media: Conceptualization, scale development and validation. *Journal of Interactive Marketing, 28*, 149–165.

Hollebeek, L. D., Juric, B., & Tang, W. (2017). Virtual brand community engagement practices: A refined typology and model. *Journal of Services Marketing, 31*, 204–217.

Howletts Zoo. (2021). *Animal encounters – Feed your favourite animal at Howletts*. The Aspinall Foundation. https://www.aspinallfoundation.org/howletts/experiences/animal-experiences/animal-encounters/. Accessed 28 Aug 2021.

IMDB. (2021a). *Any Which Way You Can* (1980) – IMDb.

IMDB. (2021b). *Every Which Way but Loose* (1978) – IMDb.

Instagram. (2021). *How do I learn more about wildlife exploitation?* Instagram Help Center. https://help.instagram.com/859615207549041. Accessed 21 Mar 2021.

IPS. (2008). *Opposition to the use of nonhuman primates in the media*. International Primatological Society. http://www.internationalprimatologicalsociety.org/OppositionToTheUseOfNonhumanPrimatesInTheMedia.cfm. Accessed 14 Feb 2021.

JAAN. (2015). *Indonesia bebas topeng monyet*. Jakarta Animal Aid Network. https://www.jakartaanimalaid.com/domesticcampaigns/free-dancing-monkeys/. Accessed 21 Feb 2021.

Jacobson, S. L., Ross, S. R., & Bloomsmith, M. A. (2016). Characterizing abnormal behavior in a large population of zoo-housed chimpanzees: Prevalence and potential influencing factors. *PeerJ, 4*, e2225. https://doi.org/10.7717/peerj.2225

Jha, L. (2019). *Producers guild and animal board to govern animal use in cinema, TV*. Mint. https://www.livemint.com/news/india/producers-guild-and-animal-board-to-govern-animal-use-in-cinema-tv-1557228357350.html. Accessed 21 Feb 2021.

Kanagavel, A., Sinclair, C., Sekar, R., & Raghavan, R. (2013). Moolah, misfortune or spinsterhood? The plight of slender loris *Loris lydekkerianus* in southern India. *Journal of Threatened Taxa, 5*, 3585–3588.

Kencana, S. (2017). *Sudah Ada Sejak Masa Kolonialisme, Beginilah Sejarah Topeng Monyet di Indonesia*. Yukepo.com. https://www.yukepo.com/hiburan/indonesiaku/sudah-ada-sejak-masa-kolonialisme-beginilah-sejarah-topeng-monyet-di-indonesia. Accessed 22 Feb 2021.

Kerr, M. (2017). *Great apes in Asian circus-style shows on rise – So is trafficking*. Mongabay Environ. News. https://news.mongabay.com/2017/04/great-apes-in-asian-circus-style-shows-on-rise-so-is-trafficking/. Accessed 28 Mar 2021.

Kitson, H., & Nekaris, K. A. I. (2017). Instagram-fuelled illegal slow loris trade uncovered in Marmaris, Turkey. *Oryx, 51*, 394.

Klook Travel. (2021). *Breakfast with Orangutans at Bali Zoo*. Klook Travel. https://www.klook.com/activity/7800-breakfast-with-orang-utan-in-zoo-bali/. Accessed 4 Sept 2021.

Knox, P. (2021). *Monkeys 'tortured' for YouTube views*. NewsComAu. https://www.news.com.au/technology/science/animals/youtube-videos-showing-monkey-abuse-slammed-by-peta-action-for-primates/news-story/6e7d28a1a61ea0d521d8c948c0f6b28d#.g6ctw. Accessed 11 Mar 2021.

Kurniawan, Y., Habsari, S. K., & Nurhaeni, I. D. A. (2013). Selfie culture: Investigating the patterns and various expressions of dangerous selfies and the possibility of government's intervention. *Proceedings of the Second Journal Government Policies International Conference*, 324–332.

LaFleur, M., Clarke, T. A., Reuter, K. E., et al. (2019). Illegal trade of wild-captured *Lemur catta* within Madagascar. *Folia Primatologica, 90*, 199–214. https://doi.org/10.1159/000496970

Leighty, K. A., Valuska, A. J., Grand, A. P., et al. (2015). Impact of visual context on public perceptions of non-human primate performers. *PLoS One, 10*, e0118487. https://doi.org/10.1371/journal.pone.0118487

Lennox, L. W. P. (1860). *Pictures of sporting life and character*. Hurst and Blackett.

Lenzi, C., Speiran, S., & Grasso, C. (2020). "Let me take a selfie": Implications of social media for public perceptions of wild animals. *Society and Animals*, 1–20. https://doi.org/10.1163/15685306-bja10023

Lopresti-Goodman, S. M., Kameka, M., & Dube, A. (2013). Stereotypical behaviors in chimpanzees rescued from the African bushmeat and pet trade. *Behavioural Sciences, 3*, 1–20. https://doi.org/10.3390/bs3010001

Mallonee, J. S., & Joslin, P. (2004). Traumatic stress disorder observed in an adult wild captive wolf (*Canis lupus*). *Journal of Applied Animal Welfare Science, 7*, 107–126. https://doi.org/10.1207/s15327604jaws0702_3

Ministry of Tourism and Sports. (2021). *International tourist arrivals*. กระทรวงการท่องเที่ยวและกีฬา [Ministry of Tourism and Sports]. https://www.mots.go.th/more_news_new.php?cid=521. Accessed 28 Mar 2021.

Molloy, C. (2011). *Popular media and animals*. Palgrave Macmillan.

Morgan, J., & Chng, S. (2018). Rising internet-based trade in the Critically Endangered ploughshare tortoise *Astrochelys yniphora* in Indonesia highlights need for improved enforcement of CITES. *Oryx, 52*(4), 744–750. https://doi.org/10.1017/S003060531700031X

Muehlenbein, M. P. (2017). Primates on display: Potential disease consequences beyond bushmeat. *American Journal of Physical Anthropology, 162*, 32–43.

Muehlenbein, M. P., & Wallis, J. (2014). Considering risks of pathogen transmission associated with primate-based tourism. In A. E. Russon & J. Wallis (Eds.), *Primate tourism: A tool for conservation?* (pp. 278–287). Cambridge University Press.

Mundy, J. (2009). *Tropical towers*. Pocket Gamer. https://www.pocketgamer.com/articles/011773/tropical-towers/. Accessed 11 Mar 2021.

Musing, L., Suzuki, K., & Nekaris, K. A. I. (2015). Crossing international borders: The trade of slow lorises, *Nycticebus* spp., as pets in Japan. *Asian Primates Journal, 5*, 12–24.

Myrtle Beach Safari. (2021). *Myrtle beach safari*. https://myrtlebeachsafari.com/. Accessed 21 Feb 2021.

Nekaris, B. K. A.-I., Campbell, N., Coggins, T. G., et al. (2013). Tickled to death: Analysing public perceptions of 'cute' videos of threatened species (slow lorises – *Nycticebus* spp.) on Web 2.0 sites. *PLoS One, 8*, e69215. https://doi.org/10.1371/journal.pone.0069215

Nekaris, K. A. I., Musing, L., Vazquez, A. G., & Donati, G. (2015). Is tickling torture? Assessing welfare towards slow lorises (*Nycticebus* spp.) within web 2.0 videos. *Folia Primatologica, 86*, 534–551. https://doi.org/10.1159/000444231

Newsable. (2018). *These 7 Hindi movies showed animal cruelty and no questions were asked!* Asianet News Network Pvt Ltd. https://newsable.asianetnews.com/entertainment/7-films-showed-animal-cruelty-n-no-questions-were-asked. Accessed 21 Feb 2021.

Norconk, M. A., Atsalis, S., Tully, G., et al. (2020). Reducing the primate pet trade: Actions for primatologists. *American Journal of Primatology, 82*, e23079. https://doi.org/10.1002/ajp.23079

Osterberg, P., & Nekaris, K. A. I. (2015). The use of animals as photo props to attract tourists in Thailand: A case study of the slow loris *Nycticebus* spp. *Traffic Bulletin, 27*, 13.

Oxford Dictionary. (2021). *Definition of ENTERTAIN*. Lexico.com. https://www.lexico.com/definition/entertain. Accessed 28 Mar 2021.

Pagel, C. D., Orams, M., & Lück, M. (2020). # BiteMe: Considering the potential influence of social media on in-water encounters with marine wildlife. *Tourism in Marine Environments, 15*, 249–258.

Peres, C. A. (1991). Humboldt's woolly monkeys decimated by hunting in Amazonia. *Oryx, 25*(2), 89–95.

Perry, D. (1999). *Ape Escape: Innovative control and traditional platform aspects merge in the PlayStation's best platformer ever*. https://www.ign.com/articles/1999/06/24/ape-escape

PETA. (2021). *Elephants abused for tourists' cheap photos and entertainment*. People for the Ethical Treatment of Animals. https://investigations.peta.org/samutprakan-zoo-animals-abused/. Accessed 28 Mar 2021.

Preuschoft, S., & Van Hooff, J. (1995). Homologizing primate facial displays: A critical review of methods. *Folia Primatologica, 65*, 121–137.

Qian, X., Tuck, B., & Elizabeth, T. (2018). *Pine County Fair: Attendee and participant assessment and economic value*. University of Minnesota.

Reuter, K. E., & Schaefer, M. S. (2016). Captive conditions of pet lemurs in Madagascar. *Folia Primatologica, 87*, 48–63.

Riddle, E., & MacKay, J. R. (2020). Social media contexts moderate perceptions of animals. *Animals, 10*, 845.

Romano, A. (2019). *You can do outdoor yoga with lemurs at this hotel in England*. Travel & Leisure. https://www.travelandleisure.com/travel-news/lemur-yoga-lemoga-wellness-england. Accessed 14 Feb 2021.

Ross, S. R., Lukas, K., Lonsdorf, E. V., et al. (2008). Inappropriate use and portrayal of chimpanzees. *Science, 319*, 1487.

Ross, S. R., Vreeman, V. M., & Lonsdorf, E. V. (2011). Specific image characteristics influence attitudes about chimpanzee conservation and use as pets. *PLoS One, 6*, e22050. https://doi.org/10.1371/journal.pone.0022050

Rule, S. (1988). *Luring apes from mist to movies*. The New York Times.

Russon, A. E., & Wallis, J. (Eds.). (2014). *Primate tourism: A tool for conservation?* Cambridge University Press.

Schmidt-Burbach, J., Ronfot, D., & Srisangiam, R. (2015). Asian elephant (*Elephas maximus*), pig-tailed macaque (*Macaca nemestrina*) and tiger (*Panthera tigris*) populations at tourism venues in Thailand and aspects of their welfare. *PLoS One, 10*, e0139092. https://doi.org/10.1371/journal.pone.0139092

Schroepfer, K. K., Rosati, A. G., Chartrand, T., & Hare, B. (2011). Use of "entertainment" chimpanzees in commercials distorts public perception regarding their conservation status. *PLoS One, 6*, e26048. https://doi.org/10.1371/journal.pone.0026048

SDNY. (1983). *Universal City Studios, Inc. v. Nintendo Co.*

Shanee, N., & Shanee, S. (2021). Denunciafauna – A social media campaign to evaluate wildlife crime and law enforcement in Peru. *Journal of Political Ecology, 28*(1), 533–552. https://doi.org/10.2458/jpe.2987

Shanee, N., Mendoza, A. P., & Shanee, S. (2017). Diagnostic overview of the illegal trade in primates and law enforcement in Peru. *American Journal of Primatology, 79*, e22516.

Siriwat, P., Nekaris, K. A. I., & Nijman, V. (2019). The role of the anthropogenic Allee effect in the exotic pet trade on Facebook in Thailand. *Journal for Nature Conservation, 51*, 125726.

Space Coast Daily. (2020). *IT'S A HOOT! See monkeys riding dogs at the space coast state fair's banana derby.* Space Coast Daily. https://spacecoastdaily.com/2020/10/its-a-hoot-watch-monkeys-riding-dogs-at-the-space-coast-state-fairs-banana-derby/. Accessed 8 Aug 2021.

Statista. (2021). *Number of social network users worldwide from 2017 to 2025.* Statista. https://www.statista.com/statistics/278414/number-of-worldwide-social-network-users/. Accessed 11 Mar 2021.

Stazaker, K., & Mackinnon, J. (2018). Visitor perceptions of captive, Endangered Barbary Macaques (*Macaca sylvanus*) used as photo props in Jemaa El Fna Square, Marrakech, Morocco. *Anthrozoös, 31*, 761–776. https://doi.org/10.1080/08927936.2018.1529360

Stiles, D., Redmond, I., Cress, D., Nellemann, C., & Formo, R. K. (2013). *Stolen apes: The illicit trade in chimpanzees, gorillas, bonobos and orangutans – A rapid response assessment.* UNEP/UNESCO.

Thạch, H. M., Le, M. D., Vũ, N. B., et al. (2018). Slow loris trade in Vietnam: Exploring diverse knowledge and values. *Folia Primatologica, 89*, 45–62.

Thaler, A. D., & Shiffman, D. (2015). Fish tales: Combating fake science in popular media. *Ocean and Coastal Management, 115*, 88–91.

Thomas, V. C. (2018). For your entertainment: Researching animal cruelty under the big top. *Michigan Bar Journal, 97*, 40–41.

Three Monkey's Photo Emporium. (2021). *Three Monkey's Photo Emporium.* Facebook. https://www.facebook.com/3monkeysphotoemporium. Accessed 14 Feb 2021.

Tyson, E. (2020). *Licensing laws and animal welfare: The legal protection of wild animals.* Palgrave Macmillan.

Wallace, B. (2012). *Why Crystal, the monkey in 'Animal Practice,' is NBC's most valuable commodity.* New York Magazine.

Walt Disney Company. (2012). *Disney's use of live animals in entertainment policy.*

WAP. (2020). *A close up on cruelty: The harmful impact of wildlife selfies in the Amazon.* World Animal Protection.

Waters, S., Setchell, J. M., Maréchal, L., et al. (2021). *Best practice guidelines for responsible images of non-human primates.* IUCN Primate Specialist Group Section on Human Primate Interactions.

Webster, J. (2016). Animal welfare: Freedoms, dominions and "A life worth living". *Animals, 6*, 35. https://doi.org/10.3390/ani6060035

WFFT. (2019). *Statement on animal welfare at Samut Prakarn Croc Farm and Zoo.* Wildlife Friends Foundation of Thailand. https://www.wfft.org/wildlife-general/statement-on-animal-welfare-at-samut-prakarn-croc-farm-and-zoo/. Accessed 28 Mar 2021.

Chapter 18
Conclusion: Twenty-First-Century Primatology

Michelle A. Rodrigues, Siân Waters, and Tracie McKinney

Contents

The future of primatology requires embracing the idea of "shared space" (Lee, 2010) and developing solutions for mutually respectful coexistence. With a growing human population, increasing demands for natural resources, and continued fragmentation of traditional habitats, primates across their range face the challenge of surviving in human-dominated ecosystems. In many cases, their survival within human-dominated and human-managed environments is contingent on human stewardship to protect their populations and manage their welfare. Finding ways to make these relationships work for all species involved is vital both for their welfare and for ours. The futures of human and nonhuman primates are intertwined, and for primatologists, the challenge is in articulating to our fellow humans the need to value, protect, and manage relationships with our distant relatives.

The chapters collected in this volume demonstrate the spectrum of possible anthropogenic influence. We can no longer divide study populations into a simple dichotomy of "wild" versus "disturbed." Rather, there exists a wide range of

---

The original version of this chapter was revised. The correction to this chapter is available at https://doi.org/10.1007/978-3-031-11736-7_19

M. A. Rodrigues (✉)
Department of Social and Cultural Sciences, Marquette University, Milwaukee, WI, USA
e-mail: Michelle.Rodrigues@marquette.edu

S. Waters
Department of Anthropology, Durham University, Durham, UK

T. McKinney
School of Applied Science, University of South Wales, Pontypridd, UK

T. McKinney et al. (eds.), *Primates in Anthropogenic Landscapes*, Developments in Primatology: Progress and Prospects, https://doi.org/10.1007/978-3-031-11736-7_18

anthropogenic influence, from the complete control experienced by animals in captivity, through to the long-term impact we have on intact forests through climate change and other biogeographical processes. The gradations between these extremes highlight the dizzying array of interactions in which humans and our relatives might find ourselves. As described by our contributors, nonhuman primates may compete for space and resources in agricultural landscapes. They may be protected and treated with reverence in religious temples while being viewed as urban pests only meters away. And primates fill a special place in our hearts and minds, as shown by their prevalence in media, zoos, and tourism settings.

Following from our recognition of the kinds and degrees of anthropogenic influence is a growing understanding of the range of habitats used by wild primates. Our contributors have explored the literature on primates living outside of primary forest and found that many landscapes can provide suitable habitat. The importance of matrix landscapes surrounding habitat fragments has been long recognized (Anderson et al., 2007; Cowlishaw & Dunbar, 2000), and our increasing knowledge of landscape supplementation, travel patterns, and ecological plasticity can help us improve the permeability of matrix and increase overall habitat availability for at-risk populations. Likewise, the chapter on regenerating forests highlights the importance of further study on this habitat type. Our review suggests that more primates can make use of regenerating forests than currently recognized and that protecting these young forests is an important strategy for primate conservation. Agricultural landscapes can also provide suitable space for foraging and travel (Estrada et al., 2012; Galán-Acedo et al., 2019). Finally, even urban spaces – which may seem the most alien of landscapes for wild primates – have been used by some genera for millennia. Our review shows that some species can not only survive but can truly thrive in urban landscapes. This coexistence may have shaped our fascination with our primate relatives, which has led us to keeping primates under human care, whether exploitative or enriching, across homes, entertainment venues, laboratories, zoos, and sanctuaries. The threads underlying our complicated relationship dynamics vary across environments, but through the chapters in this book, our authors explore both those common threads and the intricately woven local patterns embedded in cultural contexts.

Our growing understanding of the complexity of primate lives has led primatologists to increasingly diverse research questions and methodologies. Just within the pages of this volume, we have encountered primatologists using traditional field observations, conducting laboratory analyses of hormones, genetics, and other biological markers, and branching into the social sciences. We strongly encourage this mixed-methods approach in primatology. Ethnoprimatology, for example, is a research paradigm focused on the interrelations between humans and nonhuman primates, which has relevance to nearly every primatological endeavor (Fuentes, 2012; McKinney & Dore, 2018). Multi-species ethnography uses the tools of cultural anthropology to explore the many types of relationships shared by human and nonhuman primates (Malone et al., 2014; Setchell et al., 2017). Our expanding laboratory toolkit helps us identify primates' physiological stressors, genetic relationships, and dietary composition, which arms us with knowledge to better understand how our interactions and habitat modifications shape their lives. Integrating social and biological data will deliver a more nuanced understanding of the human-primate interface (Jost-Robinson, 2017; Setchell et al., 2017).

Along with a diversity of research approaches, we argue for the importance of a diverse, inclusive, and decolonial primatology. This book, with contributions from 78 authors representing 24 countries, is only a small sample of the important work with primates being done globally. We stress here the importance of widening our cultural and interpretative lens in primatology and see this as an important goal for the future of our discipline. We need to respect and learn from the work of a global community of scholars, rather than holding research from certain countries or languages in higher esteem than others. We need strong communication with policy-makers and with the public, whose values drive laws and behaviors. Finally, we need to listen to and amplify the voices of people who live with primates every day and center their knowledge and expertise in leading conservation work. Training, funding, and supporting local researchers and field staff to take on leadership roles will help ensure that conservation efforts move away from colonial conservation paradigms and make positive change for both primates and human communities. In regard to conservation success, no one size fits all; success remains highly subjective and dependent on context. The case studies in this volume illustrate the success of small scale, local conservation projects with smaller teams well-placed to establish a social connection with the people coexisting with primates, earning their support to improve the prospects of people and primates alike. With ever-increasing pressure of anthropogenic activities on primate populations, this coexistence is more important than ever.

It has been a pleasure working with such a large community of scholars to review the position of primates in anthropogenic landscapes today. It has also been a stark reminder that primate conservation and primatology as a scholarly pursuit are intrinsically linked. Conservation can no longer be viewed as a niche area for activists. Rather, it is a vital part of our field. Effective conservation means meeting people and nonhuman primates wherever they are. In some cases, this may require cross-cultural collaboration and ethnographic work to understand the social and economic factors that exacerbate tensions between humans and primates in places of conflict. In other cases, this might mean trying to understand people's desire to take selfies with wildlife or watch primate performers, to develop alternative ways of engaging audiences. In some cases, it may require engaging with colleagues who differ in ethical perspectives regarding conducting fieldwork in the landscape of a global pandemic or maintaining captive populations of primates in laboratory settings. Whatever our backgrounds, we need to take an approach that encourages compromise and encompasses diverse worldviews to deliver positive human-primate coexistence for the long term.

References

Anderson, J., Rowcliffe, J. M., & Cowlishaw, G. (2007). Does the matrix matter? A forest primate in a complex agricultural landscape. *Biological Conservation, 135*, 212–222.

Cowlishaw, G., & Dunbar, R. I. M. (2000). *Primate conservation biology.* University of Chicago Press.

Estrada, A., Raboy, B. E., & Oliveira, L. C. (2012). Agroecosystems and primate conservation in the tropics: A review. *American Journal of Primatology, 74*, 696–711.

Fuentes, A. (2012). Ethnoprimatology and the anthropology of the human-primate interface. *Annual Review of Anthropology, 41*, 101–117.

Galán-Acedo, C., Arroyo-Rodríguez, V., Andresen, E., Verde Arregoitia, L., Vega, E., Peres, C. A., & Ewers, R. M. (2019). The conservation value of human-modified landscapes for the world's primates. *Nature Communications, 10*, 152. https://doi.org/10.1038/s41467-018-08139-0

Jost-Robinson, C. A. (2017). Introduction to part III. In K. M. Dore, E. P. Riley, & A. Fuentes (Eds.), *Ethnoprimatology: A practical guide to research at the human-nonhuman primate interface* (pp. 253–256). Cambridge University Press.

Lee, P. C. (2010). Sharing space: Can ethnoprimatology contribute to the survival of nonhuman primates in human dominated globalized landscapes? *American Journal of Primatology, 72*, 925–931.

Malone, N., Wade, A. H., Fuentes, A., Riley, E. P., Ramis, M., & Jost Robinson, C. (2014). Ethnoprimatology: Critical interdisciplinarity and multispecies approaches in anthropology. *Critique of Anthropology, 34*, 8–29.

McKinney, T., & Dore, K. M. (2018). The state of ethnoprimatology: Its use and potential in today's primate research. *International Journal of Primatology, 39*, 730–748.

Setchell, J. M., Fairet, E., Shutt, K., Waters, S., & Bell, S. (2017). Biosocial conservation: Integrating biological and ethnographic methods to study human-primate interactions. *International Journal of Primatology, 38*, 401–426.

Correction to: Primates in Anthropogenic Landscapes

Tracie McKinney, Siân Waters, and Michelle A. Rodrigues

Correction to:
T. McKinney et al. (eds.), *Primates in Anthropogenic*
Landscapes, **Developments in Primatology: Progress**
and Prospects, https://doi.org/10.1007/978-3-031-11736-7

The original version of FM and Chapters 1 and 18 were inadvertently published with incorrect affiliations for authors, Tracie McKinney and Siân Waters. The book has been updated with the changes.

The updated original version of these chapters can be found at
https://doi.org/10.1007/978-3-031-11736-7_1
https://doi.org/10.1007/978-3-031-11736-7_18

Glossary

Adaptability A measure of a non-human primate species' ability to adapt to new conditions or events.

Affiliative behavior Positive social interactions that promote social bonds and group cohesion.

Agroecological matrices The set of agricultural land and patches of vegetation, natural or cultivated, that facilitate gene flow between populations and preserve biodiversity.

Agroecosystem A cultivated ecosystem that integrates the social, economic, and ecological environment.

Allogrooming One individual grooms another.

Alopecia Abnormal partial or complete absence of hair. Alopecia is often associated with high stress levels in animals.

Anthropogenic effects The indirect pressures caused by human activities creating constraints on primate ecosystems, activities, or behaviors.

Anthropogenic Resulting from human impact.

Anthroponosis An infectious disease whose source of infection is an infectious human, and can be naturally transmitted from humans to other animals.

Artificial canopy bridge An artificial connector that helps animals travel between trees, particularly for arboreal animals such as primates or squirrels, in areas that have lost tree cover.

Attitude toward primates The way in which a person thinks and feels about non-human primates. Attitudes towards animals are driven by factors such as socio-cultural background, environmental values, previous experience, and personality traits.

Autogrooming Grooming oneself.

Behavioral flexibility The ability to modify one's behavior in an adaptive way.

Biodiversity hotspot Regions rich in biodiversity and endemism that are at high risk for destruction. There are currently 36 recognized biodiversity hotspots, characterized by having at least 1500 species of endemic vascular plants and having lost at least 70% of its native vegetation, usually due to human activity.

Capacity building Developing and strengthening the skills and abilities needed for people to survive, adapt, and thrive in a fast-changing world.

Climate change Refers to changes in the climatic conditions, including changes in temperature, precipitation, and the frequency and severity of extreme events.

Climate-smart agriculture Cost-effective solutions for increasing agricultural productivity in a sustainable manner and helping to mitigate the factors contributing to climate change.

Commercial hunting An economic practice of pursuing, capturing, and killing of wildlife by people as a way of life, to harvest useful animal products and for sale to potential buyer.

Construct validity In study design, construct validity refers to whether a set of metrics actually measure what they intend to.

Cooperative A farm, business, or organization owned by and run jointly by its members, who share the profits or benefits.

Correlative species distribution models Species distribution models which are used to identify relationships between species' distributions (or densities) and environmental conditions.

COVID-19 A respiratory illness that is caused by the coronavirus SARS-CoV2.

Cross-species transmission of viruses The spread of an infectious pathogen among hosts of different species, which poses a sustained threat to public health. Also known as interspecies transmission or spillover.

Culling Selective killing to control population sizes.

Demography The study of populations in terms of age/sex composition, birth, death, and fertility rates, emigration and immigration rates, and changes in these and related parameters over time.

Direct costs Negative outcomes that are clearly linked to tourism activities, such as risk of pathogen transmission and risk of injuries.

Dispersal The movement of individuals from one location and/or social group to another.

Dormancy A dormant seed is one that is unable to germinate.

Ecological integrity The degree of forest modification as a result of the effects of human pressures and the resulting loss of forest connectivity. Low ecological integrity is often led by logging, fragmentation, crop and livestock farming, urbanization, over-hunting, wood fuel extraction, and altered fire or hydrological dynamics, resulting in deforestation.

Ecosystem services Services or goods (e.g., food, water, pollination of crops, prevention of soil erosion) provided by ecosystems that are beneficial for people.

Ecosystem-based adaptation The use of biodiversity and ecosystem services as part of an overall adaptation strategy to help people adapt to the adverse effects of climate change.

Ecotone A transition area between two biological communities that often has features of both.

Ecotourism Ecologically sustainable tourism with low negative impact on biodiversity and environment, embracing cultural sensitivity and supporting the local community.

Emerging infectious diseases (EID) Infectious diseases appearing in a host population for the first time or previously existing but rapidly increasing, either in number of new cases in a population or geographic range.

Endangered Threatened with extinction.

Endozoochory Seed dispersal by animals in which the seeds are carried inside the animal.

Enrichment Enhancing animal environments in a way appropriate for their species' natural biology and behavior to improve animal welfare.

Ethnoprimatology The study of human and non-human primate interactions and relational coexistence combining biological, ecological, and anthropological methods.

External validity In study design, external validity refers to whether causal relationships between variables are generalizable across contexts.

Fair trade certification Process by which a third party gives assurance that a product meets social, environmental, and economic standards that promote poverty alleviation and sustainable development among small scale farmers through safe working conditions, environmental protection, sustainable livelihoods, and community development funds.

Food taboo A traditional belief, systematized set of rules, and/or prohibition against consuming certain foods or combination of foods.

Free ranging exhibit An exhibit in which animals are not contained by bars or glass but rather roam freely in space that is shared with human guests.

Gender gap The difference between women and men as reflected in social, political, intellectual, cultural, or economic attainments or attitudes.

Genetic drift The cumulative and non-adaptive fluctuation in allele frequencies resulting from random sampling of genes in each generation.

Glucocorticoid Class of steroid hormones that is measured to assess physiological stress levels in animals.

Habitat fragmentation A process that occurs over time as habitat loss converts once-continuous habitat into a patchwork of isolated habitat fragments separated by matrix.

Habitat loss The conversion of habitat to an alternate landcover type, for example, from forest to pastureland.

Habitat Any area with all the necessary conditions (e.g., food, shelter, conspecifics) to sustain a species over time.

Habituation Process through which repeated exposures between primates and humans lead primates to tolerate the presence and proximity of humans; it is often undertaken for primate research and tourism activities.

Heterogeneous landscape A landscape with a variety of different habitat types (e.g., a mosaic of woodland, savannah, agricultural areas).

High conservation value forest Forest that has significant biological, ecological, social, or cultural value that makes it a high priority for conservation.

High-ranking individuals Dominant individuals that are above others in the group hierarchy.

Home range The area a primate group uses during its daily travel and activities but does not defend.

Human-animal relationship (HAR) A history of interactions between an animal and human(s) that can allow each to make predictions about how the other will behave.

Human-primate coexistence When humans and primates share habitats.

Human-primate communication Any communication cues emitted by either human or primates and perceived/interpreted by the receivers of the other species. These cues include facial expressions, gestures, or vocalizations.

Human-primate interaction Encounters between humans and primates where the behavior performed by one elicits a response from the other one. Examples are viewing, observing, photographing, provisioning, or hunting.

Illegal trade Buying, selling, or exchange of primate species locally, nationally, or internationally through unlawful channels or when forbidden by local or international laws.

Incidental primate tourism Viewing, encountering, and/or interacting with primates without having traveled to a location with the specific goal of participating in primate tourism.

Indigenous Culturally distinct groups that have ancestral ties to the land they currently occupy or have been displaced from.

Indirect costs Negative outcomes due to tourism activities that may be intangible or take longer to appear, such as ethnic conflict, financial precarity, or political marginalization.

Internal validity In study design, internal validity refers to whether there is a causal relationship between variables.

Landscape connectivity The degree to which movement is facilitated or limited between habitat fragments in a given landscape. Connectivity is determined by the amount and types of matrix that are present between the habitat fragments in question, coupled with the distance that separates them.

Landscape A heterogenous land area defined by the combination of various, interacting environmental patterns and ecological processes.

Livestock Domestic animals kept on a farm.

Living fences Trees or shrubs that are planted to delimit an area of land, such as a farm or an animal enclosure.

Management Actions taken to oversee and/or regulate primate tourism activities varying in the level of management from strict to loose.

Matrix All of the non-habitat landcover types surrounding and separating habitat fragments.

Mechanistic model A model which aims to assess the potential impacts of climate change by understanding the underlying processes that shape species-environment relationships, based on either physiological or behavioral processes.

Mestizo A person of mixed race. In Latin America, it refers to someone having Spanish and Indigenous descent.

Microbiome Community of microorganisms.

Monetary leakage Primate tourism income that escapes the local economy or system to profit nonlocal businesses or for the purchase of external goods and services.

Monoculture Agricultural area where only one crop is grown. This method is frequently used in intensive agriculture but is associated with reduced biodiversity, declining soil quality, and a higher risk of disease.

Multispecies ethnography A field in anthropology focusing on how the lives of other animals are intricately linked to human lives, social worlds, economies, politics, and cultures.

Naturalistic An exhibit (or enrichment item) that is intended to look as if it belongs in nature.

Niche construction The process by which organisms, through their own activities, alter the selective pressures operating on themselves and others.

Non-invasive surveying Surveying of wildlife by researchers in which no trapping or handling of animals occurs. This form of surveying includes approaches such as line-transect sampling, point counts, camera trapping, acoustic detectors, or others.

Observer effects The effects of direct human interaction, including observation and habituation. Such responses are those that are directly in responses to interaction or observation by humans.

Occupancy modeling Estimating the probability of occurrence via presence/absence data while accounting for imperfect detection. This form of modeling contrasts with other forms of population modeling in that rather than estimating density, researchers often focus on the animal's occurrence or spatial distribution (where, not how many). As a result, data are collected at the species rather than individual level.

One Health Collaborative and transdisciplinary effort – working locally, nationally and globally – to achieve optimal health outcomes by recognizing the interconnection between people, wildlife, domestic animals, plants, and their shared environment.

Pathogen transmission Infectious agent or organism such as bacterium, virus, worm, fungi, or other microorganism that can produce a disease and be transmitted between humans and primates.

Permaculture The creation of systems that are ecologically sound and economically viable, which provide for their own needs, do not exploit or pollute, and are therefore sustainable in the long term.

Pet An animal with a human owner, no longer living in their natural habitat, and reliant on humans for food.

Physiological indicator Measures of biological functioning such as heart rate, blood pressure, hormone levels, and immune function.

Population density The number of individuals of a given species per unit of area.

Positive reinforcement training A behavioral management technique using operant conditioning in which animals are presented with a stimulus, perform a target behavior, and receive a desired reward. Often used to allow animals to cooperate voluntarily in their own care.

Primate tourism Traveling to view captive, semi-free ranging, or wild primate populations for consumptive or non-consumptive activities.

Productive market chains The set of actors that interact in the process of supplying products to the market, including producers, intermediaries, and final consumers.

Productive reforestation The process of replanting commercially valuable trees in an area that was previously occupied by forests and woodlands.

Provisioning The deliberate feeding of wildlife by humans.

Regeneration The process by which new tree seedlings become established (naturally or artificially) after forest trees have been cut down or have died from fire, insects, or disease.

Rehabilitation A managed process in which ill, injured, orphaned, or displaced wild animals regain the health and skills needed for successful reintroduction and survival.

Reintroduction The release of captive or confiscated individuals into wild habitats from where there is population loss to supplement the numbers of an existing population, or where the species had been extirpated. Reintroduction may also include translocation or introduction outside a species' historic range.

Rescue The opportunistic or targeted removal of individuals from situations of harm including injury in the wild, confiscation from illegal trade, former pets, the entertainment industry, research facilities, or captured for translocation.

Riparian area The interface between terrestrial and aquatic systems alongside bodies of water such as creeks, streams, rivers, and wetlands.

Scale of effect The minimum size of an area at which we can observe variation in the response of an animal to changes in their environment.

Seed bank The seeds that are stored in the soil.

Seed dispersal The mechanism by which seeds are transported away from the parent plant to new sites for germination and the development of new individuals.

Seed rain The seeds that fall to the ground as a result of seed dispersal.

Sensitivity In the context of climate change research, sensitivity refers to the ways in which a species' existing biological traits make it resilient or vulnerable to ongoing changes.

Sentinel species In the context of zoonotic infectious diseases, animals that serve to detect and prevent disease risks to human health by providing an advanced warning signal.

Silvopastoral systems A system that combines tree growing with the production and maintenance of livestock within the same area.

Single-species occupancy Estimating species occurrence across a single survey or seasonal periods. The population is assumed to not gain or lose individuals via birth, immigration, emigration, or deaths.

Soil compaction The process in which soil particles are pressed together, reducing pore space between them, and consequently reducing the potential for water infiltration and drainage.

Soil quality The ability of a soil to function for specific land uses that are essential to people and the environment.

Species distribution models Quantitative models which identify areas of suitable habitat for a species by correlating its current distribution to current environmental conditions and using the obtained relationship to project which areas will be more or less suitable under future conditions. See also correlative modelling approaches.

Spillover Transmission of a pathogen from a natural primary host species, which acts as a reservoir for the pathogen, to a novel host species, leading to infection in the new host.

Stereotypy A repetitive behavior with no apparent function or goal. These behaviors are thought to be an indicator of poor welfare, caused by a lack of opportunity to express a species-specific behavior.

Sterilization Medical procedure aimed at preventing an animal from reproducing.

Strict management Structured actions taken to oversee and/or regulate primate tourism activities, focusing mainly on managing the behavior of tourists to mitigate the potential negative effects of tourism on primate behavior, health, and welfare.

Subsistence hunting A practice whereby hunters hunt strictly to provide food for self-consumption for themselves and their families.

Successional agroforestry The integration of a variety of trees and cultivated crops based on natural succession dynamics.

Sustainable agriculture certifications Process by which a third party gives assurance that a product process or service conforms to specified requirements and certification standards based on sustainable farming practices.

Sympatry Occurring in the same geographical location.

Synanthropy When a species benefits from living in close proximity to humans.

Tourism revenue Income generated by tourism activities that may be distributed to regional/national governments, local communities, or researchers.

Tout A vendor, usually on heavily touristed streets, squares, beaches or bars, aggressively selling opportunities to interact with or be photographed with captive animals.

Trade Buying, selling, or exchange of primate species, locally, nationally, or internationally.

Trait-based models These models use expert opinion and published knowledge to determine how a species is likely to respond to climate changes depending on its biological traits.

Translocation The intentional capture and release of animals from one location to another.

Travel corridors Linear feature that joins fragmented habitats to increase opportunities for animals to disperse.

Two-species occupancy modeling A form of occupancy modeling in which the probability of occurrence and/or detection for a target species is estimated while accounting for the influence of a co-occurring species.

Unmanaged tourism Sites where no actions are taken to oversee and/or regulate primate tourism activities.

Unsustainable off-take Lack of a comprehensive monitoring system of wild animal offtake or harvest levels, especially for terrestrial species.

Urban mosaic A habitat made up of areas of building density, residential human-density, anthropogenic disturbance, green areas, and linear anthropogenic structures.

Vagility The ability of an animal to move around in its environment.

Visitor attraction hypothesis In studies on the relationships between visitors and animals in zoos, the visitor attraction hypothesis predicts that animals performing active behaviors attract larger, more active, and noisier crowds.

Visitor effect hypothesis In studies on the relationships between visitors and animals in zoos, the visitor effect hypothesis predicts that behavioral and physiological changes in animals while in the presence of visitors are a direct response to stimuli from visitors.

Wildmeat Flesh of wild terrestrial mammals, birds, reptiles and amphibians hunted for human consumption.

Zoochory Seed dispersal by animals.

Zoonosis An infectious disease that can be transmitted from non-human animals to humans.

Index